Mixing Music

This series, Perspectives on Music Production, collects detailed and experientially informed considerations of record production from a multitude of perspectives, by authors working in a wide array of academic, creative, and professional contexts. Initially we solicit the perspectives of scholars of every disciplinary stripe, alongside recordists and recording musicians themselves, to provide a fully comprehensive analytic point of view on each component stage of record production. Each volume in the initial multi-authored series thus focuses directly on a distinct aesthetic "moment" in a record's production, from pre-production through recording (audio engineering), mixing and mastering through to marketing and promotions. Perspectives on Music Production, as a series, will also be welcoming monographs from authors in the field of music production. This first volume in the series, titled *Mixing Music*, focuses directly on the mixing process.

This book:

- Includes references and citations to existing academic works; contributors draw new conclusions from their personal research, interviews and experience.
- Models innovative methodological approaches to studying music production.
- Helps specify the term "record production", especially as it is currently used in the broader field of music production studies.

Download and Listen samples for chapters 6 and 9 are available on www.routledge.com/9781138218734 and on http://hepworthhodgson.com.

Russ Hepworth-Sawyer has been involved in professional audio for over two decades and, in the last, mastering through MOTTOsound.com. Throughout his career, Russ has maintained a part-time higher education role teaching and researching all things audio (www.hepworthhodgson.com). Russ is a current member of the Audio Engineering Society and is a former board member of the Music Producer's Guild and helped form their Mastering Group. Russ currently lectures part time for York St John University, has taught extensively in higher education at institutions including Leeds College of Music, London College of Music and Rose Bruford College and has contributed sessions at Barnsley College. He has written for *MusicTech* magazine, *Pro Sound News Europe* and *Sound on Sound,* as well as many titles for Focal Press/Routledge.

Jay Hodgson is Associate Professor Music at Western University, where he primarily teaches courses on songwriting and project paradigm record production. He is also one of two mastering engineers at MOTTOsound—a boutique audio services house situated in England, and now Canada. In the last few years, Hodgson has worked on records nominated for Juno Awards, which topped Beatport's global techno and house charts, and he has contributed music for films recognized by the likes of *Rolling Stone* magazine and which screened at the United Nations General Assembly. He was awarded a Governor General's academic medal in 2006, primarily in recognition of his research on audio recording; his second book, *Understanding Records* (2010), was recently acquired by the Reading Room & Library of the Rock and Roll Hall of Fame.

Perspectives on Music Production Series

Series Editors
Russ Hepworth-Sawyer
Jay Hodgson
Mark Marrington

Titles in the Series

Mixing Music
Edited by Russ Hepworth-Sawyer and Jay Hodgson

Mixing Music

**Edited by Russ Hepworth-Sawyer
and Jay Hodgson**

 Routledge
Taylor & Francis Group

NEW YORK AND LONDON

First published 2017
by Routledge
711 Third Avenue, New York, NY 10017

and by Routledge
2 Park Square, Milton Park, Abingdon, Oxon OX14 4RN

Routledge is an imprint of the Taylor & Francis Group, an informa business

Library of Congress Cataloging in Publication Data
A catalog record for this book has been requested

ISBN: 978-1-138-18204-2 (hbk)
ISBN: 978-1-138-21873-4 (pbk)
ISBN: 978-1-315-64660-2 (ebk)

Typeset in Times New Roman
by Apex CoVantage, LLC

This book is dedicated to those recently lost who contributed so much:

George Martin
David Bowie
Prince
Zenon Schoepe

Contents

Figures

Tables

Series Introduction

Perspectives on Music Production

This series, *Perspectives on Music Production*, collects detailed and experientially informed considerations of record production from a multitude of perspectives, by authors working in a wide array of academic, creative and professional contexts. We solicit the perspectives of scholars of every disciplinary stripe, alongside recordists and recording musicians themselves, to provide a fully comprehensive analytic point of view on each component stage of music production. Initially, each multi-authored volume in the series thus focuses directly on a distinct stage of music production, from pre-production through recording (audio engineering), mixing and mastering to marketing and promotions.

As a series, *Perspectives on Music Production* was designed to serve a twofold purpose. Situated within the emerging field of music production studies, *Perspectives on Music Production* aims to specify what exactly scholars and recordists alike mean by the term 'record production'. In recent research, the term is often used in simply too nebulous a manner to provide any substantive, concrete utility for researchers interested in studying specific details of the production process. In fact, both tacit and explicit definitions of 'music production' offered in recent research often bear a certain tautological resonance: *record production is everything done to produce a recording of music*, or so the argument usually seems to run. But this overly inclusive approach to defining the object of study simply doesn't withstand sustained analytic scrutiny. The production process is broad, to be sure, but it is rationalized into numerous component procedures, each of which, while holistically related, nonetheless requires its own specialized expertise(s). And this is true whether that expertise is located in a team of people or in one single individual, as the 'project' paradigm would demand. Every record production, regardless of genre and circumstance, requires at least the following procedures: pre-production (conception vis-à-vis available technology), engineering (recording and/or sequencing), mixing and mastering (even if only bouncing without any further processing) and distribution of some sort (lest the recording remains inaudible data). While record producers are indeed responsible for overseeing a project through each of these component phases—and, thus, while it may seem fair to simply refer to the totality of these phases as 'record production'—every phase has its own unique aesthetic priorities and requirements, and each of these reacts back on, and (re)shapes, the musical object being produced in turn. Ultimately, it is uncovering and understanding the broader musical ramifications of these priorities and biases that comprises this series' primary analytic concern.

Perspectives on Music Production also looks to broaden methodological approaches that currently prevail in music production studies. The place of traditional academic and scholarly work on record production remains clear in the field. However, the place of research and reflection by professional recordists themselves remains less obvious. Though music production studies tend to include professional perspectives far more conscientiously than other areas of musical study, their contributions nonetheless are often bracketed in quiet ways. Producers, engineers and recording musicians are often invited to participate in scholarly discussions about their work only through the medium of interviews, and those interviews typically follow more 'trade' oriented than straightforwardly academic lines of inquiry. We thus invite contributions from professional recordists which elucidate their own creative practice, and in whichever ways they deem most relevant to scholarly considerations of their work. Similarly, we hope the series will encourage greater collaboration between professional recordists and the researchers who study their work. As such, we invite contributions that model novel and inclusive methodological approaches to the study of record production, encompassing professional, creative, interpretive and analytic interests.

It is our sincere hope that *Perspectives on Music Production* provides a timely and useful intervention within the emerging field of music production studies. We hope each volume in the series will spur growth in music production studies at large, a more detailed and comprehensive scholarly picture of each particular procedure in a record production, as well as a general space for researchers to pause and reflect back on their and their peers' work in this exciting new area.

Jay Hodgson and Russ Hepworth-Sawyer

About the Editors

Russ Hepworth-Sawyer has been involved in professional audio for over two decades. Throughout his career, Russ has maintained a part-time higher education role teaching and researching all things audio (www.hepworthhodgson.com). Russ is a former board member of the Music Producer's Guild and helped form their Mastering Group. Through MOTTOsound (www.mottosound.com), Russ works freelance in the industry as a mastering engineer, writer and consultant. Russ currently lectures part-time for York St John University, has taught extensively in higher education at institutions that include Leeds College of Music, London College of Music and Rose Bruford College and has contributed sessions at Barnsley College. He has written for *MusicTech* magazine, *Pro Sound News Europe* and *Sound on Sound*, as well as many titles for Focal Press/Routledge.

Jay Hodgson is on the faculty at Western University, where he primarily teaches courses on songwriting and project paradigm record production. He is also one of two mastering engineers at MOTTOsound—a boutique audio services house situated in England, and now Canada, whose engineers' credits include work on projects containing the likes of Rush, Three Days Grace, The New Pornographers, Glen Campbell, Billy Ray Cyrus, The Barenaked Ladies, and numerous other household names. In the last few years, Hodgson has worked on records nominated for Juno Awards, which topped Beatport's global techno and house charts, and he has scored films recognized by the likes of *Rolling Stone* magazine and which screened at the United Nations General Assembly. He was awarded a Governor General's academic medal in 2006, primarily in recognition of his research on audio recording; and his second book, *Understanding Records* (2010), was recently acquired by the Reading Room of the Rock and Roll Hall of Fame. He has other books published, and forthcoming, from Oxford University Press, Bloomsbury, Wilfrid Laurier University Press, Focal Press and Routledge.

Notes on Contributors

Chapter Authors

Gary Bromham is a songwriter, producer and mix engineer. He has worked with numerous successful artists including Sheryl Crow, George Michael, Graham Coxon of Blur and Editors. He was signed as an artist to EMI Records in the mid-1990s as part of his band 'The Big Blue'. More recently, he has acted as a guest speaker for Apple and is a visiting lecturer at several universities in the UK and abroad. He is currently a PhD student at Queen Mary University of London researching retro aesthetics in digital audio technology.

Andy Devine studied the creative applications of music technology and composition at Bretton Hall and completed his master's in music technology at the University of York. Devine now teaches music production at both undergraduate and postgraduate levels at York St John University. Along with academic pursuits, Andy has produced electronic music both for pleasure as well as for client commissions and has been a DJ nationally, presenting a bi-weekly radio show showcasing the best in electronic music, new and old.

Ruth Dockwray is a senior lecturer in Popular Music and the program leader for the BA Music and BA Popular Music Performance courses at the University of Chester, UK, where she teaches historical, critical and analytical studies of pop music. Her research areas include analysis of video game soundtracks and musicology of popular music production.

Phil Harding joined the music industry at the Marquee Studios in 1973, engineering for the likes of The Clash, Killing Joke, Toyah and Matt Bianco by the late 1970s. In the 1980s, Phil mixed for Stock, Aitken & Waterman tracks such as 'You Spin Me Round' by Dead or Alive followed by records for Mel & Kim, Bananarama, Rick Astley, Depeche Mode, Erasure, Pet Shop Boys and Kylie Minogue. In the 1990s, Phil set up his own facility at The Strongroom with Ian Curnow. Further hits followed with productions for East 17 (including 'Stay Another Day'), Deuce, Boyzone, 911 and Let Loose. Recent projects include the book *PWL from the Factory Floor* (2010, Cherry Red Books) and mixing Sir Cliff Richard's 2011 album *Soulicious*. Harding has recently worked for Holly Johnson, Tina Charles, Samantha Fox, Belinda Carlisle and Curiosity with his new production team PJS Productions. Phil is also Vice Chairman of JAMES (Joint Audio Media Education Services) and is soon to complete his PhD with Leeds Beckett University.

Christopher Johnson launched his career as a singer-songwriter tinkering with a Tascam PortaOne to demo songs for his band. He turned more decisively to production in order to stretch a publishing advance as far as it could go by setting up a better tracking studio, thereby saving on future recording costs. As a multi-instrumentalist producer, he is known in the UK progressive rock scene for his studio and live work with Halo Blind, Mostly Autumn and Fish. Chris is currently a lecturer of Music Production at York St John University. His research focuses on the musician's self-view of their own authenticity.

Alex Krotz has been surrounded by music since a very young age and has always had a talent and undeterred passion for creating music. He has worked with a wide range of artists, including some of Canada's largest acts (Shawn Mendes, Three Days Grace, Drake), as well as many up-and-coming bands. He currently is an engineer at Noble Street Studios.

Dylan Lauzon is a graduate of the University of Western Ontario's Music Administrative Studies Program and professionally is a musician and songwriter based out of London, Ontario. His current project, Nikki's Wives, has been a feature act at the Leather and Laces 2016 Victoria's Secret Super Bowl Pre-Party in 2016 and is currently touring as direct support for Grammy Award–winning artist CeeLo Green. In the last few years, Dylan has written songs with producers such as Alastair Sims (Rush, Three Days Grace) and Gavin Brown (Billy Talent, Metric), as well as having playing featured on records by Die Mannequin and the New Pornographers.

Mark Marrington is an academic specializing in a number of areas, including musicology of record production, popular music studies, electronic music history/practice and the creative uses of music technology. He is currently a Senior Lecturer in Music Production at York St John University, UK, where his teaching is primarily focused on the musicology of record production, critical theory and DAW-based composition.

William Moylan is professor and coordinator of Sound Recording Technology at the University of Massachusetts Lowell. He has been active in music and recording technology communities for over thirty years, with extensive experience and credits (including several Grammy nominations) as a record producer, recording engineer, composer, author and educator. Audio production professionals, scholars and audio programs worldwide have used Moylan's writing. His *Understanding and Crafting the Mix: The Art of Recording* (Focal Press, 2015) informs much of this chapter. His new book *Recording Analysis: How the Record Shapes the Song* (Routledge, 2017) builds on these foundations.

Dean Nelson is a producer and engineer mixer and professor of Music Production & Pro Tools at Ontario Institute of Audio Recording Technology

(OIART). Nelson began assisting for producers Neal Avron, Mark Trombino and Ethan Johns. He later moved on to Chalice Studios in Hollywood, then Ocean Way Studios. Nelson assisted Jack Joseph Puig for over a decade, but began a five-year mentor relationship with Jack that eventually grew from assistant to engineer. During his tenure with Jack, Dean assisted/engineered projects for the Rolling Stones, Fergie's Grammy-nominated hit, 'Big Girls Don't Cry', Mary J. Blige and U2. In addition to his engineering duties, Dean also assisted Jack and Waves with PuigChild and PuigTec. Dean was later offered a job as Beck's engineer. In this role, he recorded and/or mixed an array of projects including Stephen Malkamus and the Jicks's latest *Mirror Traffic*, Sonic Youth founder Thurston Moore's album *Demolished Thoughts*, 'The Record Club' projects, Charlotte Gainsbourg's *IRM* and *Stage Whisper*, Jamie Lidell's *Compass* and the Beck/Bat For Lashes Twilight collaboration *Let's Get Lost*. Upon relocation to London, Ontario, he produced and mixed several songs off Buck 65's *Neverlove*.

Justin Paterson Associate Professor of Music Technology at London College of Music | University of West London, where he leads the MA Advanced Music Technology. He is also a music producer and author of *The Drum Programming Handbook*. His research has an international profile, and in 2015, he developed a new format of music release featuring interactive playback together with Prof. Rob Toulson – a project that at present has a patent pending based on intelligent audio crossfades. Current research is around interactive 3-D audio for virtual reality (VR), collaborating with record label *Ninja Tune* and VR company *MelodyVR*. Justin is also co-chair of the Innovation in Music conference series.

Martyn Phillips graduated in Engineering Science from Oxford University in 1981. Martyn then went on to produce, engineer, program, write, perform, mix and occasionally master audio recordings in a wide range of genres. A pioneer of now-ubiquitous techniques such as retuning vocals, time-stretch remixing and re-splicing drum-loops, he produced a number of hits, including number 1 records from Erasure, Londonbeat, Jesus Jones and Deva Premal. Lecturing in Music Technology at the London College of Music, University of West London since 2009, he now plays and sings in his band Mpath in Germany.

Joshua D. Reiss is a reader with Queen Mary University of London's Centre for Digital Music, where he leads the audio engineering research team. He has investigated music retrieval systems, time-scaling and pitch-shifting techniques, music transcription, loudspeaker design, automatic mixing, sound synthesis and digital audio effects. His primary focus of research, which ties together many of the above topics, is on intelligent signal processing for professional sound engineering. Reiss has published over 160 scientific papers. He was nominated for a best paper award by the IEEE and received the 134th AES Convention's Best Peer-Reviewed Paper Award. He co-authored the textbook *Audio Effects: Theory, Implementation and Application*. He is a former governor of the Audio Engineering Society and co-founder of the start-up company LandR, providing intelligent tools for audio production.

Matt Shelvock is completing his PhD at ABD in Ontario, Canada. Shelvock possesses diverse research and musical interests including teaching, session work, producing, mixing and mastering. As a guitarist, he has had the pleasure of working with multi-national successes such as Skip Prokop (Lighthouse, Janis Joplin, Mike Bloomfield/Al Kooper), Josh Leo (Emerson Drive, LeAnn Rimes, Reba McEntire) and many others. Matt's research interests were inspired by time spent recording and gigging professionally, and his work aims to explain how different various recordist methods—such as tracking, mixing and mastering—constitute distinct musical competencies.

Alastair Sims is passionate about music and has devoted his life to helping bands and artists recognize their dreams. Working with some of Canada's largest bands (Rush, Walk Off The Earth, Three Days Grace) in many of the greatest studios (Noble Street, Revolution), he has been able to work alongside and learn from some of the best engineers and producers. Continuing his work with up-and-coming Canadian artists along with well-established bands, Alastair is cementing himself as a strong presence in Canadian music.

Rob Toulson is Professor of Commercial Music at the University of Westminster. Rob is also an active music producer, sound engineer, songwriter and performer. His research focuses on the relationship between technology and creativity in commercial music production and he has developed a number of innovative software applications in this area. Rob is the inventor of the iDrumTune iPhone app, which assists percussionists with tuning their instruments and has been adopted by musicians all over the world. In 2015, he led the AHRC-funded Transforming Digital Music project, which developed immersive mobile applications for connecting artists more intimately with their fans.

Robert Wilsmore graduated with a doctorate in composition from Nottingham University in 1994 and has since written on subjects such as prog rock group Yes for the Routledge journal *Parallax* and on Kraftwerk for Continuum Press as well as many other publications that explore collaboration and philosophy in popular music. His current philosophical conceit that 'we're all just writing one big song' manifests itself through the concept of the group The And and their singular work 'The Song of a Thousand Songs' as well as through the 'real' group The And Ensemble, which he directs, composes for and produces with Christopher Johnson at York St John University, where Robert is Head of School for Performance and Media Production.

Chapter 10: Interviewee Biographies

Pierre Belliveau (Gone Deville) is a Producer/DJ based in Montreal. He is currently working on a full-length album. Previous releases of note include an appearance on Toolroom's *Ibiza 2014* poolside compilation and releases for Mile End Records. Pierre's production style has led him to collaborate with (among others) Raekwon from Wu-Tang Clan, Hector Couto, Martin Roth and NOTV.

Craig Bratley's debut album was released on Tsuba in 2014, following a number of releases on labels such as Bird Scarer, Is It Balearic, Robsoul and Foto, all of which received critical acclaim. Craig also heads up the Magic Feet record label. Alongside regular DJ sets across the UK and Europe and at festivals such as Glastonbury, Festival Number 6, Electric Elephant and Low Life, his plans for 2016 include releases and remixes on Claremont 56, Throne of Blood and Magic Feet.

Rick Bull (Deepchild/Acharné/Concubine) is a Berlin-based Australian artist who has earned himself a reputation as a groundbreaking producer and DJ though a dynamic approach to House and Techno. Rick has performed live at many of the most respected electronic music institutions, such as Berghain and Tresor.

Ryan Chynces (Rion C) makes techno, basically!

Andy Cole (LuvJam) is a UK-based DJ/Producer/Label Boss and Graphic Designer; 2014–15 saw Andy touring across Germany, Belgium, Sweden, Norway, Hungary, Spain and Romania in addition to maintaining a residency with We Love Space and performing at Bestival for the tenth year running. Andy also runs the much-revered vinyl arts projects *Blind Jacks Journey, Crow Castle Cuts and Nip7*. His new project *The Legend of Gelert* launched in April 2015 and, as with his other projects, is developing a cult following.

Phil France (Cinematic Orchestra/Phil France) is a Manchester-based musician/producer. Phil is the principal collaborator (alongside Jason Swinscoe) in the Cinematic Orchestra, which entails co-writing, arranging and producing on their albums such as *Everyday*, *Man With The Movie Camera*, *Ma Fleur* and also the triple-award-winning soundtrack for *The Crimson Wing* nature documentary. Phil has recently released his first solo album, *The Swimmer*, to great acclaim.

Adam Marshall (Graze/New Kanada) was born and raised in Toronto and is now based in Berlin. Adam has been active as a DJ/Producer and record label boss for ten years.

Noah Pred is a Berlin-based Canadian DJ/Producer whose *Thoughtless* imprint is now on its hundredth release. The past two years have seen Noah perform extensively throughout North America and Europe alongside recent releases on *Cynosure, Highgrade* and *Trapez LTD*. Noah has collaborated with a diverse collection of artists including Tom Clark, Pablo Bolivar and Tim Xavier.

TJ Train (Room 303) signed his first release under the Room 303 brand to the UK's Love Not Money label in 2011. He has since gone on to sign tracks to other esteemed labels including On the Prowl Records (New York), Suruba Records (Spain) and Subtrak Records (Berlin). TJ has also performed alongside Miguel Campbell, Damian Lazarus, Infinity Ink, No Regular Play, Tone of Arc, Nitin and James Teej, to name a few.

Acknowledgments

The authors would like to thank, first and foremost, all those at Routledge Press who shared our vision for this project and who were instrumental in seeing this series to print.

We must, however, extend our thanks to those who have contributed their research, their conclusions and of course their patience as we've encouraged, edited, disheartened (perhaps) and edited again. The work in this book, and the subsequent series already in full flow, is wide ranging. We'd like to thank everyone who has contributed. We'd also like to thank those who put in a call and were not selected for this book—perhaps we'll see you later in the series. Thank you for your support and shared vision for the project.

Finally, we'd like to thank our families for their patience, too, as we write and pull together another book.

Russ Hepworth-Sawyer and Jay Hodgson
www.hepworthhodgson.com

Introduction

Russ Hepworth-Sawyer and Jay Hodgson

Welcome to Perspectives on Music Production:
Mixing Music

As you will have learned from our series introduction, this is the first in a series of planned multi-authored books exploring academic interests in music production from a range of *perspectives*. This approach enables us, as a discipline, we believe, to study any topic from differing angles. Whether that be an angle of perspective, such as the method of approaching a mix, or the way in which you look at or listen to a mix, or to consider the prism through which you analyze a mix, Perspectives on Music Production is a broad church.

Much of our writing to date has been conceptualized and organized within the adopted term of *the production process*. For example, *From Demo to Delivery: The Process of Production* (edited by Hepworth-Sawyer, 2010) involved multiple authors discussing phases of that exact staged production process. From the outset of Perspectives on Music Production, our proposal to Routledge, and the invitations to the contributors here, we were clear what the titles of each of the books would be. True to form, they are based upon broad segments of the production process. The output, however, would dispel any such theory. We have welcomed this. A researcher will take the book out of the library most applicable to their current investigation. What transpires is a number of discussions that mixing music cannot be pigeonholed into one specific stage of the music production process. The mix is within the concept, the seed of the initial composition, or idea. Reconceptualization can occur again later through the production of a track, which could of course be equally applied before going into the studio and after the recording in the mixing stage. Expect, therefore, to read about recording, production, pre-production and even mastering within this book, all in reference to the creation of a product. The focus of Perspectives on Music Production: *Mixing Music* therefore is the culmination, the moment if you like, of a discrete number of tracks or sources coming together to make an exciting, cohesive whole.

Many might query why the first book in a series loosely based around the production process should launch with *Mixing Music*. As we researched and interviewed for our forthcoming book *Audio Mastering: The Artists* (due out at a similar time to the release of this book), we noted how little research had actually been carried out on the mix. We were equally alarmed that outside of the popular music technology press, there was little ethno-graphic, or experiential, discussion of the mix. Additionally, we felt that the prism by which the mix is viewed should also be considered further.

Contributions have been submitted from both experienced academics and early career researchers. Contributions have also been accepted from professionals discussing their work. In addition, interviews have been transcribed and are presented to you here for your own conclusions. You will find some pieces based upon discussions and experiences, too. We welcome these *perspectives*.

The relatively recent emergence of Music Production as an academic topic has dawned at a time that coincides with open access journals, the Internet and considerable change within the audio industry itself. These have shaped both the academic activities we engage in, but also the indus-try we write about today. Music Production, we believe, as a field of study is therefore part history, part present and part future.

Zenon Schoepe, the late editor of *Resolution* magazine, wrote in one of his last editorials in the March/April 2016 edition:

> As an industry we are populated with personalities. Individuals who represent a notion, an idea, a set of values, maybe even an era that we all know has gone but curiously still hanker for . . . [When they are gone] what we are left with is what they stand for . . . I think it is all the more alarming for us because it is unlikely that we will ever see their kind again.

At that time, one presumes he was writing about the recent loss of Sir George Martin, or perhaps David Bowie. He could not have known that Prince was about to join the reference. However, those of us currently studying the world of music production can empathize with Schoepe. We dedicate ourselves to music production's 'part history'. Through the lens of part history, we explore and dissect the historical developments and undertakings in studios around the globe and judge, compare and root our knowledge of the present.

As a form of study, our history is actually not that old. The industry based upon the distribution of recorded audio has only clocked up around 100 years or so, yet there is still so much knowledge still to catalog, ana-lyze and discuss.

We feel that much present work should be extracting or formalizing historical or tacit knowledge held within current professionals. As an aca-demic community we should continue to be noting this living history as best we can, ensuring the skills, techniques and anecdotes are kept for future researchers. We hope to capture and analyze as much as is possible of what Schoepe referred to as 'what they stand for' in this and the forth-coming books in the series.

Chapter Introductions

In the process of constructing this book, we received a number of abstracts from potential contributors internationally. As we considered each contri-bution upon its merit, we also considered the tone and flow of the whole book as an item. We wanted to launch the book with a historical account of the mix. Martyn Phillips, in chapter 1, brilliantly discusses many of the pivotal milestones that have enabled and expanded the mix to the quality we persistently expect today. Later in the chapter, Phillips explores perspectives through which the mix is analyzed and understood. This is a fantastic introduction to perspectives of mixing music and for what enters later in the book. One can appreciate that he has considered the mix as the *artifact* for some time and the ways in which is it appreciated, viewed and consumed. Phillips's considerable professional experience is drawn upon, although not overtly pressed upon you, although is best demonstrated by his section on mixing guidelines, something that is also touched upon by Phil Harding later in this book. Phillips has introduced an analysis into the process of the mix, and in doing so has introduced some office practice theory we never thought we'd see within a music production book in the form of Edward De Bono's *Six Thinking Hats*. This is discussed from the perspective of being productive and is also considered alongside Brian Eno's *Oblique Strategies*. This chapter explores the landscape of mixing music from the prism of part-history, citing important landscapes, additionally analyzing part-present working practices, but concluding with some consideration for the future of mixing practice.

To understand music mixing, it is important to understand what we hear. Listening from either a consumer or professional standpoint needs to be considered throughout the process of creating a mix. Many contributors in this book discuss or refer to William Moylan's work. In chapter 2, Moylan joins us discussing 'How to Listen, What to Hear', where he expands upon his renowned work on interpreting the mix. Moylan investigates the problems and technicalities for listeners to become attuned to listening to the mix and how to hear it. He begins by reinvestigating his earlier work and looking at the need to refine listening skills and the challenge to achieve this. Moylan later analyzes the mix on separate levels, which he incidentally calls 'levels of perspective', through which mix analysis should take place. Through a number of figures and tables, Moylan explores the mix and how we listen to it, exploring aspects that set a foundation for chapters later in this book. Moylan discusses the terminology we all use to describe the mix. This is a theme picked up and considered later in the book, too. The sound stage is introduced within Moylan's chapter and is once again another theme explored later within the book when describing the mix landscape with the more ethnographic contributions. The chapter concludes with an apt observation on the mix and its interpretation.

Building upon Moylan's area of investigation, Ruth Dockwray explores sonic spacialization in 'Proxemic Interaction in Popular Music Recordings', chapter 3. Using a number of tracks for discussion, Dockwray explores the use of space and intimacy in mixing, using tracks spanning rock to pop over a number of decades. Focusing on vocal placement compared with the musical backing and the sound space, Dockwray explores the singer's persona in relation to the environment, or 'zones'.

There is of course the 'part-present'. Later in the book, professionals who have tacit knowledge discuss their craft, such as Phil Harding and Gary Bromham, who are both working towards their PhDs after considerable professional careers, and those professionals whose interviews have been transcribed by contributors featured within this book.

Capturing the part-present first is Phil Harding in chapter 4, where he provides an ethnographic insight into how he continues to professionally employ the 'Top-Down' compared to his native 'Bottom-Up' mixing principle. Citing his many chart hit mixes as examples, Harding demonstrates the nuances and differences between the two methods. To achieve this he has developed a twelve-step program he follows to achieve the necessary, and expected, professional standards. Harding expresses that this methodology can be effective equally on the SSL G-Series console, upon which many of his professional successes were mixed, or by using Pro Tools, as is cited here within his chapter. What Harding leaves you with for the rest of the book is an insight into how he approaches a mix, and the quality level to which a 'mix standard' is judged.

Just as Harding explores a widely debated topic of 'how to' approach mixing, Justin Paterson in chapter 5 introduces discussions that are a cause of modern debate around mixing within the computer Digital Audio Workstation environment, compared to the once-traditional 'out of the box' variation using a mixing console and outboard processing. Paterson also begins our part-future perspective of mixing music. He elucidates to the future, discussing possible working practices and structures in the future beyond simply working 'in the box' or out of it.

Both chapters 4 and 5 are important at this juncture of the music production evolution or journey. They notate the nuanced changes in mixing methodology and practice over the past twenty-five years and inform the part-future element of this book.

Chapters 6, 7, 8, 9 and 10 represent the opportunity for the researcher to read either discussions or thought pieces with DJs, editors, engineers and producers about their craft and the industry within which they reside.

Chapter 6 begins with an interview with Alastair Sims by Jay Hodgson discussing primarily digital editing as a craft. Editing has not been a new phenomenon in mixing, but has in the past two decades become more of an expectation rather than a once-in-a-blue-moon desire to amend performance errors. There is, of course, the argument that the historic technology prevented us editing in such a non-destructive way, and that had it have been available, the editing 'phenomenon' would have been more prevalent. Alastair Sims offers both comment and audio (via the book webpage on www.hepworthhodgson.com) to discuss the art of editing in modern music production. In addition, Sims describes the communication and professional interplay with all those involved within the production process.

Dylan Lauzon in chapter 7 discusses his view of pre-production within mixing. His argument, well formed, explains that the concept or idea of the song has a 'mix' attached to it as it is written, or developed. Many of these characteristics will either remain as elements within the final mix or set the blueprint for its completed style. Lauzon describes his methodology when writing a song or, we believe, when writing a production. The best analysis he comes to is that he's 'writing a mix'.

As introduced earlier by Justin Paterson, significant moves have been made within the industry from the analog recording studio (and console) to the digital audio workstation. Dean Nelson's discussion piece (chapter 8), aptly titled 'Between the Speakers' begins by expresses his thoughts on modern mixing and the facilities provided to us. He asks the question whether our 'most superior' equipment has led us to create world standard records. The chapter, however, grows into an ethnographic discussion of the process of mixing per se. In considerable detail, Nelson explores rationale and his response to many aspects of mixing, including examples of his experience working with the likes of Beck, Jamie Lidell and Bat For Lashes, as well as many other artists.

Chapter 9 is an interview with Alex Krotz by Jay Hodgson. Krotz is staff engineer at Noble Street Studios and has worked with the likes of Shawn Mendes, Three Days Grace and Drake. In his interview, titled 'Mixing for Markets', Krotz discusses the process of mixing from the perspective of the deal, the communication and management of the project. He discusses what's involved and how the mix concept is approached differently depending on the occasion, or market (genre). In contrast to Phil Harding, Krotz describes his adoption of the bottom-up method for his mixing. Ethnographically this interview is an important addition to Perspectives, addressing the part-present working practices this series aims to disseminate. Krotz closes by expressing thoughts on the future of mixing.

Andrew Devine and Jay Hodgson host a mixing roundtable discussion in chapter 10 that includes many prominent electronic music producers and DJs. In this chapter, the interviewers are keen to extract key differences between mixing for dance music compared to the oft-covered pop and rock balancing. Conclusions are drawn about the processes that work and the equipment utilized. One conclusion is that mixing is not a distinct stage of the production process in electronic dance music. As expected, it concurs with Dylan Lauzon's argument about pre-production and mixing.

Matt Shelvock explores mixing within hip hop in chapter 11. Through a number of sources, Shelvock explores the differences and methods of mixing music within the genre. As with the previous chapter, many references support the fact that mixing is an integral part of the music making often within hip hop. The discussion expands to encapsulate the creativity and individualistic nature of the production and therefore mixes. This perspective is quite different from mainstream popular music, where it could be argued there is a formulistic method to the production of music. Shelvock explores some of the specific mix techniques employed to create this individualism by citing many of hip hop's core artists.

In chapter 12, Robert Wilsmore and Christopher Johnson discuss ontological issues within the mix. They explore how different perspectives of the mix reveal themselves as multiple authenticities. Wilsmore and Johnson explore the meaning of the mix, whether it is singular or a multiple identity. 'For function and as well as out of pretentiousness' they have employed the word 'themix' to express the singular multiple expression of the mix. If this confuses you, it is worth exploring the full argument. Through areas of philosophy and real live practice with their experimental act 'The And', this chapter explores the map that is the mix, concluding that 'it's only rock n roll but we like it'.

Mark Marrington continues the ontological theme in chapter 13 by evaluating concepts bestowed upon the art of mixing. Marrington focuses first on the role of the mix engineer in relation to the production process and the people within the process. He later addresses the work the mix engineer is undertaking on the music in front of them. Marrington discusses the importance of the mixer 'speaking on behalf of the artist' through the medium of the mix. Marrington explores other aspects of this rhetorical art, discussing the constraints mixers are placed within because of expectation or genre.

Jay Hodgson completes the ontology section with chapter 14, titled 'Mix as Auditory Response'. Hodgson explores the concept of the mix becoming the one article comprising the multiple multitrack elements.

Joshua Reiss, building upon his work on intelligent systems in music, has contributed chapter 15 covering the recent emergence of algorithmic approaches to mixing multitrack content. Using his experience of researching in this area, this contribution explores the current thinking and practice in intelligent systems that could, in time, transform the workflow of the mix engineer completely. Reiss explores the factors in achieving a form of automated mixing, but is keen to express that the creativity in audio production would not be lost. The intelligent system cannot, at this stage, replace the artistic decisions. Whilst the end product (if there will ever be an end) would perhaps remove the monotonous tasks for the sound engineer, Reiss is keen to point out that it will allow musicians to concentrate on the music more.

As a professional engineer, Gary Bromham considers how academic practice can inform mixing in chapter 16. Bromham notes that, within music production, there is a natural thought that the flow of information about mixing should be from practitioner to academia. Throughout the chapter, Bromham discusses the potential flow of knowledge and consideration that could be diverted from academia to the professional. Through a series of interviews and personal experience, Bromham moves to look at areas of the mix and how the 'flow' could benefit both parties. The chapter does not only elaborate upon the mix and the mixing, but also the mixer as a 'sonic trend' with a 'sonic signature'.

In the final contributed chapter 17, Rob Toulson considers the final aspect of mixing—the handing of the audio to the mastering engineer. Through a series of interviews with mastering and mixing engineers, Toulson explores the factors, problems and issues that are faced in this often-silent transaction between the mix and the mastering of an artifact. The chapter also explores the more modern phenomenon of mixing and mastering as 'a single process'. Toulson argues that the possibility to 'cut corners' and put off decision-making until later in the production process may have led to the processes of mixing and mastering becoming more 'porous' over time.

This series is also partly about the future. As described above, Perspectives on Music Production: *Mixing Music* contains a number of papers and interviews that discuss the route map for mixing as we develop technologies and skills. For example, Joshua Reiss expands on his work on intelligent systems in mixing. Developments such as landr.com have, despite robust industry concerns of quality, opened up the starting acceptance of automation. Never before has an automated system been accepted by a portion of the music production community. Of course we have presets within, say, Logic Pro, which claim to process a bass drum, but due to the fact that every bass drum and every studio is different (let alone the drummer), the settings will all need tweaking to get close to what's required. Automated services such as landr.com offer, at the time of writing, very little in the way of honing. It is, however, only a number of years before substantial portions of our workflow will be automated in mixing, opening up a whole new debate. The final contribution from Russ Hepworth-Sawyer in chapter 18 explores the future perspectives in mixing music.

1

Exploring Potential of the Mix

Historical Milestones and Expanded Perspectives

Martyn Phillips

Introduction

Working beyond the plentiful suggestions from experts, copying others or trusting their art to happenstance, how can mix engineers find their own unique way of deepening their own craft? Is it possible to find new approaches? This article looks at conceptually what is possible in this regard and suggests some practical ways of achieving this.

Historical Context

The level of possibility and complexity available to the mix engineer has increased in steps since the role came into being. To appreciate the significance of the increasing finesse that can be applied to the process of accessing and manipulating component parts of an audio piece, some historical perspective is beneficial.

Originally, the mix was intimately connected with the performance, both of these associated with the immediate, and transitory, fading with the physical sound vibrations to reside only in the memory of the listener. The mix was confined to the arrangement and guidance of the parts, a job sometimes employed by a conductor.

This was changed in 1853, when an endurable artifact that represented audio was created, committed to a medium of soot on paper. Parisian Édouard-Léon Scott de Martinville recorded an incomprehensible squawk that is likely a human voice. It would be seven years later, on April 9, 1860, before he was able to record something intelligible on his apparatus which he now called a phonautograph, a recording of someone, possibly himself (MacKinnon, 2012), singing 'Au Claire de la Lune', amidst a sea of noise. Scott de Martinville never was able to play back his recording or even appeared to contemplate the possibility, and it would take 147 years before the development of a virtual stylus by Carl Haber and his team at the Lawrence Berkeley National Laboratory in Berkeley, California, would enable it to be heard for the first time and open the door for an endurable mix, in this instance between the signal and noise, something that can be addressed with current technology (Rosen, 2008).

Edison's sound recordings of a snippet of Handel's *Oratorio* in 1877 enabled performances to be played back or, for the first time (Rosen, 2008), recalled for later examination. This necessarily would have brought with it a new self-awareness, an aural mirror, which gave recording musicians a new ability to refine their own performances. In addition, others could now access and manipulate the recorded artifact with a degree of autonomy from the performers.

Around 1920, a young German composer, Stefan Wolpe, created a Dada provocation by simultaneously playing Beethoven's fifth symphony at different speeds on eight separate gramophone players. This was a conceptual development as, for the first time, a piece that combined several separate previously recorded elements was created. Wolpe possibly missed the opportunity of exploring the mixing of entirely different recordings, this being fulfilled three decades later in 1951 by John Cage's piece for twelve radios (Ross, 2013).

Cage was in step with the times as by then, the jazz guitarist, inventor and legend Les Paul was laying the groundwork for combining separate recordings of musically related parts. This was a significant paradigm shift as, for the first time, it became practical to consciously, and sequentially, combine the component parts of a piece.

Les Paul initially achieved this by the process of Sound on Sound recording. As early as 1949, Paul was getting results by switching off the erase head of a tape recorder to overlay parts on same piece of tape. An alternate method involved recording a performance to an acetate (later tape) and then playing along with it while recording to a second one. Paul replaced this technique by using a modified Ampex 200 tape recorder with a fourth head, enabling this process to be done using one, more portable machine (Snyder, 2003). One mistake in the performance or extraneous noise, however, and the process would need to be started from the beginning; 'How High The Moon', recorded in 1951, had to be recorded three times as the first two recordings were ruined by first a siren and then an overflying airplane (Buskin, 2007).

Les Paul also explored moving beyond representations of live acoustic performances, for instance, by recording his electric guitar at half speed so that, on replay at the correct speed, it would be one octave higher. Recording multiple vocal parts inspired both controversy and inspiration. Richard Buskin (2013) reports a conversation with Bruce Swedien, destined later to become a highly acclaimed mix engineer, about How High the Moon: "Up to that point the goal of music recording had been to capture an unaltered acoustic event, . . . (it) left no room for imagination, but when I heard 'How High the Moon', which did not have one natural sound in it, I thought, 'Damn, there's hope!'"

Sel-Sync (or Selective Synchronous recording), also conceived by Paul in 1953 (Petersen, 2005) and fabricated by Ross Snyder at Ampex in 1955, opened up a realm of new possibilities, although initially not being considered important by Snyder (Petersen, 2005). While monitoring previously recorded parts via the record head (henceforth known as the sync head), additional recordings could be made it time with them. Earlier recorded parts could be individually re-attempted or patched up with reference to what had been originally later recorded ones; with the Sound on Sound technique, subsequent recordings would have been lost.

The development of Sel-Sync development had another crucial significance: recordings could now be to be separately manipulated in a number of ways, such as attenuating or equalizing. Parts could be re-contextualized or even discarded.

The process of deferment of the final mix from the recording of the parts starts with Les Paul and the forward planning of his Sound on Sound recording. In Les Paul's case, this might be the time it took his wife, Mary Ford, to cook macaroni cheese for their dinner but can now span to decades as has been the case with the Beatles' 'Fly Like a Bird' (Roger, 1994). This ability to be being able to examine the work at different times is not trivial; perspectives change as the listening environment does, also with the weather, the time, the cultural context. A mix then can make sense in many different environments and at different times has the potential to resonate with a greater number of listeners.

In more recent decades, the use of automation has enabled the engineer to refine a process in a number of different ways, to go beyond relying on manual control or the time constants of outboard processes to effect dynamic changes over time and to apply a number of simultaneous changing processes to a single part.

Up until the advent of digital, there was a one-to-one correspondence between all links in the audio chain: sound pressure, microphone capsule displacement, electrical current, capacitor charge, tape magnetization and speaker cone displacement could all be mapped to each other by some bijective function. The introduction of digital techniques in audio, such as random access, digital filters and other mathematical techniques, enabled this correspondence to be broken. A raft of new processes of manipulation could now be employed. For example, Melodyne Editor plug-in's Direct Note Access, created by Peter Neubäcker, is able to change individual notes within polyphonic audio, something that would have been undreamed of just a few years ago (Celemony, n.d.). iZotope's RX5 software editor can access individual harmonics or areas of the spectrum within complex audio for processing (iZotope, 2016). Other processes that go beyond the traditional techniques of sound manipulation include Sound Radix's Surfer EQ plug-in (Sound Radix, 2015), which can track harmonics within a melodic monophonic part, and Pi plug-in, which can access hitherto opaque phase relationships between different parts to create a more phase-coherent mix (Sound Radix, 2015).

Artificial intelligence is now increasingly being used, not only for composition, but also for audio manipulation and can be utilized in the span of the recording process from reducing spill on drum kit microphones with Accusonics' Drumatom (Accusonics, n.d.) to the mastering of the final stereo mix with LANDR (LANDR, n.d.).

The scope of what is possible in the mix is practically unlimited. Given sufficient artistic permission, any audio result can be produced. It has not been unknown for re-mixers, the author included, to have work accepted that has nothing of the original recording left, a metaphorical spade with a new blade and a new handle. Whether it was his intention or not, the new possibilities that Les Paul created for the mix engineer have developed sufficiently so that the job is comparable to that of the performer.

The Platonic Mix

Where is all this increased sophistication leading to? Is there some perfect solution that can be aspired to? Although this term is often used, is it actually possible to create a 'perfect mix'?

Plato might have proposed that, like his eponymous solids, the ideals of which only exist in the transcendent realm of Forms, there exists in there a divine exemplar of the completed mixed work, a perfect piece of mixed music that physical reality aspires to but can only approximate (Banash, 2006).

Music experienced in altered states of consciousness, dreams or through near-death experiences (NDEs) suggests that there might indeed be something perfect that can be accessed. The latter music has been described as "transcendental, unearthly harmonic beauty, angelic, sublimely beautiful, exquisite harmonies, heavenly, a celestial choir of angels, a tone so sublimely perfect, joyous and beat-less melody, an orchestra of voices" (Williams, 2014). Pieces in such states can have the impression of appearing complete, seemingly before any apparent human input has been done to create it. The issue of where this music arises—in the mind of the listener or pre-formed elsewhere and witnessed—opens up fundamental questions on the nature of consciousness.

Incompleteness

Can these revelations of sublime music be authentically considered perfect? Is the concept of a perfect mix useful?

A possible solution to whether perfection is in fact possible comes from the Austrian-born logician, mathematician and philosopher Kurt Friedrich Gödel, whose two incompleteness theorems, which discuss the limitations of mathematical systems, have implications beyond the purely theoretical. Smith (2013: 3) reports that "Gödel's first incompleteness theorem shows that the entirely natural idea that we can give a complete theory of basic arithmetic with a tidy set of axioms is wrong". He goes on to summarize the second theorem thus: "nice theories that include enough basic arithmetic, can't prove their own consistency" (Smith, 2013: 6).

Recent research at University College London on the spectral gap (Cubitt et al., 2015) has demonstrated that Gödel's theorems do indeed have real-world effects, that certain material properties are 'undecidable'—they are neither true nor false (Knight, 2015).

Gödel's proofs have also been extended beyond the realm of mathematics to the theory of mind, along with the support of Alan Turing's 'Turing Machine' thought experiment. One major corollary is that the human mind can always find some aspect that cannot be contained within any ideal (machine in this context) (Anon., n.d.).

Gödel's first theorem is of particular interest, as it might be rephrased as stating that a self-consistent axiomatic system cannot be completed. A mixed piece of music might be considered such a self-consistent, or at least self-referential, system as meaning is derived from within the context of the combined elements. It might thus be postulated that there is, in fact, no perfect mix, or indeed any piece of art, because as soon as it is produced, a new perception can be applied to it, which negates its perfection.

Artistic Exemplar

So why could it be that the music experienced in other states of consciousness come across as so perfect if they are not actually so? Williams's NDE experience and following research (2014) leads him to comment,

> In the spirit realm, gardens sing and colors can be heard. It is a realm where light and sound, color and geometrical patterns are all combined into a totality of harmonic perfection. This is music that is on a level that is beyond hearing.

These experiences are rich in synesthesia, the involuntary stimulation of one sense by another and comparable links between the sensory experience and emotional responses. Perhaps these strong emotional links persuade the witness that what is being heard is on a more sublime level that anything in the physical realm. Alternatively, could it be that the emotions being experienced are more truly represented by the melodies, arrangements and textures experienced in these rarefied states of consciousness than by most of what can be created from everyday consciousness?

Wherever this music may be generated, some paragon of a completed work, even if not perfect, with associated emotions, does appear to be accessible, apparently instantaneously, to the human mind. By acknowledging its existence, even if not readily available, the mix engineer can view the process of mixing as a process of discovery as opposed to one of creation. Without entering an altered state of consciousness or having the refined discipline to mentally construct a complex audio artifact, how might this objective be revealed?

Unlike the musician, who must distill what they can of their lifetime's worth of art into the moment's performance, the mixer's art does not need to be similarly condensed. The same piece of work may be sequentially revisited, thanks to Les Paul's legacy, from a multitude of perspectives and interrogated using different mindsets, intuitions or emotions, here collectively termed as perceptions. It is not necessary to simultaneously be aware of all the various aspects of the artifact, but each dynamic can be sequentially addressed.

Holding the idea of what the target artifact might be will generate responses on mental, emotional or physical levels, and these can enable appropriate decisions to be made that may get closer to it. The perspectives chosen against which to assess the mix should be pertinent to the song, the artist, the genre and the age. The more perceptions that can be considered, the more refined the potential. Once dealt with, other previously addressed perspectives should then be checked to see if they are being adversely affected. There are not-insignificant dangers of dilution of what might be considered the essence of the piece and a resultant homogenized product that is a jack-of-all-trades and master of none.

It is tempting to view this as chipping away at the edifice of the work, as if working the facets of a crystal, incrementally closing in and gradually revealing the final product. What materializes, however, can end up surprisingly different from that envisaged at the outset. Why should this be?

Illusion

It is a part of human self-image to believe that we are witnessing the world as it happens. In 1870, the German physiologist Hermann von Helmholtz made measurements of the speed of the signals in nerves demonstrating that the observation will always lag the event (Zimmer, 2009). Furthermore, the awareness of our decisions lags the implementation of the effect of them; we act and then have the thought to do so (Haggard and Eimer, 1999). Our interpretation of the world is also entirely conditioned and filtered through our makeup and experiences so that we do not experience reality in any real sense. Many religious perspectives observe this, for instance in early meditations in *A Course of Miracles* such as "I have given everything I see in this room . . . all the meaning that it has for me" and "I do not understand anything I see in this room" (Schucman, 1996: 4–5).

The illusion of our senses includes musical or audio constructs. We cannot truly experience what is really there in its pure form—what we experience is more than just colored by our individual makeup, experiences, prejudices and tastes; it is entirely a mental fabrication, only a cartoon representation of reality.

Imbalance

The human trait of exploring that which is off the path of harmony and venturing into distortion and dissonance, may, at least in part, be attributed to imbalance in the functions of the two brain hemispheres. Far from being equal and opposite, there is a considerable variation in aptitude between them, the dominant left hemisphere being less adaptable, which has been demonstrated when it ceases to function properly and the right can take over (Gynn and Wright, 2008: 4–8). The two hemispheres of the brain can, in fact, operate independently from each other, as patients having undergone corpus callosotomy, a surgical procedure for the treatment of epilepsy, have demonstrated (Yonekawa et al., 2011). The imbalance between hemispheres creates a flow or conversation. Glynn and Wright (2008: 11) suggest that music helps restore the symmetry and reduce the internal chatter. Is the seeking of an appropriate sonic balance a reflection to find meaning in the hemispheric imbalance, in the skewed relationship with the harmony of natural processes?

Strange Attractor

The imagined pure form of the mix, the practically perfect mix, exhibits characteristics of a strange attractor in chaos theory. This can be viewed as a goal that, as has been shown, can never be reached, only orbited around. These particular orbits are chaotic, unlike the neat classical models of the planets orbiting the sun in fixed orbits. This was demonstrated in 1971 by Edward Lorentz, who showed that with only a few (three) degrees of freedom, a never-repeating infinite number of such orbital paths around such an attractor could be created (Gleick, 1988:139–140).

In considering the mix as such an orbit around the idealized, if not perfect, objective, the point at which it is deemed complete, the process of orbiting the target is stopped, sampled in what is known as a Poincaré, or return, map. This is the final mix, abandoned in its orbit rather than completed, the artistic moment crystallized.

Although sometimes close to the objective, sometimes the point of return, the mix outcome, can be radically different to that initially envisaged. The choice of processing used or the inability to match it with the desired intent may have created unforeseen consequences, but even minor changes in the choices made in the initial conditions or during the process can result in a significantly different outcome to that expected: this is like the popular idea derived from chaos theory of the butterfly flapping its wings causing a hurricane on the other side of the world. Professor David Pérez-García, co-author of 'Undecidability of the Spectral Gap' (Cubitt et al., 2015) comments, "the results show that adding even a single particle to a lump of matter, however large, could in principle dramatically change its properties" (Knight, 2015).

Limitations

There are, at any particular time, limitations placed on how deep or profound a perception can be achieved in both the technology and the human use of it.

Technical limitations are imposed by the quantization of audio data. Although large, the number of possible solutions to an audio problem using a digital format is finite. This may be familiar to engineers emerging from working with analog and finding certain sonic subtleties absent in digital. Advances, particularly with higher sample rates and bit depths, have ensured that this obstacle is being steadily eroded.

The limits of the current toolset is another limit to possibilities, although each new process generates an exponential increase in possibility as methods multiply.

The improvement in deepening perceptions that can elevate what can be communicated within a mix is perhaps being held back by the mix engineer's adhesion to old ways of working, not consciously engaging in seeking out new possibilities and applying them. Although the 'what' and 'how' are often well accounted for, interviews with mix engineers often reveal little of the reasons, the 'whys', that drive their decisions beyond something along the lines of 'because that gave the sound I was looking for'. Can intentionally focusing on a greater range of perspectives not only help illuminate the reasons for decisions but also enrich the final product, producing more facets on the metaphorical crystal of the mix?

Mixing Guidelines

There have been a number of attempts to create clear perspectives to establish the basics of a balanced mix.

Sherman Keene (Keene, 1981) sets out eight aspects that warrant attention: three frequency bands that need separate attention; appropriate and moderate effects; dimension—a sense of depth; motion—intelligent use of panning; at least one true stereo track; and some acoustic information.

In addition to getting these basics right, how may potentiality be improved?

Six Thinking Hats

Edward de Bono suggests using six different 'thinking hats' as an aide to better decision-making (Bono, 1990b).

The blue hat provides focus, defines the issue and sets out how to approach it; how the other thinking modes will be organized.

The white hat can help in setting the context with its area of facts and figures. What is the market for the mix? How much revenue can be made from it? How much time is there? In addition, facts about the piece belong here. What are the frequency ranges of the tracks? What is the song or piece about?

The black hat is concerned with what is wrong. Advised by white-hat thinking, it is dominant in the initial sorting of the piece. The tidying up, editing, compiling and related tasks are the focus here. In addition, the unweaving of tangled frequencies by reducing frequency masking and separation in the stereo picture belong here. Comparisons with other pieces, possibly in the genre, are also appropriate, as are previous successful strategies.

The yellow hat is concerned with positive assessment, "a mixture of curiosity, pleasure, greed and a desire to 'make things happen'" (Bono, 1990b: 110). The yellow hat provides the motivating force to get the job done. Alternative choices, processes or direction can be considered at this time.

The red hat deals with emotional responses, such as 'Hate it, get rid of it!' Although termed a hat, the choice of acceptance or rejection originates in the inductive grey matter of the gut, so reactions in that part of the body should be noted. This hat is also about shifting emotions: for instance, one man's spill is another man's ambience. Degrees of emotions can be managed with the red hat where language is often inadequate in the task.

The green hat is worn for creative thinking. This is the realm of 'lateral thinking', the creation of fresh possibilities, unexpected leaps of perception. This mode of thinking is also concerned with one of the most important ingredients in a mix: that of humor, as created by the quantum jump from one viewpoint to another. Stylistic references can be thought of in terms of this, as the listener is carried from one musical stream to another. It could well be argued that green-hat thinking is a part of what differentiates the engineer from the producer.

Time and budget pressures in the mix will often dictate that the problem-solving, black-hat thinking dominates the mix process. Beyond this, black- or red-hat thinking advises what balance needs to be addressed. This is done with additional yellow-hat thinking. Green-hat thinking opens up new considerations that can then be examined in the same way.

Chance

In aleatoric music, composition is to some degree determined by chance events; the same concept can be applied to sonic choices within the mix.

Gibson (2005: xxvi) suggests randomizing all settings on processing equipment to create a radically different solution which can provide unexpected insights.

Po

As a part of the aforementioned green-hat thinking, Edward de Bono created the word 'po' to describe the process of random juxtaposition, selecting a word at random and considering its relationship to the subject in question to throw up new perceptions (Bono, 1990a).

Eno's Oblique Strategies

The process of chance is also incorporated in the 'Oblique Strategies' record cards, originally created with Peter Schmidt and Brian Eno in 1975. These offer 115 different instructions of meditations designed to be selected at random and applied, or at least considered in relation to, the creative process. The set could be considered related to the Tarot, I-Ching or Runes, where contemplation of a seemingly random selection of their elements always appears curiously relevant to the subject in question.

Although used extensively by David Bowie (Hendrikse, 2013) and others, these suggestions are incomplete; indeed Eno intended others to make their own card systems and not just use his. However useful, a finite set of cards can only take the process so far. An analysis of the cards can help extend these processes.

Choice of Polarity

It can be seen that many of the Oblique Strategy cards deal with finding an appropriate level or instructing one of some parameter. This can also be considered as dealing with the balance of a duality, opposite or polarity or the appearance or state of a monistic aspect.

The Oblique Strategy card instructions that deal with polarities suggest working with the following: center, accretion, level of structure, completion, personality, cleanliness, extravagance, cascades, courage, decoration, self-indulgence, desire, time distortion, activity, ease, differences, repetitions, flaws, ghost echoes, comfort, humanization, glee, gradations, nobility, humility, credibility, intonation, absence, mechanization, idiosyncrasy, wholeness, uniqueness, change, heroism, ambiguities, consistency, radio-friendliness, insignificance, novelty and note density. This collection is perhaps an extension on Keene's nine principles for balancing, but obviously not an exhaustive list of parameters that can be considered.

In fact, as de Bono suggests with using the word 'po', random words selected for the dictionary can open up a galaxy of other parameters for consideration that could be argued to offer comparable insights to Eno's suggestions. For example, in a test run of twenty random words selected for this writing (eleven were discarded) the following words were picked:

1) Reluctance
2) Gratify
3) Extremist
4) Flock
5) Histrionic
6) Pusillanimous
7) Insomnia
8) Detraction
9) Distend

In the creative process, these could generate related instructions in the spirit of the Oblique Strategy cards, such as:

1) "Postpone the hook and replace"
2) "Start with the hook"
3) "Boost at all frequencies"
4) "Triple track"
5) "Show off"

6) "The small quiet voice"
7) "Repeat and increment"
8) "Remove
9) "Over arrange, extend and mute"

In the mix process, the same random selection of words could be reframed as polarities to be considered within it, for instance:

1) How much is being held back?
2) How much is being delivered / how satisfying is each section?
3) How balanced are the balances? (Should there be more shocks?)
4) Mix density
5) Histrionics
6) Small sounds
7) Boredom
8) How much has been removed?
9) How much overall information flow?

It is apparent that this process can be considerably further extended with other random selections.

Micro / Macro Focus

Focus may be shifted in order to seek more appropriate balances.

Parts may be considered self-referentially using obvious parameters such as level, dynamics, panorama, frequency and timbre. There should be caution applied, as examination of individual elements separates the listener from the whole and it is impractical to guess how a part will interact with others on its own. Plato appropriately commented, "The concentration of the mind on a single object, or on a single aspect of human nature, overpowers the orderly perception of the whole" (Jowett, n.d.). Phase relationships with co-incident sounds will modify the part's appearance in the overall mix, and aspects of it may be masked by other sounds with similar frequencies. Therefore, the relationships between parts, and between parts and any effects introduced, are of greater importance than examination of the components. The same parameters can be considered with particular regard to phase relationships and frequency masking.

Furthermore, the assembled piece's adhesion to the genre needs to be considered. What will satisfy the target audience is of great importance. The timelessness of the techniques and of the sound of the piece can be examined here; how closely are musical archetypes represented or musical memories recalled? Also, there should be awareness of the level of forward thinking. Can the mix be sufficiently ahead of the fashion to appear cutting edge and important without being so far ahead to appear unmusical?

Ultimately, the relationship that is the most important to consider is that between the listener and the music. The mix engineer is practiced in playing the role of the recipient; however, this can be examined in greater detail. How much attention should the mix ask of the listener? This will vary between genres from meditation music to avant-garde. How familiar should the mix feel? Too much or too little and it will not inspire a need to purchase it (if that is a source of funding for the project). A sufficient distance from what chimes with the listener's experience can inspire a need to own the product, to include it in their life and inspire others to follow suit. This is a way to create a hit and an income stream for those involved in the production.

Calculus

In addition, calculus may provide further insight. By considering the first derivative, rates of change are illuminated: for instance, how fast the piece is developing or the relationship between transients and sustains. The second derivative reveals acceleration, such as tempo changes, the intensification of the lead into the final choruses or use of breakdowns. Integral calculus can aid in the consideration of the accumulation of parts or spectral density.

Some analysis of groove can be achieved by considering calculus; however, considerably more needs to be addressed to optimize this. Discussion thus far has been limited to reactions created through mind, emotions and gut, but much music with a rhythmic component should address far more of the human system and, ideally, all body parts should be satisfied. There should be an awareness of, and even skill in, dance forms that might be associated with the genre, otherwise the result will not function in that form. If the dance forms are undefined, there should arguably still be some appropriate movement inspired in the mixer.

What Is Possible?

Are these dualities exclusive; is it indeed possible to separate the differing balances or do they affect each other? For instance, considerations of criteria such as warmth, distortion, timbral balance and weight will likely not be independent and the change in any parameter may have an impact on the groove. Can the dualities be considered new dimensions or finer gradations of the same ones or, in other words, how many degrees of freedom are available? Considering that the audio is physically only represented by number streams with as many degrees of freedom as there are channels at any moment, it can easily be concluded that there are indeed more degrees of freedom available; that is, it is possible for changes made in consideration of one perspective to be made without compromise of another. Potential limits imposed by the human auditory system would appear to not have been reached.

The quest for a set of criteria that have the greatest independence from each other will give us the greatest control. In geometric terms, orthogonal parameters will offer this. By considering the aspects of a sound in terms of width (X or the panorama), height (Y or frequency) and depth (Z or field), these orthogonal dimensions can be acquired. Changing the perspective from one set of coordinates to another reveals a different set of variables, for instance from Cartesian to polar. In this example, the distance from a central listening position could be considered instead. The use of Mid/Side processing can assist in this perception shift in the X-Y and the question of where the listener's 'home' position in the spectrum lies when considering if this shift to polar can also be applied on the Z-axis.

Numerology

How can the process of balancing polarities be further extended in finding the best solution in terms of the mix? Contemplation of other numbers and their significance might help in developing deeper insights. An examination of the integers up to nine with the esoteric art of numerology opens up further perceptions.

With the number one, opportunity, drive or inspiration can be thought of. What is about to take form? What are the possibilities?

In addition to the previously discussed dualities, sharing or partnering can be examined with the number two, such as with the classic dialogue between the main electric guitar and lead vocal of rock music.

Consider the three-fold concept of subject–verb–object of an action. What is the verb and how strong is it? How easily do ideas evolve from one space to another? What are the goals? Gibson considers three levels of dynamics of sound images: individual placement, patterns and movement (Gibson, 2005:152). Practical consideration of this number in connection with the work is that of the three desirable qualities, that of being cheap, good and on time; it might be considered that only two of these can ever be achieved for any particular project.

Four-fold processes can reveal the cyclical natures, such as those coded in seasonal changes. Consider what is repeating, where there is growth and decay. Also the 'elements' of form: earth, air, fire and water, or in Shakespearean terms: the humors of blood, phlegm, choleric and melancholy might be examined here. For instance, the element of earth can be concerned with foundations and stability.

With number five, the spirit, aether, quintessence (Greek) or akasa (Sanskrit) can be added to the four physical elements. Where is the life in the mix? Also, what gets added to a repetition? Gabriel Roth's 5Rhythms is a dance-orientated process that can explore five-fold processes including this concept (5Rhythms, n.d.). Five physical senses can be considered.

A six-fold process can be regarded as the balance of opposing processes, as symbolized by the two upward- and downward-pointing triangles that make up the Star of David. Both Eno and de Bono offer paths to insight by reversing the normal flow of the work. Harmony and communication can be considered here.

Contemplation of the number seven reveals the full sequence that completes an octave jump an addition of one octave or a doubling of frequency. This number invites consideration of how musical notes are spread.

Number eight can attend to contradictions, consequences and also relationships in time, such as what is changing or unchanging, movement.

Number nine offers an insight into what is catalytic in a process, what arrives and transforms one form into another, such as a bridge section to a chorus. What is concluded, reaches a completion?

Paul Carr (2012) suggests prioritizing the following ten 'elements' of music: melody, harmony, lyrics, form, texture, tempo, meter, timbre, dynamics and mix. As an example, he points out that for Bob Dylan, the lyrical content could figure higher than the mix, implying that the latter should be less obtrusive.

Gibson (2005) identifies eleven components of a great recording: concept, melody, rhythm, harmony, lyrics, density, instrumentation, song structure, performance, equipment quality and the mix, and suggests creating a hierarchy for them for a particular piece.

The Major Arcana of the Tarot, which is developed from numerology, goes further with describing processes or situations up to twenty-one. The Minor Arcana is perhaps a less useful tool, though not to be ignored.

Conclusion

Inviting in new perceptions on how to assess the mix increases the level of sophistication available to the mix engineer to refine performances. The mixer can then more accurately account for imbalances in the human condition and, by more accurately resonating with them, connect with the listener sufficiently to inspire them to make an investment into the product.

This new power calls for those with the purse strings to allow for new approaches to the mix, such as those suggested here, to be more fully explored and not to be satisfied with just the antecedent working methods. The investment in this part of the production process should result in a return as the audience engages more fully with the product. Holistic awareness from mix engineers is also called for so that they may be able to effectively engage in considering new methods.

Bibliography

5Rhythms. (n.d.). 5Rhythms | *Gabrielle Roth's 5Rhythms*. [online]. Available at: http://www.5rhythms.com/gabrielle-roths-5rhythms/ [Accessed: 22 December 2015].

Accusonics. (n.d.). Achieve the Drum Sound of Your Dreams. *Drumatom: The World's First Drum Leakage Suppression Tool*. [online]. Available at: http://drumatom.com/ [Accessed: 6 January 2016].

Anon. (n.d.). *Minds, Machines and Gödel*. [online]. Available at: http://users.ox.ac.uk/~jrlucas/Godel/mmg.html [Accessed: 3 December 2015].

Banash, D. (2006). *Plato's Theory of Forms*. [online]. Available at: http://www.anselm.edu/homepage/dbanach/platform.htm [Accessed: 28 November 2015].

Bono, E.D. (1990a). *PO: Beyond Yes and No*, new edition. London: Penguin Books Ltd.

Bono, E.D. (1990b). *Six Thinking Hats*. London: Penguin Books.

Buskin, R. (2007). *CLASSIC TRACKS: Les Paul & Mary Ford 'How High The Moon'*. [online]. Available at: http://www.soundonsound.com/sos/jan07/articles/classictracks_0107.htm [Accessed: 11 November 2015].

Buskin, R. (2013). The Clash 'White Riot' Classic Tracks. *The Clash 'White Riot' Classic Tracks*. [online]. Available at: http://www.soundonsound.com/sos/oct13/articles/classic-tracks-1013.htm [Accessed: 23 September 2014].

Carr, P. (2012). The Elements of Music: A Good Start to Basic Analysis. *Paul Carr*. [online]. Available at: http://paulcarr.org/2012/01/27/the-elements-of-music-a-good-start-to-basic-analysis/ [Accessed: 10 December 2015].

Celemony. (n.d.). *Celemony | What is Melodyne?* [Online]. Available at: http://www.celemony.com/en/melodyne/what-is-melodyne [Accessed: 6 January 2016].

Cubitt, T.S., Perez-Garcia, D. and Wolf, M.M. (2015). Undecidability of the Spectral Gap. *Nature* 528 (7581): 207–211. [Online]. Available at: doi:10.1038/nature16059.

Gibson, D. (2005). *The Art of Mixing: A Visual Guide to Recording, Engineering and Production*, Second revised edition. Boston, MA: artistpro.com LLC.

Gleick, J. (1988). *Chaos: Making a New Science*. New York: Penguin.

Gynn, G. and Wright, T. (2008). *Left in the Dark. Kaleidos Press*. [online]. Available at: https://books.google.co.uk/books?id=vOj_GZ2–1dEC.

Haggard, P. and Eimer, M. (1999). On the Relation between Brain Potentials and the Awareness of Voluntary Movements. *Experimental Brain Research* 126 (1): 128–133. [online]. Available at: doi:10.1007/s002210050722.

Hendrikse, W. (2013). *David Bowie—The Man Who Changed the World*. London: New Generation Publishing.

iZotope. (2016). *RX 5 Audio Editor: Noise Reduction & Audio Repair Software*. [online]. Available at: https://www.izotope.com/en/products/audio-repair/rx/ [Accessed: 6 January 2016].

Jowett, B. (n.d.). *Ion, by Plato*. [online]. Available at: http://www.gutenberg.org/files/1635/1635-h/1635-h.htm [Accessed: 11 November 2015].

Keene, S. (1981). *Practical Techniques for the Recording Engineer*, Second edition. (s.l.): MixBooks.

Knight, G. (2015). Gödel and Turing Enter Quantum Physics. *UCL-CS*. [online]. Available at: http://www.cs.ucl.ac.uk/computer_science_news/article/goedel-and-turing-enter-quantum-physics/ [Accessed: 23 December 2015].

LANDR. (n.d.). LANDR: Instant Online Audio Mastering Software. *LANDR*. [online]. Available at: https://www.landr.com/en [Accessed: 6 January 2016].

MacKinnon, E. (2012). Edison Voice Recording Is Old, but Not Oldest. *LiveScience.com*. [online]. Available at: http://www.livescience.com/24317-earliest-audio-recording.html [Accessed: 6 January 2016].

Petersen, G. (2005). Ampex Sel-Sync, 1955 | *Mixonline*. [online]. Available at: http://www.mixonline.com/news/facilities/ampex-sel-sync-1955/367111 [Accessed: 15 December 2015].

Roger, C. (1994). *'Free as a Bird' Emerging from Beatles' Studio Reunion—Tribunedigital-Thecourant*. [online]. Available at: http://articles.courant.com/1994–04–05/features/9404050734_1_lost-lennon-tapes-homing-bird-mccartney-and-harrison [Accessed: 21 November 2015].

Rosen, J. (2008). Researchers Play Tune Recorded Before Edison. *The New York Times*, 27 March. [online]. Available at: http://www.nytimes.com/2008/03/27/arts/27soun.html [Accessed: 20 November 2015].

Ross, A. (2013). Beethoven Dada. *The New Yorker*. [online]. Available at: http://www.newyorker.com/culture/culture-desk/beethoven-dada [Accessed: 20 November 2015].

Schucman, H. (1996). *A Course in Miracles Combined Volume*, Second edition. New York: Viking.

Smith, P. (2013). *An Introduction to Gödel's Theorems*. Cambridge: Cambridge University Press.

Snyder, R. (2003). Sel-Sync and the 'Octopus': How Came to be the First Recorder to Minimize Successive Copying in Overdubs. *ARSC Journal* 34 (2): 209–213.

Sound Radix. (2015). Sound Radix. *Sound Radix*. [online]. Available at: http://www.soundradix.com/ [Accessed: 6 January 2016].

Williams, K. (2014). *Music and the Near-Death Experience*. [online]. Available at: http://www.near-death.com/science/research/music.html [Accessed: 21 November 2015].

Yonekawa, T., Nakagawa, E., Takeshita, E., Inoue, Y., Inagaki, M., Kaga, M., Sugai, K., Sasaki, M., Kaido, T., Takahashi, A. and Otsuki, T. (2011). Effect of Corpus Callosotomy on Attention Deficit and Behavioral Problems in Pediatric Patients with Intractable Epilepsy. *Epilepsy & Behavior* 22 (4): 697–704. [online]. Available at: doi:10.1016/j.yebeh.2011.08.027.

Zimmer, C. (2009). The Brain: What Is the Speed of Thought? | DiscoverMagazine.com. *Discover Magazine*. [online]. Available at: http://discovermagazine.com/2009/dec/16-the-brain-what-is-speed-of-thought [Accessed: 21 December 2015].

2

How to Listen, What to Hear

William Moylan

Introduction

Perhaps nothing could be more important to *Mixing Music* than listening and hearing. To accurately hear the dimensions of the mix and the many unique sonic qualities of recordings we need to develop refined listening skills.

Learning to listen does present significant challenges, though. Listening is a highly personalized activity—internally processed in isolation, individualized by experience and physiology, shaped by perception and cognition. Add in the absence of a vocabulary for sound and being confronted with the learning of unimaginable experiences, the challenge grows.

Still, listening for mixing is a process that has been learned by others and that we can learn. The listening process can be developed with knowledge, awareness, intention and diligence. The listening process is one of active awareness and concentration; one of being open to possibilities of the unknown at any moment; searching with intention for new information or holding sounds in the focus of attention. Identifying the purpose for listening guides the process and informs the activity.

Successful listening brings accurate hearing. In successful listening, the elements of sound, levels of perspective and functions of the materials are recognized. Nearly all of these qualities are unique to audio recordings and the mix; they do not exist in the same way in nature or do not exist at all in natural acoustics. Thus, we must learn what these qualities are (or what might be present in the mix) in order for these qualities to be experienced. We will explore these qualities in detail and learn what they contribute to the mix. This all allows for successful hearing and will bring an understanding of the mix, in all its subtleties and complexities, dimensions and sonic qualities.

Learning What to Listen For . . .

As beginners, our first mixes invariably fall short sonically in a great many ways. We persevere and redo them, or create others, and those next novice attempts also typically fail, often miserably. Sometimes we recognize the shortcomings in frustration; sometimes we are simply overwhelmed and confused. Each time we listen it sounds different, but definitely not better—and we can't determine *what* about the sound is 'wrong'. We don't know what to hear, or how to listen to begin to hear. We just know the mix is not sounding right; is not working for the music. There is so much to hear in mixing music; so much to keep in mind, so many dimensions of sound to shape—most of which is beyond our prior experience.

In our first experiences listening to mixes, both our own or those of others, we instinctively use our prior listening methods. Our attention is drawn within the music, or to the performance, or to following the story of the text—but rarely toward hearing the mix. The mix itself shapes the music and performance in a great many ways, but its sounds are largely unheard by the vast majority of music listeners. It is understandable—beginners have not experienced the subtleties of the mix within the sounds of records, but that must change. To mix music, we must learn to hear the mix's dimensions and sound qualities; we must learn what they are, and we must experience them.

To understand mixes, and to craft our own mixes, our listening process engages sound in many different ways—some of which are counterintuitive and some contradict prior listening experiences. In a sense, we need to re-learn how to listen in order to hear all of these dimensions of the mix.

Our listening process is most naturally drawn to delineating the mix by the vocals and musical instruments. So we can productively begin preparations for the listening process by recognizing all of the sound sources that are within the recording, that are within the mix. In the listening process that ensues, we will seek to hear:

- How the sound sources relate to one another,
- How the sound sources relate to the whole, and
- The many qualities of each individual sound source.

These three concepts are critical to remember to understand the nature of the music mix. These relate to levels of perspective, or the level of detail at which one is listening. All instruments or voices fit into the mix at various levels of perspective, and function differently within different levels (see Table 2.1; Moylan, 2017).

All instruments or voices also fit into each of the different dimensions of the mix. These dimensions of sound are represented as artistic elements and provide the mix with great richness. Some of these dimensions are extensions of the traditional musical materials of pitch, dynamics and rhythm, while the elements of sound qualities and spatial properties often bring sound relationships and concepts not found in nature. A summary of artistic elements appears in Table 2.2. These are important dimensions that we will come to recognize and learn to hear, through learning to listen in new ways (Moylan, 1992).

As we experience these elements, our perception and understanding awakens. As we learn to navigate listening at various levels of perspective, we become aware of the richness of the mix and of how sounds and

Table 2.1 Levels of perspective and related levels of detail

Level of Perspective	Level of Detail
Highest Level, Large Dimension Overall texture	Overall, primary shape Overall character and characteristics Big picture, overview Experience of all individual sounds coalescing into one blended sound
Upper Level, Middle Dimension Composite texture of the activities of all individual sound sources	Relationship of all sound sources as being equal Detachment from individual sound sources Level where prominence and balance relationships can be accurately judged
Middle Level, Middle Dimension Individual sound sources in the mix	Overall characteristics of individual instruments or voices and their musical materials Typical perception, surface-level detail of life experiences Relationship of self to the world, and to others Perceived relationship of sound source to the rest of the mix inaccurately places prominence on the source
Lowest Levels, Small Dimension Characteristics of individual sounds Small-scale activities of elements	Detail below surface level of sound sources and musical materials Detail is heightened, exaggerated Focus on the characteristics of sounds Microscopic details

Table 2.2 Artistic elements as dimensions of the mix

Artistic Elements

(Unique to or More Prominently Shaped in Music Recordings)

Pitch Levels and Relationships—register, range, pitch areas, pitch density, timbral balance

Dynamic Levels and Relationships—reference dynamic level, program dynamic contour, musical balance, dynamic contour/shape of musical ideas, dynamic contour of individual sounds

Rhythmic Patterns and Rates of Activities—tempo, rhythms, patterns of durations in all elements

Sound Sources and Sound Quality—timbral balance, pitch density, arranging, performance intensity, performance techniques; sound qualities and timbre of sound sources, component parts of timbre/sound quality

Spatial Properties—stereo location, surround location, phantom images, moving sources, distance location, sound-stage dimensions, imaging, environmental characteristics, perceived performance environment, space within space

elements work on various levels. We begin to understand how to balance and delineate sounds, not only by loudness but also by shaping all of the elements; we learn about prominence and significance, intense focus and open listening (Moylan, 2015).

From this brief examination of what we are trying to hear, let's move forward to explore how to listen.

How to Listen

The following sections are for learning to listen. Building our listening skills will lead us towards hearing—perceiving and recognizing—the elements that create dimension in the mix. With diligence, we will learn to listen deeply. In thoughtfully learning to listen, we can learn much about ourselves, and our unique relationships to sound.

Listening Is Personal . . .

Listening is a highly personalized activity. Many aspects of the listening process are unique to each individual. With a cursory glance at another, one identifies the individual's unique physique, experiences, knowledge, cognition, perception and more.

The unique physical qualities of individuals shape sound and its perception. Beginning with the dimensions and shape of the head, shoulders and pinnae, sound is modified by the body of the listener in subtle ways that are different from all other people. The sound qualities reaching the hearing mechanism in one individual are not precisely the same as those reaching others. These differences proceed in becoming greater as we factor in differences between the two ears and other physical attributes of our hearing mechanisms (B. Moore, 2012; Deutsch, 2013).

In addition, we all come to listening with our own set of experiences. We each have prior listening experiences that uniquely shape our ability to engage sound and musical materials. We each have our own set of experiences, brought by the great many but uniquely defined events and activities in our lives (Clifton, 1983; Thompson, 2015). Our listening skill levels improve with more experience and increasing exposure, especially once the process has been informed, guided and directed with the knowledge of what might be present to be 'heard' (Polanyi, 1962).

These experiences also shape how we process sound and perceive sound. We each have our own relationships with sound, our own ways of making sense of sound (beyond language) and of understanding or identifying the qualities of sound. We all gauge pitch in our own way, we all tend to recognize some pitch levels more deeply than others whether or not we have pitch recognition (Levitin, 2006; Roederer, 2008; Thompson, 2015). We can all calculate time relationships accurately by using our individual memories of the tempo of songs we have found meaningful (Butler 1992; Snyder, 2000). We all can recognize thousands of timbres, many unique to our personal lives (Schaeffer, 1966; Moylan, 2015). Further, we react physiologically to sound in ways common to all and in ways that are subtly unique to each of us (Levitin, 2006; Thompson, 2015).

We can learn to access our unique ways of perceiving and understanding sound. Through this, we can establish a listening practice that is both effective and relevant.

Learning to Listen . . . Is Uniquely Complicated

Learning requires a willingness to fail and is often uncomfortable. Learning requires admitting to not knowing what is to be learned and being receptive to something new. Adding to this, learning to listen requires a commitment to trust that what is not known is actually present—that this sound quality that has never been perceived, never been experienced, actually exists.

If you have never experienced a certain sonic dimension, how do you know what to listen for? How might you come to hear it? Where do you take your attention to recognize a sonic dimension that is beyond your imagination?

We all know it is entirely possible to listen to a recording and not hear certain sound qualities that are very pronounced to a skilled listener. Some faith that something is present that has yet to be experienced needs to be present for learning to begin. Leaning in to the unknown needs to be cultivated, as well as allowing oneself to fail safely and constructively (this is especially important in beginning work).

This whole process of developing a new skill is particularly difficult for learning listening. We are challenged to discover aspects of sound that defy previous experience, to use a new and specific type of attention while seeking dimensions only known by descriptions. And our language on sound challenges us further. How can we describe the aspects of sound we are learning to hear? To do so with clarity and precision is not a simple task.

We do not have a vocabulary for describing sound and its specific dimensions. It is simply not part of our language. Instead, our custom is to describe sound by using analogy and by using the terminology of the other senses. We resort to words such as 'warm', 'dark', 'smooth', 'mellow', 'edgy', 'bright', 'crisp' and a *great* many others—imprecise at best, and typically grossly inadequate and ineffective; often misleading, and commonly merely meaningless jargon. If one has never heard a sound or sound quality, it is nearly impossible to accurately imagine it by having it described with such a vocabulary. Still, we try, and certain terms become widely used, though they mean different things to different people—though they provide the illusion that the user has some inside knowledge.

As one example (of many), the term 'punchy' is currently used in many contexts in a variety of styles of music and recordings. The term might communicate a general sense of energy and intensity, but tells of nothing about sound. Still, most budding recordists would be quick to say they know what a 'punchy sound' is—though each would define it differently, given the means and skills to accurately describe their idea of the qualities of a 'punchy' sound. They have experienced recordings that others have described as 'punchy' so they believe they know what 'punchy' is—though the person who used that term in the first place may not have a clear idea of why they used the term.

Objective and more functional communication about sound is possible, however. It does take some knowledge of the physical dimensions of sound and some focused practice and skill in identifying the characteristics of the dimensions. Meaningful communication is possible, but it takes attention to develop and some discipline not to resort to old patterns (Moylan, 1992).

The states and activities of the physical dimensions of sound can be described, as they travel through the duration of the sound. A dynamic envelope can be described by its contour and levels against time; spectral content can be addressed in specific terms of what partials are present when, and the contours and levels of their individual amplitudes. All other aspects of sounds can be addressed with equal precision and detail by describing their unique dimensions and activities against time (Moylan, 2015).

This brings us to recognize: meaningful communication about sound—with sufficient accuracy and depth to be of use in learning to listen, in understanding recordings, and in audio production work—requires considerable skill and knowledge. Thus, talking about sound in a meaningful way can only be accomplished if one is sufficiently prepared (Schaeffer, 1966).

Aids for Learning to Listen

Certain background studies will prove invaluable for recognizing what is heard. This provides resources to draw upon to understand the concepts, bringing greater ease toward ultimately being able to experience the sound qualities of recordings. For learning the process of listening, with the goal to hear the unique qualities of audio, one will be well served with some background in

- Acoustics
- Psychoacoustics
- Dimensions of sound in audio recordings
- Music studies

A background in the first three establishes a framework to talk objectively about the physical characteristics of sound, taking into account how the sounds are transformed within our perception. Sounds can be conceptualized by what is physically present in the sound, as one perceives it. This is an experience all listeners will have in common, for most aspects of sound. Though we will likely not be very accurate at first, our descriptions will share objective information of the actual sound and its characteristics—not transferring the sound experience to touch, taste, sight or smell terms, or worse. Different levels of accomplishment in the above areas will bring different levels of detail to communication and commensurate precision to the descriptions (Moylan, 2015).

This way of describing sound can be applied to acoustic sounds, and then later be directed toward recognizing the unique qualities of recordings.

How music studies can prove useful should become clear in the following section.

Beginning to Hear . . .

Learning to hear these sound dimensions will require some new listening skills. These listening skills will be directed toward identifying:

- Frequency and pitch levels and activity
- Amplitude and dynamic levels and changes of level
- Time judgments and rhythmic activity

Frequency, amplitude and time considerations are separate and distinct from pitch, loudness and rhythm. With some experience in these six dimensions, timbre and spatial aspects can be engaged.

A number of resources are available for learning frequency and amplitude recognition, including Dave Moulton's *Golden Ears* (1995) exercises. A comprehensive set of exercises and guidance from *Understanding and Crafting the Mix* (2015) by this author provides a sequential resource for the physical dimensions of frequency, amplitude and time, and also the music-related percepts of pitch, dynamics and rhythm; related studies in timbre and spatial dimensions are also included, as well as guidance in improving musical memory.

Drawing upon one's previous experiences can also support the development of these skills, and others. For instance, many of us have learned to perform an instrument by mirroring performances on recordings; many others have learned pieces of music from listening to recordings. These activities develop and refine a sense of pitch, rhythm, dynamics and expression (timbre shaping, etc., from performance intensity) that can be applied to the process of understanding sound qualities and relationships. This activity supports the development of many listening skills, as it can be a resource in understanding subtleties of many dimensions of sound, as well as bringing facility to traditional musical materials.

Similarly, if we have experience with traditional music dictation, that skill will prove valuable in this context of learning to hear the unique dimensions of recorded sound. Music dictation can pull the listener into learning how to follow sound changes over time and to write down that activity. Further, the act of taking music dictation develops memory for musical materials, which will lead to developing memory applicable to all other listening procedures that occur in the mix. Importantly, the act of music dictation engages writing down what was heard, making it available for reflection or for execution at a future time.

Of course, this skill utilizes musical notation and has limited application to sound in audio and to the mix process. Another notational system for writing the dimensions of sound, and for documenting dimensions of recorded sound and the mix, need to be found. X-Y graphs can be adapted for this purpose. Observations of many aspects of sound can be plotted as X-Y graphs. This can be helpful to track how an individual dimension of sound might exist over time. In Figure 2.1, we can observe the component parts of a synthesizer timbre against time. This is the Moog synthesizer glissando that appears at 0:12, ending the introduction to The Beatles' 'Here Comes the Sun' (*Abbey Road*).

X-Y graphs can now be used to support descriptions of sounds, supplementing objective language with visualizations of data. Describing sound might thus become more direct, articulate, meaningful and effective. Verbal description becomes more focused and universally understood, especially when supported by the graph's plotting of the sound's characteristics.

The X-Y graph is highly adaptable for all of the elements of sound we might seek to teach. Further, it can be used at any level of perspective, from showing the smallest detail to depicting the overall shape of any element's material. Following is a table of graphs that are central to this approach; each is dedicated to a single element and focused at a specific level of perspective (see Table 2.3; Moylan, 2015).

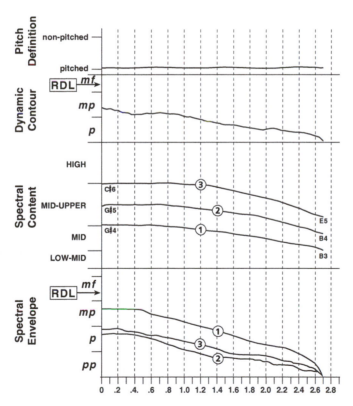

Figure 2.1   Sound quality evaluation graph of a Moog synthesizer sound from 'Here Comes the Sun' (Moylan, 2015)

Table 2.3 Primary X-Y graphs for artistic elements at various levels of perspective

Element	Perspective	Graphs of Mix Dimensions
Pitch	Individual Sound	Pitch Area [Graph]
	Individual Source	Melodic Contour
	Individual Source	Pitch Density
Dynamics	Composite Texture*	Musical Balance
	Composite Texture	Musical Balance vs.
	Overall Texture	Performance Intensity
		Program Dynamic Contour
Rhythm	*All levels*	*Implied in all graphs*
Sound Quality	Individual Sound	Sound Quality (Timbre)
	Overall Texture	Timbral Balance
Spatial Properties	Individual Source	Environmental
	Composite Texture	Characteristics
	Composite Texture	Stereo Location
	Composite Texture	Surround Location
	Overall Texture	Distance Location
	Overall Texture	Sound-Stage Dimensions
		Perceived Performance
		Environment

* Composite Texture of the activities of all individual sound sources.

Creating these graphs draws one into the listening process directly, and deeply. It brings detailed discoveries into what makes each mix unique, and how they are shaped, as well as developing listening skills.

Engaging Listening, Listening Deeply

The process of listening begins with intention. When we bring intention to listening, we have a sense of purpose and direction for our attention and for what we are seeking to find; this can be a sense of what to focus upon, or a sense of how to approach searching for information. We learn to bring awareness to specific musical materials or ideas, to specific elements or dimensions of sound, to bring our focus to specific levels of detail, to conceptualize a level of perspective where sounds can be perceived as equal in prominence in order to gauge their relationships correctly, and so much more.

Listening with intention allows one to plan what information is sought, and to take one's attention to that task solely. Certainly, a great many types of information can be extracted from what is heard at any moment; intention allows us to recognize that all of the information cannot be extracted all at once, and that each listening has a specific purpose and may bring attention to any specific aspect of sound.

Listening for one set of information is especially important at the beginning. This provides the luxury of listening to only one aspect of sound, ignoring all others. Limiting the listening to one topic will more quickly and directly improve recognition of that element, and it will develop the disciplines of focus and attention. It also embraces the reality that it is not possible to hear several things simultaneously. Instead, there will be repeated hearings of the material. Listening repeatedly to the same material produces new results, as results are checked and observations refined, or as the listener's intention shifts to another element.

We will begin each listening experience with an agenda of intention; the more specific the intention, the more effective the effort. This involves directing attention correctly at every stage of the process. Attention can be directed to a specific level of detail, a specific type of information and specific aspects of sound. This is part of defining intention.

We listen and process information at various levels of detail. At each level, sounds have certain characteristics, coalesce into a unique texture and have certain types of relationships to other sounds. Certain qualities might appear at one level and be significant, and not be relevant at others. Identifying a specific level of perspective for each listening is important for defining its intention. The level of detail might be most readily approached, especially during early learning, as 'middle dimension' levels and 'large dimension' and 'small dimension' activities (LaRue, 1992; Moylan, 2014). Table 2.1 provided some definition to these levels.

As skill develops, the process becomes more involved, and we engage and contrast critical listening and analytical listening issues. We listen for several types of information, related to the context of the material. Some information will be related to the message of the recording (musical materials and their relationships) and some information will be unrelated to context (such as the sound characteristics of a microphone or its placement). Knowing that a sound's qualities are being examined in isolation (out of context) or in relation to other sounds (in context) brings the learner deeper understanding and often a more direct way to listen to the material. Listening with intention continues. This alternation of critical versus analytical listening practice can strengthen the sense of intention and focus in the developing listener, particularly once control is becoming established. Until control of material is gained, listening with intention brings greater awareness of each aspect studied, with its dimensions and level of perspective, and develops the listening discipline to not be distracted by or pulled to other aspects of sound (Moylan, 2015).

As a product of this process, sound memory is improved. Encouraging listening as a separate process from writing brings us to remember what is heard. Further, it is simply not possible to write and listen simultaneously. Separating the two deliberately improves memory and brings greater understanding to the material.

Once one is able to track each element of sound accurately, it is possible to move to engaging several elements. This process does not seek to hear the qualities of several elements simultaneously. Instead, a deliberate alternation between elements brings control to the process and greater awareness to the material. After gradually improving and engaging more sophisticated alternations of materials and levels of perspective, a time will arrive when one can have an open listening field, scanning and listening to discover unique or important sound qualities or sound relationships. Ultimately, the listening process is one of either controlled exploration or of specific focus, depending on the situation and purpose.

Listening Without Expectations . . .

The process of listening requires an open mind; a mind ready to engage as relevant all that comes along, without expectations or personal desires. This includes not having preconceived ideas for what will be heard and of what we expect to hear in the recording. When listening to determine what we believe to be present, we can imagine it when it is not present; worse, we usually miss what actually *is* present.

It is common to listen *for* something, and to miss what actually occurs. We can do this in conversation, when we hope to hear someone say what we want to hear so intensely, that we hear it, although it is not stated; or do not hear it when it is not stated precisely as we had hoped or expected. In music this can happen similarly, as our minds find diversion in twisting musical lines in ways not actually present in the song. In audio we can miss aspects of sound that are unexpected as well; we anticipate a certain sound quality and can imagine it present even if it is not, we can misplace our attention and not perceive certain sound characteristics, the mind can take the listening experience in unimaginable directions of misperceptions based on our prior listening experiences, and so much more (Handel, 1993).

These all work to distort the listening experience, often substantially for beginning listeners (and not-so-beginning listeners). This problem can show itself prominently in accurately hearing the mixes of existing recordings and in the many dimensions of the mix process itself.

How do we cultivate listening with a purpose, while listening without expectations?

The successful listener is open to the possibility that any sound, sound quality or sound relationship might take place, at any moment; that this could be in any aspect of sound, and at any level of perspective. They are aware that this sound material might be or might contain something they have never before experienced—in any manner of form. They listen without expectations to all sound material, even what might not be the most important aspect of sound, as it is often still significant. In our listening, we use this concept of equivalence to establish awareness that any aspect of sound, at any level of perspective, may contain some quality or relationship that is significant and that is worthy of attention (Tenney, 1986).

It is further understood that significance does not relate to importance; the least important aspect of sound to the mix may still contain qualities that are significant in one way or another (such as a barely audible sustained synth chord, or the rattle in a rarely played tom drum). Likewise, significance is not interrelated with prominence, as a significant sound does not need to dominate the listener's perception.

Listening without expectation allows one to be receptive to the unexpected and the unknown, ready to engage sounds or aspects of sound never before experienced just the same as those known and expected. It allows for accurately hearing and understanding something new, something unexpected—or recognizing and evaluating something known. This openness to all aspects of sound will allow things to be detected and embraced that otherwise might not have been heard. From this position, new creative ideas or solutions might be discovered, or minute degradations in the signal path detected; from this position, a sense of engaging all dimensions of the recording will be held.

From this position of equivalence, we engage listening to all levels of perspective with equal attention, to all dimensions of sound as equally worthy of our attention, to all critical and analytical information as potentially significant in their contributions to the whole.

Listening Without Hearing . . .

Listening, that is, the deep listening required of audio, is hard work. It demands one's undivided attention, and it requires focused concentration. We expend energy to keep the mind engaged. As time progresses, attention is quick to suffer.

Remaining focused on listening is not easy. Our minds wander—a lot. We can learn to bring awareness to this difficulty by reminding ourselves very regularly and frequently to remain focused while listening. While we can be quick to point at the many distractions of modern life making paying attention more difficult now than ever before, difficulties in keeping focused attention have always existed as part of the human condition (Nhất Hạnh, 2015). Indeed, it is part of our human nature and conditioning to constantly shift our attention; we do this in a great many ways, and this often serves us well—as we suddenly remember to do something important, just in time.

Here, let us remember that sound is a memory.

While we *listen* during the passage of time, we *hear* backwards in time, considering what has happened. We do not know the duration of a sound until it stops, we do not know the shape of a melodic line until it has cadenced, we do not know the message of a song until it is over. Accurate listening engages not judging or reaching conclusions until having listened completely and considered fully what has happened. All of this happens as a by-product of engaging an awareness of what is happening *now*, and retaining (making a memory of) that experience to form larger experiences and to compare one experience to others (Snyder, 2000; Moylan, 2015).

The wandering mind steps in the way of this process—in a great many ways.

Often, our attention strays to switching between listening, music concerns and the production process. This blurs our perception, focus and the intention of listening—and greatly reduces our productivity. A few common examples are

- Musical materials that are particularly interesting, or distasteful
- Sounds one recognizes from other contexts

- Sounds or relationships that are particularly interesting or attractive
- Trying to figure out how a sound, or a mix (etc.), was created
- Being absorbed by the music or performance
- Being drawn to a certain line, such as bassists listening to the bass part
- Appearance of unique sounds or sound relationships (such as shifts in the mix)

The process of listening has many distractions. We seem to have no limit to things that can pull our listening off task and bring something off topic into the center of our attention.

In an instant, the focus of attention can shift from what is being heard to what to have for lunch; from the timbre of the snare drum, to getting our car's oil changed. Random thoughts and daydreams drift through unnoticed as you listen intently to the edit points in your comp track—and miss them, for the fourth time. Remembering to pay attention, we work through fatigue and distractions to stay on task. And we fail—quickly and consistently; we fail to keep the focus of our attention. So, we should learn to recognize this will happen, that this is normal. We can learn to recognize when it happens and simply start over. If we acknowledge that this *will* happen, we are more apt to recognize it when it *does*.

Now, let's examine, and begin to hear, the dimensions of a mix.

What to Hear: Hearing the Dimensions of the Mix

When we are learning to mix, we are learning to balance prominence. We bring the focus of our listening attention to individual sounds, and begin to compare those sounds to others. Some skilled mixers will begin with all instruments and voices present, then begin adjusting; many skilled mixers will begin with specific sounds and build their mix; hybrids of each are used by various individuals as appropriate to the project. However it is approached, the mix unfolds in this process of bringing balance to the prominence of instruments and voices.

When we listen to a mix, some sounds or musical ideas are more noticeable than are others. Our attention is drawn to them. Or perhaps they stick out. For one reason or another, they are prominent. Prominence is perhaps what is conspicuous, or perhaps it is the thing that draws us to hold it in the center of our attention. What is prominent is not necessarily most important, or most significant. What is prominent is simply dominating our attention in some way. Prominence can certainly be created by loudness, but any other element of the mix can also bring prominence to a sound, to an instrument, to a musical idea, and so forth.

When mixing, we are also making a fundamental decision of the amount of clarity and blend of instruments and voices. The clarity of a sound in the mix can provide prominence, and the blending of instruments and voices might bring them to be more-or-less equal within the texture and in our perception.

We hear the mix at various levels of detail (perspective). In each, we balance the prominence of sounds, but in different ways. 'Balance' often brings to mind loudness, but the mix crafts balance within all elements. Shaping the smallest details of individual sounds contributes to clarity and blend in a great many ways and begins the pyramiding of perspective relationships.

Details Within Individual Sounds

At beginning stages of the mix, the sound qualities of individual instruments and voices are shaped. Shaping the qualities of sounds begins within tracking and is finalized in the mix. Shaping of sounds often takes place at the lowest level of perspective, and builds upwards through various levels of perspective, as the mixing process adds sounds until all are present. It might prove helpful to refer back to Table 2.1.

Individual sounds are shaped for their timbres in very exacting ways. To hear this material, we need to allow ourselves to enter into the sound to hear the components of timbre individually, to hear the dynamic envelope, spectral content and spectral envelope of the sound. We have been taught from birth to hear timbre as an overall quality, blending of all these elements; it is unnatural for us to try to enter deeply into a sound, to bring our attention to the qualities within its timbre. It is possible to perform this, however, and the seasoned mixer can hear the minute aspects of timbre qualities of instruments and voices, and follow them as they are combined with other sounds and placed spatially during the mixing process.

The Moog synthesizer that ends the introduction to 'Here Comes the Sun' was presented in Figure 2.1, above. At this level of detail, we can perceive and recognize all aspects of the dynamic envelope, the frequency of each partial of spectral content, and the dynamic contour of each partial over the duration of the sound. Bringing our attention to the physical dimensions of the sound allows us to hear the subtle details and allows us to calculate them precisely in terms of frequency and in dynamic relationships within the sound. This process of hearing timbral detail will prove vital to controlling a number of mix elements (Moylan, 2015).

Distance of sounds is perhaps the most important element explicitly determined by timbral detail. Instruments and voices appear to be located at a distance from the listener (the person hearing the finished mix) by the amount of low-energy timbral information present in the sound. The higher the amount of subtle low-level information, the closer the sound appears. In this way, sounds can be crafted to sound astonishingly close to the listener, or infinitely far (Moylan, 1992, 2012).

While we might instinctively reach to add reverberation to create distance, timbral detail alone is responsible for this illusion. When reverberation brings perceived distance to increase it is because timbral detail is masked; without the loss of timbral detail, the instrument or voice will simply sound like it is in a different, more reverberant room but at the same perceived distance (B. Moore, 2012).

To understand the distance placement of sounds, we bring the focus of our attention to the level of timbral detail—those aspects just described.

We then begin to use our knowledge of the sound to attempt to locate the sound based on our previous experiences. We use our sense of self, our sense of occupying personal space, to help determine if the sound is within our area of proximity (where we might be able to touch the instrument or slightly beyond our grasp), or a bit farther but still near to us (perhaps in the same room), or perhaps at a location of considerable detachment, far or well beyond our sense of space. Levels of timbral detail bring clarity to these changes of relationship (Moylan, 2012).

Focus your attention on the distance cues of the McCartney's lead vocal in 'Eleanor Rigby'. Hold the vocal in the center of your attention and concentrate on the amount of timbral detail; notice how the vocal shifts location as low-level spectral information appears or diminishes. Some of these changes are very subtle, and some will be very pronounced. You may hear little or no difference at first, but once you are able to experience this quality and then hold this element within your awareness, these qualities and their changes will unveil themselves. Listen carefully to the vocal in the introduction, the first verse and the first chorus; track where McCartney is in relationship to you, the listener. Once you begin to accurately perceive this dimension, the substantial distance changes between sections will become clear, and you will become aware of the subtle changes within the verses. Listening to the two stereo versions (1987 and 2009) will reveal subtle differences of distance, comparing them to the original mono mix will create another contrast that should become noticeable (and ultimately quite obvious) in its differences with repeated listenings. Remember to keep awareness of timbral detail to determine distance; be aware not to confuse loudness levels, or the amount of reverb present, with the distance of McCartney's vocal to your listening location.

Perspectives of Individual Lines and Aggregate Texture of All Sources as Equals

Two distinct levels of detail comprise the middle ground of perspective. A level up from individual sounds, sound sources (individual instruments or voices) exist as individual lines, with their own characteristics in all elements. These sources have their own shapes and characters that can be heard and crafted individually and can be related to other, specific sources.

One level higher is the perspective of the aggregate texture; here all sources are present and exist as equals. At this level, sounds can be held with equal attention and compared as equals. Only on this level of perspective can sounds be accurately balanced; it is at this level that sounds can be compared without one being more prominent by misconceived focus of attention. All or any number of instruments/voices might be compared, or all might be perceived simultaneously listening in this way. With this equal attention, the mix can be crafted so one sound (or instrument, or musical idea) might purposefully be made more prominent by the use of some dimension of the mix, or all could be made equal in a blended sound.

Crafting musical balance is where this concept is critical. Musical balance is balancing the loudness levels of sound sources. It is natural to hold one instrument in the center of one's attention and attempt to compare that loudness to others; this very act distorts the relationships. A shift to holding a sound within attention can happen quickly, such as when a new instrument enters, and especially when the lead vocal appears. When we hold a sound in the center of our attention, it 'appears' to be loudest, because it is occupying our awareness; for loudness to be balanced, sounds need to be held equally in our awareness (Moylan, 2015).

We confuse loudness with prominence easily. We must remember loudness is simply loudness—the perception of the sensation of the amplitude of the waveform. Loudness is not prominence or significance; it is not distance or timbral complexity; it is not necessarily what is in the center of our attention; numerous other perceptions are easily confused with loudness. Of course, something can be both loudest and prominent, etc., but the point is the most prominent, closest, most complex, widest, most interesting, or most significant sound (etc.) is also not automatically the loudest. One productive way of improving skill in musical balance is to plot out simple dynamic contours and dynamic relationships of several instruments—individually, then against one another, then as an aggregate sound; this draws the listener deliberately into the appropriate level of perspective (Moylan, 1992).

Figure 2.2 provides the musical balance of the opening measures of 'Strawberry Fields Forever'. Listen to these opening measures and bring your attention to the Mellotron, then follow it throughout the example;

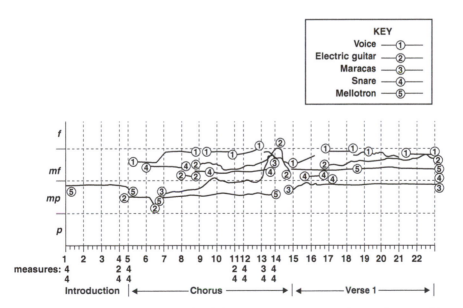

Figure 2.2 *Musical balance graph of the opening measures of 'Strawberry Fields Forever' (Moylan, 2015)*

note how its loudness is shaped, and notice how your attention is diverted (perhaps momentarily) when the vocal enters. Repeat this several times and notice how you can obtain greater clarity in perceiving the changes and subtleties of characteristics. Now try to shift your attention to the perspective where all sounds are held equally in your perception to notice the musical balance (or loudness balance) of the sounds. Next, follow John Lennon's vocal for dynamic shape, notice carefully the changes between sections; listen again, but shift perspective to observe the vocal's loudness in relation to the instruments of this section. Remember to remain focused on loudness contours, or on loudness relationships and balance; practice one skill at a time.

Timbres of the instruments and voices that we embraced above are also present and functioning in this middle dimension. The mix also joins all sounds by their overall timbre, which is the result of their inherent sound quality and timbral alterations created by the expression and level of energy of their performance intensity. The timbre of sounds sets up a pitch density in the mix, whereby certain frequency/pitch ranges are emphasized in the mix dependent upon instrument selection and the characteristic timbres of those sounds. Some find it easiest to conceptualize pitch density as a vertical dimension to the mix, where low to high pitch/frequency is conceived bottom to top (A. Moore, 2012). Pitch density exists at this dimension of the individual sound source, and again in the aggregate mix; it is later reconceived as 'timbral balance' in the overall texture dimension. How timbres are combined in the mix can provide blend or clarity to sounds and musical ideas, and more. This is a rather advanced concept that is typically learned after control of loudness and spatial dimensions.

Beginning mixers are quick to recognize loudness and lateral locations of sounds in the mix. In stereo, phantom images can span 90 degrees from a point 15 degrees outside each loudspeaker, when the speakers are at a 60-degree relationship to each other. Instruments and voices are placed in the stereo field during the mix process, and certain conventions have arisen concerning where certain instruments and the lead vocal are placed, in certain types of music or to be most effective. The sound field is not comprised of placement alone. Images also have width. The width of images coupled with the location of images in relation to others will play very important roles in the prominence, clarity and blend of sounds. Image sizes can range from a very precise point source to occupying the entire width of the sound stage; obviously most images fall between these extremes.

Figure 2.3 plots the stereo location of the primary sound sources in the opening of 'A Day in the Life'; two separate tiers allow us to see the sounds more clearly. Note the different widths and locations of the sounds, and how the acoustic guitar and piano images change in width at certain points during the passage. Bring your attention to Lennon's vocal, and follow it as it moves gradually across the sound stage, and how its width changes, sometimes subtly and sometimes markedly. Listen carefully to the placement and widths of the percussion sounds, and notice how they relate to and complement those sounds on the graph.

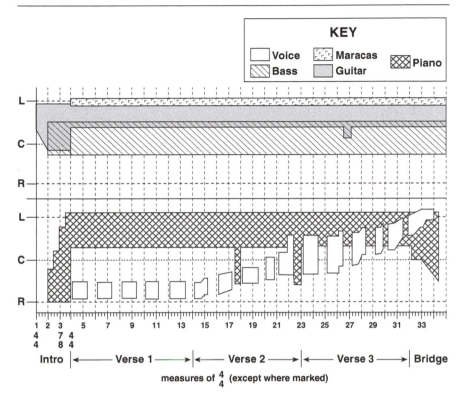

Figure 2.3   Stereo location graph of 'A Day in the Life' (Moylan, 2015)

Conceptualizing the Interactions of Elements in the Aggregate Texture

At the middle dimension, it is possible to observe how two or more elements interact. Hearing these relationships is obviously a more advanced skill, but it is well worth acquiring, as it leads to engaging more advanced mix concepts and much greater control of crafting mixes.

We can take our attention to a higher level of perspective and equally hold two elements in our attention simultaneously. For instance, we can perceive the interactions of pitch density information and stereo location. In doing so, we can hear and recognize how the imaging that distributes instruments/voices across the sound stage will also provide a location to their frequency content—their pitch densities. Figure 2.4 provides a graph that can allow this information to be plotted. In this graph, we can clearly recognize how instruments occupy the same, different or similar frequency regions, and how their lateral placements bring them to be separate or overlapping with others. Attention to the distribution of pitch material across lateral space can prove an effective means of localizing sounds to provide clarity or blend, as desired (Moylan, 2012).

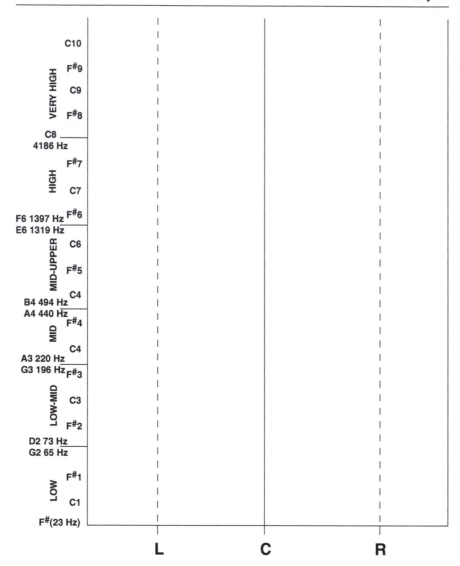

Figure 2.4   Sound source frequency content against stereo imaging

Allan F. Moore (*Rock: The Primary Text*) has devised a way of charting three elements simultaneously. His 'soundbox' plots pitch density (vertical), stereo location (lateral) and depth (distance) simultaneously with a quasi-three-dimensional box. The soundbox can bring visualization to sections of a mix with a clarity that many have found helpful, and it has proved a useful tool in both studies of mixes and in practice of planning and executing mixes.

A recording's sound stage is the area that encompasses the lateral and distance locations of all of the instruments/voices of a mix. It establishes and defines the left/right and front/back boundaries of the illusory stage from which the 'performance' that is the recording emanates. The imaging of a mix locates each source in relation to the listener and at a specific place on the sound stage. The sound stage is established by the interaction of the elements: stereo location and distance location.

Figure 2.5 presents a sound-stage diagram with instruments and voice placements. Viewing this diagram one can recognize the locations of sounds, both in their placement on the sound stage and in their relationships to the listener. The diagram can be useful in planning a mix or in evaluating an existing recording's sound stage.

Let us remember: phantom images have width, and therefore span an area from one point to another. They may be very narrow or very wide. But whatever their width, all of our human experiences have brought us to simply identify where a sound is, and react accordingly. We have never needed to try to process the width of a car horn, but merely localize where it is and quickly determine the proper course of action. Determining the width of sources requires bringing attention to the edges of the images to determine where they cease to occupy space; this requires some practice.

The other dimension of the sound stage is depth. Remembering that distance is determined by timbral detail allows for the correct perception of distance cues and distance placement on the sound stage. Distance location allows us to identify the sound's level of intimacy or level of detachment from the listener. While phantom images occupy an area of lateral width, these sounds do not occupy a depth area. We hear sounds at specific locations of distance from the listener; if there is a sensation of depth present, it should not be confused with a width of depth. Rather, it is the result of the listener conceptualizing the placement of the instrument/voice within

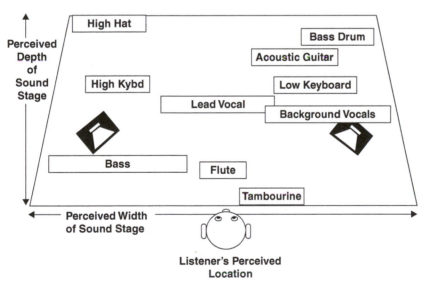

Figure 2.5 Sound stage and imaging, with phantom images of various sizes and at different distances (Moylan, 2015)

an appropriate environment. While the sound of an instrument/voice can be sensed as having depth from the environment, it is conceived as existing at a specific spot within that environment. The sound source and the environment it is performing in remain fused and inseparable in the listener's understanding—this happens in nature, so the listener expects it and accepts it in recordings (Moylan, 2015).

This fusion of source and environment has the potential to shift markedly in surround sound mixes—along with all other dimensions of the sound stage and the relationships of the mix to the listener.

The Sound Stage in Surround Sound

Surround sound can deliver a stunning experience to the listener, in ways that are very different from stereo recordings—ways that have the potential to be significantly richer. The listener's relationship to the music and to the performance has the potential to be strikingly different from stereo; it can be an experience entirely transformed.

In stereo recordings, the sound stage is in front of the listener, as if the recording is a concert of sorts. The size of the stage can vary from sources being tightly grouped in the center to a full spread of 90 degrees in front of the listener. The listener is an observer of the recording, a member of a conceptual audience. They may be very close or very far from the sound stage, but they will be detached from the stage of the recording and an observer (Moylan, 2015).

In surround, there is no telling where the sound stage will be, or the scale of its size. Every song, and potentially every moment in a piece of music, can have a unique sound stage and might change the listener's relationship to the music and its ensemble. This relationship of the listener to the music and its message, to the musical materials of the song, and to the sound sources/performers of the recording is both physical and conceptual. These are areas where surround can be vastly different from stereo, and deliver a vastly different experience. Surround can also enhance stereo with more realistic ambiance, or environmental characteristics; as such, it enriches the stereo recording without substantially changing the listener's relationship with the sound stage (Moylan, 2012).

The listener's relationship to the surround sound stage is a significant characteristic of the recording. This relationship may be examined most directly through considering (1) the listener's location either as being located within the sound stage or detached from it as an observer, (2) if detached, the conceptual or perceived distance of the listener to the sound stage, and (3) the location of the sound stage relative to the listener (Moylan, 2015).

In surround, listeners can find themselves located within the sound stage, and listeners may also find themselves surrounded by the sound stage, but detached from it. The listener within the sound stage is distinctly different from sound sources being merely wrapped all around the listener location. It is possible for the listener to get the impression they are within the music, within the ensemble, perhaps part of the band (Holman, 2008).

The sound stage can be located anywhere around the listener. Its size can encompass any portion of the 360 degrees around the listener. The position of the sound stage relative to the listener might be reduced to (1) the extent to which the sound stage resembles traditional stereo, (2) the placement of ambiance and the perceived performance environment and of the individual sources, and (3) the presence of an additional sound stage or of sounds outside of the sound stage.

Figure 2.6 plots the opening of 'I Want to Hold Your Hand' (from *LOVE*). Here the sound stage is in front, as in stereo but wider, and the ambience is in the rear. This is a common approach that is related to conventional imaging,

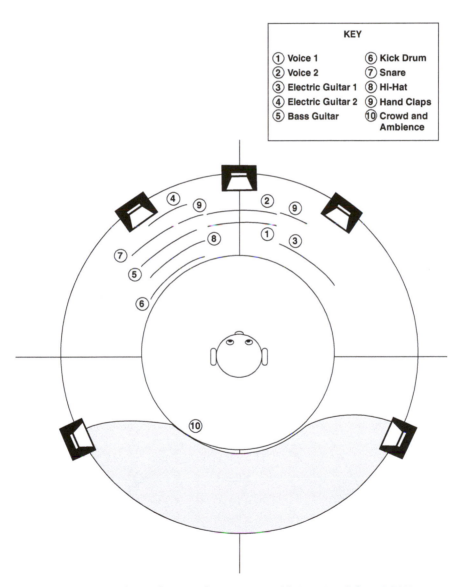

Figure 2.6   Surround sound stage of 'I Want to Hold Your Hand' from *LOVE*

with added control of the environment in the rear. In practice, the front-center channel may change imaging matters considerably, as can be heard in this example's pronounced clarity to the individual musical lines in contrast to the original mono version. This recording displays a width of sound stage that extends slightly extending beyond the 90-degree spread of stereo, though sources are largely grouped as if performing together live. The environment cues are placed to the rear or to the sides of the listener, and are joined by screams of crowd noise to depict some semblance of a live performance.

Imaging in surround sound is vastly more complicated than in stereo. In stereo, phantom images are established by the interaction of sound emanating from two loudspeakers; the resulting images may appear anywhere between the two speakers or up to 15 degrees beyond. 5.1 surround-sound imaging substantially changes the concepts of phantom image size and placement, image stability and listener's relationship to the music and the ensemble.

As seen in Figure 2.7, there are five primary phantom image locations existing between adjacent pairs of speakers in surround. These images tend to be the most stable and reliable between systems and playback environments.

Many secondary phantom images are possible as well. These can be created and appear between speaker pairs that are not adjacent. These images contain inconsistencies in spectral information and are inherently less stable. Implied are different distance locations for these images, as the

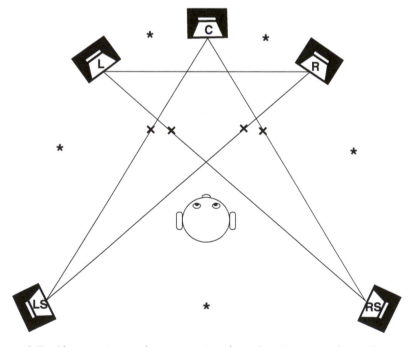

Figure 2.7   Phantom images between pairs of speakers in surround sound

trajectories between the pairs of speakers are closer to the listener position. These closer locations do not materialize in actual practice. The distance location of these images are actually pushed away somewhat by the diminished timbral clarity of these images. This can create contradictory and confusing sonic impressions.

When we consider locations caused by various groupings of three or four loudspeakers, placement size and location options for phantom images get even more complex—and less predictable.

This wealth of opportunity for mixers creates a complex set of possibilities for imaging in surround and generates great potential for other speaker-combination approaches to imaging. The format for 5.1 surround provides the opportunity for as many as twenty-six possible combinations of speakers. Table 2.4 lists the potential phantom images in surround that may be established by the interaction of any of the five loudspeakers, in any combination (Moylan, 2017).

As with imaging in stereo, the edges of phantom images are more decisive to defining source locations and size than are their centers. The edges of images allow us to understand and identify widths of images and where and how they might overlap. With this awareness of the edges of images, we can recognize how sounds are separated in space or blended. The separation of source images in surround can be very marked, and may grow more pronounced as the width of the sound stage increases.

When using non-adjacent speakers for imaging, sounds blend and fuse differently. Image placements and size can be unstable and vary quickly with any listener movement. Sounds can both separate themselves from others or become ill-defined, almost transparent sonically and largely masked. When the imaging occurs from speakers on either side of the listener, the image can be confusing, as it localizes within the conceptual listener location, yet there is the impression of detachment due to diminished

Table 2.4 Possible combinations of speakers for generating phantom images in 5.1 surround sound; combinations exclude potential inclusion of the subwoofer

Number of speakers	Number of combinations	Combinations
2	10 pairs	Adjacent pairs: L-C, C-R, R-Rs, L-Ls, Ls-Rs Non-Adjacent pairs: L-R, L-Rs, C-Ls, C-Rs, R-Ls
3	10 groups of three	L-C-R, L-C-Ls, L-C-Rs, L-R-Ls, L-R-Rs, L-Ls-Rs, C-R-Ls, C-R-Rs, C-Ls-Rs, R-Ls-Rs
4	5 groups of four	L-C-R-Ls, L-C-R-Rs, L-C-Ls-Rs, L-R-Ls-Rs, C-R-Ls-Rs
5	1	L-C-R-Ls-Rs

timbral detail. Thus, phantom images from non-adjacent speakers complicate and may confuse distance perception and depth of sound-stage mpres-
sions. These images exhibit peculiar characteristics. The interaural cues of
sounds produced from pairings of left- or right-surround and one or two
non-adjacent front speakers place the sounds close to the listener's head,
but without heightened timbral detail. This is potentially confusing for the
listener, and it tends to make the sound stage unstable—or simply provide
conflicting information (Moylan, 2017).

Examine the imaging in the surround sound stage of Figure 2.8. In
this *LOVE* version of 'Eleanor Rigby', the performance and the ensemble

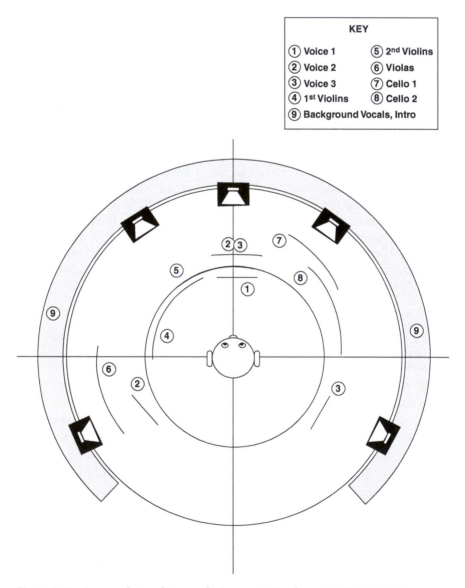

Figure 2.8 Surround sound stage of 'Eleanor Rigby' from *LOVE*, 0:00–1:09

surround the listener. During the Introduction, soft background vocals appear as broad phantom images and shift locations at each appearance. These shifting background vocals are represented in the shaded area around the listener and are at a considerable distance from the listener; while the distance location of these vocals shifts, all activity takes place within the shaded area. During this first 1:09 of the song, the double string quartet and vocals surround the listener, while a sense of being detached from the song and the performance remains. This mix is significantly different from the original and the re-mastered versions. It would be instructional for the reader to compare this mix to those of the original mono version and a stereo mix of the song—perhaps with the more contemporary timbral balance of the re-mastered stereo version from 2009.

Another important aspect of surround-sound imaging is the various ways environments might appear.

Unique to surround, the environments of individual sources can be separated from the source itself. In effect, the environment can become another source with its own location, width and character. The fusion of sources and their environments that occur in stereo (and in nature) may also happen in surround, but it can also be deliberately counteracted by physically separating the two sufficiently. The result is an impossible sound in our world, with the unnatural separation of source and its environment, which may or may not work in the mix.

The ambient sound of a perceived performance environment (PPE, to follow next) can be separated from the sound stage as in the *LOVE* version of 'Strawberry Fields Forever'. It might also approximate a live performance experience by being subtly blended within and extending the sound stage, and then filling the rear of the space. Countless variations are possible.

Highest Level of Perspective

Hearing the mix at the highest level of perspective, we engage the overall sound—the sound of the mix as one, blended entity. Final shaping of this level of perspective often is the role of the mastering engineer, though the mix engineer fundamentally establishes these qualities and relationships. The primary elements of this perspective (the overall musical texture) are timbral balance, program dynamic contour and perceived performance environment.

We hear one overall environment for the mix as the PPE. This is the 'performance environment' of the sound stage, or the world within which the performance that is the mix takes place. The PPE can be crafted for the mix, but it is most often perceived through the interactions of the environments of the mix's instruments and voices. Its dimensions shape the entire recording and bind all individual sound sources and their spaces into a single performance area (Moylan, 1992, 2015).

Timbral balance might be considered the frequency response of the mix, and it represents the recording's 'spectrum'. It is the distribution and density of pitch/frequency information of the mix and is the combination of all of the pitch densities of all of the mix's sounds.

At this highest level of perspective, all loudness levels of the mix coalesce into a single sensation of loudness. From beginning to end, the mix unveils a program dynamic contour: a single dynamic level of the composite sound that varies throughout the song, thus establishing a contour or high-level dynamic envelope. This is the dynamic shape of the mix, as a single variable loudness level.

Listening at this highest level of perspective requires one to focus on the single sensation created by all of the sounds of the mix. The single sensation will further be delineated to bring attention to only the overall environment, to only the timbral balance, or to only the overall dynamic contour of that overall sound. Just as we explored earlier, we must set and hold a clear intention to listen to this overall sound in order to keep from being drawn into details of a myriad of sources.

Conclusion: Shaping the Mix to Complement the Music

In *Mixing Music*, it can be meaningful to remember: the recording *is* the performance; the record *is* the song. The musical ideas and materials of the song are given extra dimension and are thereby enhanced by the recording, by the mix. The artistic elements described in previous sections provide the raw materials for this enhancement. They can also add substance and musical expression in themselves. This brings us to recognize that the mix engineer is a creative artist: composing the mix, performing the recording.

The mix can contribute greatly to the musicality and musical materials of the song—conversely, the mix can also be quite transparent. This points to production philosophy and shaping the mix to be most appropriate for the individual song, the musical style and the artist. The mix can have a widely varied degree of influence on the song; no matter its prominence, that influence is always present.

The recording process becomes a musical platform for shaping, enhancing, creating and presenting music. The song and the mix become inseparable. *Mixing Music* is making music.

As the song unfolds, so does the mix. The mix effectively delivers the song by shaping its dimensions to match, complement or enhance the character of the song, its message and expression. The mix is crafted to most appropriately deliver the story of the song, the drama of the music, the expression of the performance. As the mix unfolds and changes, it reflects the characteristics of the song and adds substance and enhancements. The mix and the music become entwined in synergy.

Above in 'What to Hear', we examined the artistic elements of recordings—the dimensions of the mix—and how those elements function on three levels of perspective. We considered briefly how elements interact and form a complex tapestry of relationships that become woven into the song. These discussions were overviews of topics that are very rich in detail and vast in their influences on the mix and on the music. How the mix, or more fully how the recording, transforms music is examined deeply in the author's new book: *Recording Analysis: How the Record Shapes the Song* (Routledge, 2017).

The relationships between the dimensions of the mix and the creative process of crafting the mix are also examined throughout the author's *Understanding and Crafting the Mix: The Art of Recording* (Focal Press, 2015). A significant portion of the book is dedicated to developing listening skills; some of which were introduced above. *Understanding and Crafting the Mix* also contains greater detail on the materials presented in this chapter, and applies it to analyze mixes for artistic elements and much more. It explores mixing as a creative process, in other words.

To conclude, let us remember that *Mixing Music* begins with listening. That listening is a focused activity to be learned; it is a skill that can be developed. Listening is paying attention, and remembering to pay attention. Listening is bringing attention to a specific aspect of sound, or holding an open awareness for what might arrive. Listening can be seeking specific information, and can bring awareness to any level of detail and any element of sound. We listen without preconceived ideas or expectations, and hold the possibility of the unknown arriving in the next instant.

Listening also requires knowing what to listen for—what to seek to hear. Listening and hearing work in tandem. Listening takes place during the passage of time, an event unfolding moment by moment in the present. Hearing takes place in memory; considering what has happened, we hear backwards in time.

If hearing is perceiving and recognizing, hearing only really arrives with knowledge and experience; these inform and refine perception to bring realization and understanding. It is for this reason that we carefully examine the unique qualities of recordings and refine our skills at hearing subtle details—so we might become aware of how those details are shaping our recordings, and so we might embrace mixing as a creative process.

We can learn to conceive and compose our own mixes, and establish our own musical voices—once we are in control of the recording and mix processes, and once we know what to hear, and learn how to listen.

Bibliography

Butler, David (1992). *The Musician's Guide to Perception and Cognition*. New York: Schirmer Books.

Clifton, Thomas (1983). *Music as Heard: A Study in Applied Phenomenology*. New Haven, CT: Yale University Press.

Deutsch, Diana (2013). *The Psychology of Music*, Third edition. Orlando, FL: Academic Press.

Handel, Stephen (1993). *Listening: An Introduction to the Perception of Auditory Events*. Cambridge, MA: The MIT Press.

Holman, Tomlinson (2008). *Surround Sound: Up and Running*, Second edition. Boston, MA: Focal Press, Elsevier.

LaRue, Jan (1992). *Guidelines for Style Analysis*, Second edition. Detroit: Harmonie Park Press.

Levitin, Daniel J. (2006). *This is Your Brain on Music: The Science of a Human Obsession.* New York: Plume.

Moore, Allan F. (2002). Rock: *The Primary Text.* Aldershot: Ashgate.

Moore, Allan F. (2012). *Song Means: Analysing and Interpreting Recorded Popular Song.* Burlington, VT: Ashgate.

Moore, Brian C.J. (2012). *An Introduction to the Psychology of Hearing,* Sixth edition. London: Elsevier Academic Press.

Moulton, David (1995). *Golden Ears: Know What You Hear.* Sherman Oaks, CA: KIQ Production, Inc.

Moylan, William (1992). *The Art of Recording: The Creative Resources of Music Production and Audio.* New York: Van Nostrand Reinhold.

Moylan, William (2012). 'Considering Space in Recorded Music.' In *The Art of Record Production: An Introductory Reader for a New Academic Field*, eds. Simon Frith and Simon Zagorski-Thomas. Surrey: Ashgate Publishing Limited, pp. 163–188.

Moylan, William (2014). 'Pathways Through Recording Analysis.' Delivered to the *137th Audio Engineering Society International Convention*, Los Angeles, October 9. Paper Proceedings.

Moylan, William (2015). *Understanding and Crafting the Mix: The Art of Recording,* Third edition. Burlington, MA: Focal Press, Taylor & Francis Group.

Moylan, William (2017). *Recording Analysis: How the Record Shapes the Song.* New York: Routledge, Taylor & Francis Group.

Nhất Hạnh, Thích (2015). *Silence: The Power of Quiet in a World Full of Noise.* New York: HarperCollins.

Polanyi, Michael (1962). *Personal Knowledge: Towards a Post-Critical Philosophy.* Chicago: University of Chicago Press.

Roederer, Juan (2008). *Introduction to the Physics and Psychophysics of Music,* Fourth edition. New York: Springer.

Schaeffer, Pierre (1966). *Traité des objets musicaux.* Paris: Editions du Seuil.

Schaeffer, Pierre and Guy Reibel (1966). *Solfège de l'objet sonore.* Paris: Editions du Seuil.

Snyder, Robert (2000). *Music and Memory: An Introduction.* Cambridge, MA: The MIT Press.

Tenney, James (1986). *Meta ≠ Hodos and META Meta ≠ Hodos.* Oakland, CA: Frog Peak Music.

Thompson, William Forde (2015). *Music, Thought, and Feeling: Understanding the Psychology of Music,* Second edition. Oxford: Oxford University Press.

Discography

Beatles, The. 'A Day in the Life', *Sgt. Pepper's Lonely Hearts Club Band*, EMI Records Ltd., 1967, digitally re-mastered 1987. CDP 7 46442 2.

'Eleanor Rigby', *LOVE*, EMI Records Ltd., 2006. 0946 3 79810 2 3/0946 3 79810 9 2.

'Eleanor Rigby', *Revolver*, EMI Records Ltd., 1967, digitally re-mastered 1987. CDP 7 46441 2 0.

'Eleanor Rigby', *Revolver*, EMI Records Ltd., 1967, stereo digital re-mastered 2009. 0946 3 82417 2.

'Eleanor Rigby', *Revolver*, EMI Records Ltd., 1967, mono digitally re-mastered 2009. LC 0299 509999945823.

'Here Comes the Sun', *Abbey Road*, EMI Records Ltd., 1969, digitally re-mastered 1987. CDP 7 46446 2.

'I Want to Hold Your Hand', *LOVE*, EMI Records Ltd., 2006. 0946 3 79810 2 3/0946 3 79810 9 2.

'I Want to Hold Your Hand', *Meet The Beatles!* EMI Records Ltd., 1964, re-mastered 2004. CDP 7243 8 66878 2 1.

'Strawberry Fields Forever', *LOVE*, EMI Records Ltd., 2006. 0946 3 79810 2 3/0946 3 79810 9 2.

'Strawberry Fields Forever', *Magical Mystery Tour*, EMI Records Ltd., 1967, digitally re-mastered 1987. CDP 7 48062 2.

3

Proxemic Interaction in Popular Music Recordings

Dr. Ruth Dockwray

This chapter discusses sonic spatialization and the notion of proxemics in recorded tracks. Spatialization, or rather the spatial characteristics and positioning of sounds within a track, can directly influence the way a listener can formulate their own interpretation. Through the analysis of proxemic zones within the context of the 'sound-box', their impact in terms of interpersonal distance and listener engagement will be discussed along with potential meanings.

First, the sound-box, within which the sounds appear, needs to be explained. The 'sound-box' (Dockwray and Moore, 2010) is a heuristic model of the way sound source location works in recordings.[1] The sound-box acts as a virtual spatial field within which sound sources can be located at any single moment as existing in a three-dimensional virtual space: laterality of the stereo image, perceived pitch-height of sound sources and perceived proximity to a listener. Second, 'personic environment'. This is a term coined by Moore (2005), which delineates between the persona evidenced in a recording and the musically coded environment, in which that persona is virtually located. This environment is normally communicated through three factors: location of sound sources within the sound-box, harmonic vocabulary, and form and narrative. Finally, the application of proxemics. Proxemics is used to refer to sonic, or rather vocal, characteristics that are audible on the track and more specifically, the relation between the listener and the persona and personic environment.

The term *proxemics* originated with the anthropologist Edward Hall, who used it to refer to "man's perception and use of space" (Hall, 1968: 83). Hall's theory of the communicative use of space can be separated into four zones: intimate, personal, social and public, as shown in Figure 3.1. For each of these zones, Hall lists key 'dimensions', such as postural, sociofugal, kinaesthetic, touch, retinal combinations, thermal, olfaction and voice loudness to describe the types of interpersonal behavior that may be identified. Each of these zones or distances relates to levels of comfort that people may feel with another person within that distance. Within each zone, Hall also notes a close and far phase, which helps to determine the type of interaction observed in that zone. However, the near and far phases seem only applicable to the intimate zone (see tables in

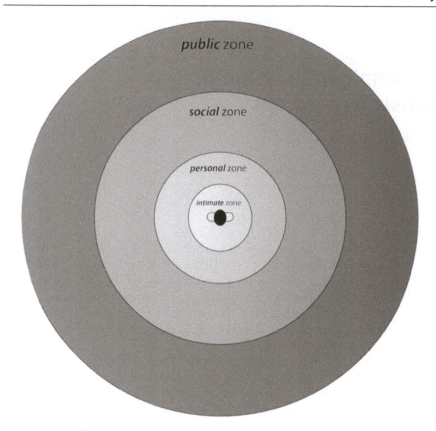

Figure 3.1   Hall's proxemic zones

Hall, 1969: 126–127).[2] Indeed, it is worth noting at this point that each type of interaction as discussed by Hall that takes place within each zone is not universal and, as he states, "proxemic behaviour . . . is culturally conditioned and entirely arbitrary" (Hall, 1969: 115).

With Hall's zones, he offers a specific way of discussing space and different types of distances, which can be applied to recorded song. More specifically, the interpersonal relationship and zone location that is focused on is that between the listener's perceptual position and the position of the song's persona, as modified by its (personic) environment (see Moore, 2005).

The use of Hall's theory enables the listener to determine elements and qualities of each zone through the concept of 'personal space' (the individualized space that surrounds a person) and the concept of 'interpersonal distance' (the related distance between two persons). To identify the perceived distance between the listener and the persona, the listener needs to engage in the overall scene of the track, which offers aural clues as to spatial placement and perceived proxemic distance.

Drawing on Hall's bodily dimensions, some important qualities refer directly to the analysis of recordings ('aural' and 'oral') and others are merely implied ('vision') when articulating the zone placement of the persona. The oral and aural aspects of Hall's receptors in proxemics perception can be directly applied to recordings, such as whispering and soft voice in the closer zones and loud voice and full speaking voice in the distant zones. These vocal distinctions can be identified as 'qualifiers' and form part of the paralinguistic features that Lacasse (2010) discusses in his work on phonographic staging, which enable the aural interpretations of the interpersonal distance and placement of the persona to be made by the listener. Lacasse also discusses 'alternants' (vocal sounds that can include "inhalations and exhalations and gasps"), which as he mentions are often used in a musical context, but when highly audible may also be identified as being part of the intimate or personal zone.

While these paralinguistic features may go some way in articulating the persona, the loudness or intensity of the voice needs to be taken into account and is again relative to the personic environment. As Moylan (2007: 140–142) discusses, the "perceived performance intensity" and the "dynamic level" are both important in the listener's perception of the persona's proxemic placement and the relationship between the persona's performance environment. The perceived performance intensity is the original level of the sound source, which may be altered in the overall dynamic level of the track. This allows, for instance, a vocal whisper to be heard at a higher level, relative to the rest of the sonic sources in the recording. While the perceived loudness has been manipulated, the sonic properties of the whispered voice have not changed and therefore give the listener clues as to perceived proxemic location of the persona. In this sense, the loudness can serve to enhance the interpersonal distance between the listener and the persona, particularly when analyzing the intimate zone. In terms of the intimacy that can be created between then listener and person, Nicola Dibben (2009) discusses the effect of the voice within the mix and the importance of vocal characteristics in the articulation of the persona and represented emotions, in terms of "performing intimacy" (p. 319). In this sense, amplification "creates intimacy between listener and singer, and communicates the 'inner' thoughts of the song character and/or performer" (Dibben, 2009: 320).[3]

Kylie Minogue's track 'Slow' presents an example of this type of intimacy, which can be perceived by the listener. Kylie's persona is presented in an intimate zone as evidenced by her audible vocal *qualifiers* and *alternants*, such as her whispered voice and clarity of breath intakes. Her intimacy and close proximity to the listener is also made evident through the lyrics that also situate the persona in the intimate zone. Hall's close-intimate phase descriptors highlight the potential of physical contact and "the high possibility of physical involvement is uppermost in the awareness of both persons" (Hall, 1969: 117). What is also interesting about this example is how Kylie's persona occupies a wide area of the sound-box, which suggests closeness to the listener and emphasizes Hall's 'vision' aspects of each zone. Vision, in terms of head size and peripheral vision, can be perceived in terms of sonic placement within the sound-box; however, in order to illustrate this notion, the concept of the 'framing variable' needs to be discussed.

The relationship between the perception of interpersonal distance and the framing variable considers the notion of *para-proxemics*, a term coined by Joshua Meyrowitz (1986), which focuses on the perceived or, rather, relative size of an object in order to judge its distance from the perceiver. Visual techniques, such as close shots, bring the main object into focus, and analogies can be drawn with audible contexts. In this context, the framing variable refers to the relative 'size' of the persona (as manifested aurally) in comparison to other environmental sounds that suggests which of the four zones are operative. So, as in the case of Kylie's track, her vocals seem to be relatively larger than the rest of the sonic elements in the track, occupying a larger space within the sound-box. Musically, the persona can be said to be in focus, referring to the increased definition of a sound source and increased sound-source width.

The qualities of spatial elements, such as sound-source width and its relationship to the environment, are examined by Moylan (2007, 2009) and are important aspects in the consideration of interpersonal distance perception and framing variable. The process of considering the aural clues to spatial placement can be described as 'auditory scene analysis', which Albert Bregman (1990) used as a way of analyzing individual sounds, through the process of segmentation, segregation and integration. While Bregman's approach focuses on isolated sounds, Francis Rumsey (2002) offers another means of classifying auditory processes that also considers the spatial environment. By focusing on not only the attributes of the sound sources but also the spatial environment (attributes of the particular space), the perception of spatial placement becomes clearer and enables what Rumsey terms a "scene-based understanding" both of where sounds are placed (with relation to each sound) and also of the type of space in which they appear.

Rumsey argues for three categories: width, depth/distance and envelopment, which are applied to individual sounds and environment. Of these, the width and depth of individual sound sources, the width and depth of the spatial environment, and the sense of envelopment remain valuable for our purposes. The difference between depth and distance is subtle: *distance* is the distance between the 'front' of a sound source and a listener, while *depth* acknowledges that there can be a 'back' to that sound source, and the difference between front and back delivers depth. By acknowledging these attributes of sounds, the proxemic positioning of the persona can sometimes be more clearly discussed.

Taking into account Hall's zone dimensions, Rumsey's approach to scene analysis and Meyrowitz's para-proxemics, it is possible to present an initial tabular summary of the aural dimensions that are common to each proxemic zone (see Table 3.1). The table focuses on the distance of persona to listener/degrees of intervention, the persona and personic environment (including envelopment, degree of separation and location of persona within the sound-box) and the articulation of persona.

One particularly effective example of different proxemic zones audible on recorded tracks is 'All My Life' by the Foo Fighters. In the opening section, the persona can also be identified as being located in the intimate zone. However, it is different from the intimacy of the previous Kylie track. Lead singer Dave Grohl's persona can be perceived as being situated in what Hall classifies as the far phase of the intimate zone. His close proximity to the listener is immediately evident by the vocal quality

Table 3.1 Proxemic zones

Zones	Distance of persona to listener/degrees of intervention	Persona/environment	Articulation of persona
INTIMATE	-Very close to listener (i.e., touching distance) -No intervening musical material	-Persona set in front environment -Normally high degree of separation between persona and environment -Vocal placed at front of sound-box and abuts the boundary of the sound-box	-Close-range whisper -Clarity of vocal sounds (coughs, breath intake) -Lyrical content suggests intimacy/ potential physical contact and addresses interpersonal relationship between two people
PERSONAL	-Closer to listener (within arm's length) -Possibility of intervening musical material	-Persona in front of environment -Still a certain degree of separation but less than in intimate zone -Vocal not a forefront of sound-box, set back from boundary	-Soft to medium vocals -Less clarity of vocal sounds - Lyrical content addresses two or three people
SOCIAL	-Medium distance from listener -Intervening musical material	-Persona within the environment -Little separation and more integration -Vocal placed within the center of sound-box	-Medium to loud vocals -Few, if any, vocal sounds heard -Lyrical content addresses small/ medium group of people
PUBLIC	-Large distance from listener -High degree of intervening musical material	-Persona engulfed and towards the rear boundaries of the environment -High degree of integration -Vocal placed towards the back of sound-box	-Full, loud vocals, shout/semi-shout -No vocal sounds heard -Vocals address large group of people

(semi-whisper) and audible vocal breaths due to the close miking of the sound source. His lyrics do not suggest intimacy, but rather are centered on himself and his search for something. The lyrics are also indicative of his close proximity, as he makes clear that "nothing satisfies but I'm getting close, closer to the prize at the end of the rope". Additionally, the guitar in the introduction also plays a key role in the proxemic positioning of Grohl, as it provides a point from which the listener can perceive the vocal's relative distance from the listener and its distance from the guitar. The wide sound source of his persona, which spans the entire scene width and high definition of the vocals, brings Grohl into focus and perceptually closer to the listener.

The ensemble width, or rather, the perceived width of the main group of sound sources, allows for the differentiation between other sound sources that are spatially detached from the rest of the ensemble. Source separation and increased source definition can also provide the listener with an increased perception of the degree of distance the persona is from the rest of the ensemble and from the listener. The sonic attributes of each sound source provide aural clues to spatial positioning; for instance, the lower dynamic and narrow source width of the guitar immediately places the guitar behind the vocals.

Half a minute into the track 'All My Life', the group width is increased by the entering of the rest of the sound sources and now seems to encompass the entire scene width. The persona is no longer detached from the environment and is enveloped by the individual sound sources of the ensemble, reducing its definition from being in focus and detached to appearing as part of the ensemble and less in focus. The difference here is that the guitars panned to each side appear to be closer to the listener, and the persona has changed in terms of vocal delivery (now shouting) and is dynamically louder in an attempt to be heard over the rest of the ensemble (or personic environment)—placed in a social zone.

Interestingly, the speed with which the persona moves from one zone to another highlights the possible discomfort Grohl feels about being in an intimate zone and in such close proximity, disclosing personal facts. His need to quickly come out of that zone into a more social zone suggests there has been an invasion of personal space, which he clearly is uncomfortable about. He later returns to the intimate zone before the chorus, albeit briefly.

Changes of proxemic zone are perhaps among the most interesting uses of space within recordings. Leona Lewis's cover of Snow Patrol's 'Run' is one such example and is representative of the gradual movement from one zone through the others. The track begins with Lewis in an intimate zone, backed by a resonant piano very much in the background. The piano is joined at 45" by mid-range sustained strings, providing a carpet of sound. At 1'08", both strings and Lewis achieve an upper range in terms of register: while the degree of reverb has not altered, it is as if she has stepped back, as the greater range opens her words to a wider audience. By 1'38", she is no longer foregrounded by her environment, but the whole texture has receded into a social space, or rather the environment has moved towards us. In the second verse at around two minutes, the drum kit (which had previously consisted simply of on-beat ride cymbals, which had been crucial in bringing the environment to the fore) falls into a conventional pattern, and any remaining illusion of intimacy is lost. (She now seems proud of her profession of constancy.) By 3'20", both she and the strings appear with full force and are subsequently joined by a full gospel choir, any sense of restraint lost, as if she no longer cares who hears. Lewis's persona is fully enveloped by the environment. Thus, the increased envelopment of the persona by the environment clearly marks the change in proxemic zone from intimate to public space.

Moore (2005) discusses how the persona can be articulated through the lyrics and the importance of lyrics in determining proxemic placement and the degree of interpersonal distance. As Table 3.1 illustrates, the lyrical content also gives the listeners clues as to the person or persons being addressed and the location of the persona with a particular proxemics zone. One such distinguishable feature is the use of pronouns, which articulate who is being addressed. The combination of vocal attributes, lyrics and personic environmental features enable the listener to identity the different zones, as evident in this next example.

With regard to the social and public zones, the perception of proxemic placement and interpersonal distance relies not only on the level of personic environmental envelopment but on lyrical modes of address. Hall's descriptors for the social and public zones state that other people are seen and are important in peripheral vision, alluding to the notion of co-presence and group communication, as opposed to the intimate zone, where communication is to an individual. Group communication relies on public modes of address, increased voice level and the use of full voice. While the voice level is "a significant variable in judging distance" (Hall, 1963: 1016), it is the lyrical content and public mode of address in this next example that places the persona in a more social/public zone.

In the rock anthem 'We Will Rock You' by Queen, the listener locates Mercury's persona as being in a social space. The full-voice vocals, a medium distance away from the listener, are evidently communicating to a group of people through Mercury's persona being situated within the environment consisting of multiple hand claps, suggesting co-presence. The social zone location remains throughout the track, with little difference between the verse and chorus in terms of Mercury's vocals. The only change occurs with the personic environment, which thickens in texture during the chorus due to the addition of multi-backing vocals, thus widening the overall width of the sound source and engulfing the lead vocals. The public mode of address as evidenced by use of the pronoun 'we' and the interchangeable use of 'you' from the singular to plural serve to emphasize the social/public proxemics placement and the increased interpersonal distance from the listener. Such is the nature of rock anthems that they rely on group participation and a sense of co-presence. Rock anthems essentially induce gestural and vocal participation, which creates a sense of community through this interaction with the listener and the audience in live contexts (Dockwray, 2005).

There are examples of rock anthems where several different zones can be heard across the track. The different audible zones are particularly significant for rock anthems as they act as a cue for audience participation, which is typically during the chorus. In another of Queen's tracks, 'We Are the Champions', the verses situate Mercury in a personal zone, characterized by the soft to medium vocals, close proximity of the listener near the front of the sound-box and the clarity of the vocals from the bass guitar and piano, offering a clear separation in terms of spatial placement. The chorus, on the other hand, situates Mercury in a public zone, characterized by the full loud vocals and his persona being engulfed by the sound sources and in particular the multi-backing vocals that occupy a wide selection of the sound-box.

Spatial movement can portray a spatial relationship, particularly in the case of rock anthems, where gestural participation, such as raised arms (clapping and swaying) is a visual indicator of audience inclusion. The gestures of reaching out with the arms as part of the audience/band interaction is a means for both groups to connect and reduce the distance between each group, metaphorically speaking. The reaching-out gesture can be seen as a form of proxemic behavior, which endeavors to reduce the interpersonal distance, albeit in a psychological sense. While the social/public zone is implicit on the recording, the proxemic interaction translates from an audible to a visual interaction within a live context.

The examples in this paper aim to demonstrate the ways in which proxemics offers a theoretical perspective that contributes to the potential way in which a listener may experience a track and provides a language that begins to acknowledge spatial placement and the perceived location of the persona, in relation to the listener.

Notes

1 See Moore (2002: 120–126) on the concept of the sound-box; Moore and Dockwray (2008) and Dockwray and Moore (2010) for the uses of the sound-box and taxonomies of sound-box uses (Moore, 2012).

2 The application of proxemics is used in the work by Maasø (2008), which focuses on the mediated spoken voice in a film sound context. Interestingly, he notes that the near-far phase seems applicable in a mediated voice context, as opposed to recorded tracks, where the different levels of intimacy can be differentiated, through vocal qualities and lyrics.

3 For more analysis on the application of proxemics in terms of intimacy and the intimate singing voice, see Dibben (2013).

Bibliography

Bregman, A.S. (1990). *Auditory Scene Analysis: The Perceptual Organization of Sound*. London: MIT Press.

Clarke, E.F. (2005). *Ways of Listening: An Ecological Approach to the Perception of Musical Meaning*. New York: Oxford University Press.

Dibben, N. (2009). 'Vocal Performance and the Projection of Emotional Authenticity.' In *The Ashgate Research Companion to Popular Musicology*, ed. D.B. Scott. Farnham, Ashgate, pp. 317–334.

Dibben, N. (2013). 'The Intimate Singing Voice: Auditory Spatial Perception and Emotion in Pop Recordings.' In *Electrified Voices*, eds. D. Zakharine and N. Meise. Göttingen, Niedersachs: V & R Unipress, pp. 107–122.

Dockwray, R. (2005). *Deconstructing the Rock Anthem: Textual form, Participation and Collectivity*. PhD thesis, University of Liverpool.

Dockwray, R. and Moore, A.F. (2010). Configuring the Sound-Box 1965–1972. *Popular Music* 29 (2): 181–197.

Hall, E.T. (1963). A System for the Notation of Proxemic Behaviour. *American Anthropologist* 65 (5): 1003–1026.

Hall, E.T. (1968). Proxemics. *Current Anthropology* 9 (2–3): 83–108.

Hall, E.T. (1969). *The Hidden Dimension*. London: The Bodley Head.

Lacasse, S. (2010). 'The Phonographic Voice: Paralinguistic Features and Phonographic Staging in Popular Music Singing.' In *Recorded Music: Performance, Culture and Technology*, ed. A. Bailey. Cambridge: Cambridge University Press, pp. 225–251.

Maasø, A. (2008). 'The Proxemics of the Mediated Voice.' In *Lowering the Boom: Critical Studies in Film Sound*, ed. J. Beck and T. Grajeda. Urbana, IL: University of Illinois Press, pp. 36–50.

Meyrowitz, J. (1986). 'Television and Interpersonal Behaviour: Codes of Perception and Response.' In *Inter/Media: Interpersonal Communication in a Media World*, ed. G. Gumpert and R. Cathcart. Oxford: Oxford University Press, pp. 253–272.

Moore, A.F. (2002). *Rock: The Primary Text*. Aldershot: Ashgate.

Moore, A.F. (2005). The Persona-Environment Relation in Recorded Song. *Music Theory Online* 11: 1–21.

Moore, A.F. (2012). *Song Means: Analysing and Interpreting Recorded Popular Song*. Farnham: Ashgate.

Moore, A.F. and Dockwray, R. (2008). The Establishment of the Virtual Performance Space in Rock. *Twentieth-Century Music* 5 (2): 219–241.

Moore, A.F., Schmidt, P. and Dockwray, R. (2009). A Hermeneutics of Spatialization for Recorded Song. *Twentieth-Century Music* 6 (1): 83–114.

Moylan, W. (2007). *The Art of Recording: Understanding and Crafting the Mix*. Burlington, MA: Focal Press.

Moylan, W. (2009). Considering Space in Music. Proceedings from the 2008 Art of Record Production Conference. *Journal on the Art of Record Production* 4. (October). [Online]. Available at: http://arpjournal.com/considering-space-in-music

Rumsey, F. (2002). Spatial Quality Evaluation for Reproduced Sound: Terminology, Meaning, and a Scene-Based Paradigm. *Journal of the Audio Engineering Society* 50 (9): 651–666.

Discography

Foo Fighters. (2002). "All My Life," *One by One*. CD, RCA 82876 50523 2.

Lewis, Leona (2008). "Run," *Spirit* (deluxe edn). CD, J Records 88697 46016–2.

Minogue, Kylie (2003). "Slow," *Body Language*. CD, Capitol CDP 7243 595645 0 0.

Queen. (1977). "We Will Rock You" and "We Are the Champions," *News of the World*. CD, EMI Parlophone—CDPCSD 132 (1993, Remastered Reissue).

4

Top-Down Mixing—A 12-Step Mixing Program

Phil Harding

This chapter is an ethnographic reflection of my working practices as a producer, engineer and mixer in the 1990s. In what follows, I suggest that the methods I used then are still useful in the context of modern popular music mixing. To do this, I conceptualize mixing in two different ways, specifically as a 'top-down' and a 'bottom-up' creative practice. 'Top-down' in this concept refers to starting a mix with the lead vocals and then working 'down' through the arrangement to the drums. 'Bottom-up' mixing refers to the opposite. Bottom-up mixing begins with the drums and ends with the vocals. The latter method, in my experience, has been the traditional routine in rock, pop and dance music genres since the 1970s.

"It will be all right at the mix" and "it's all in the mix" (Cauty and Drummond, 1988, p.117) are phrases I've heard in recording studios since I started making records in the mid-1970s. By this time, a multi-track recording, requiring a final stereo mixdown session, was already a very well established part of the production process. The mixing process is barely touched upon in *The Manual (How to Have a Number One the Easy Way)* (Cauty and Drummond, 1988) by Jimmy Cauty and Bill Drummond of The KLF, first published in 1998, and to be honest I do not go into the 1980s SAW mix process in any great detail in my own book *PWL From the Factory Floor* (Harding, 2010). William Moylan (2015) notes that "people in the audio industry need to listen to and evaluate sound". I would agree with Moylan and other academics that prolonged periods of critical listening "can be used to evaluate sound quality and timbre" (Moylan, 2015:186). What is timbre in music? Zagorski-Thomas (2014) notes that timbre is "a function of the nature of the object making the sound as well as the nature of the type of activity" (Zagorski-Thomas, 2014, p. 65). Ultimately, the mixing process is something that needs to be performed, practiced and mastered by every creative in this time of diversity. Composers and musicians, as well as engineers and producers, will find themselves in the mixing seat due to budget or time constraints in an age where budgets for all types of recorded music and audio have fallen by 25% or more since the 1990s. The following 12-step program still serves me well after 40 years of experience as an industry practitioner and is a framework for others to experiment with.[1]

It was during my boy-band production period in the1990s that I stumbled upon the idea of starting a mix with the vocals. Possibly it was the large production-based projects that Ian Curnow and I created with East 17, in particular their 1994 Christmas #1 single 'Stay Another Day'. I was mixing this track in the summer of 1994. It was the day before going on a family holiday, and a huge mixing task lay in front of me with over 50 vocal audio tracks together with multiple keyboards, drums, bass and a full orchestral arrangement, programmed by Ian Curnow, including the Christmas tubular bells. All of this was across 48 tracks of analog tape, so some sub-mixing had already been done in Cubase Audio in our production suite at the Strongroom Studios. Nevertheless, there were at least 10 to 16 tracks of vocals to be dealt with, as quickly and efficiently as possible, so that I could get home at a reasonable time for our holiday journey the next day. Apart from the usual lead and harmony vocals from vocalist Brian Harvey, there were also counter chorus vocals from band member Tony Mortimer and four-part chorus harmony vocals, all double-tracked, from each band member in the chorus. Then there were a large quantity of chorus harmonies, each quadruple-tracked by session vocalist Tee Green, plus verse harmonies and answers from the band members and Tee. Finally, there was a backing vocal counter melody on the outro of the song, again quadruple tracked by Tee Green.

As the family holiday deadline loomed, I had a 'light bulb' moment and decided that I would start the mix with Brian's lead vocal, supported by the main song pad synthesizer for some musical perspective. From that day on, this has been my adopted mixing method for every record I have mixed regardless of genre. I believe this does not just work for boy bands or pop music; it can work for any mix that contains vocals. One may need to adapt the starting null-point for other digital audio workstations (DAW) and software, but this has worked very well for me on hardware such as the SSL G-series and Pro Tools software. The following is my '12-Step, Top-Down' mix routine for either software or hardware.[2]

Step 1. Main Lead Vocal and Master Fader

For my first step, I would set the main lead vocal fader to zero—that can be the group master or the individual fader if I have more than one lead vocal track. Also, I set my master fader to zero. If the vocals are well recorded, this will be a good start point and hopefully a good null-point setting for the mix overall.

Step 2. Song Pad or Guitar

Next, I would bring in the song pad keyboard or whichever instrument plays most of the way through the song and supports the main vocals—this could be an acoustic guitar for a singer/songwriter track or the main electric guitar for a rock band. I balance whatever this choice is on the track well behind the lead vocal in a supporting role, making sure that it is not fighting for space with the vocal; it could always be raised later in the mix process.

Step 3. Main Acoustic Guitar Accompaniment

At this point, I consider turning that instrument into a stereo soundscape, if it is not already. The reason for this is to fill out the stereo picture behind the vocals. If the keyboard pad is mono, I put a simple stereo chorus on it (set as subtly as possible) via an auxiliary send. Same for the guitar if it is a mono signal; later in the mixing process I remove that from the acoustic guitar. Currently, when recording a main acoustic guitar for a typical singer/songwriter, I generally record three signals into a DAW:

a. A feed from the internal guitar pick up via the direct injection (D.I.) box.
b. A small diaphragm condenser microphone over the acoustic hole, angling in from the fretboard—typically a Neumann KM84.
c. A large diaphragm condenser microphone on the body of the acoustic guitar, below the guitarist's strumming hand but out of the way of being hit by that hand—typically a Neumann U87 would do this job well, as would an AKG C414.

Step 4. Lead Vocals

Now I would start processing the lead vocal. Below is my standard set of vocal mix techniques:

a. Insert a vocal compressor starting with a 3:1 ratio and the threshold set so that the gain reduction meter is only active on the louder notes.
b. Set up an auxiliary send to a vocal plate reverb with a decay of about three seconds and a high-pass filter (HPF) up to 150 Hz.
c. Set up an auxiliary send to a crotchet mono delay effect with around 35% feedback and 100% wet.
d. The equalizer settings are entirely dependent on how the vocals sound, but typically if the vocals were recorded flat (and I will not know unless I have recorded them), then I would boost a few decibels (dB) around 10 kHz or 5 kHz and consider a 4 dB cut at 300 Hz to 500 Hz and also an HPF up to 100 Hz, provided this doesn't lose the 'body' of the vocal sound. If the vocal is already sounding too thin, then I would try a boost from around 150 Hz to 250 Hz, but no lower than that, as I would want to save any boost of 100 Hz downwards for kick drum and bass only.
e. My final 'go to' on a lead vocal, and I apply this later in the mix, is the Roland Dimension D at its lowest setting—Dimension 1. The Universal Audio plug-in virtual copy of this piece of hardware is a good replacement and again this would be on an auxiliary send, in addition to keeping the original signal in the stereo mix. The effect on the Dimension D is still set to 100% wet and balanced behind the lead

vocal to create a stereo spread, but strangely it has a wonderful effect of bringing the voice forward, hence this is best used later in the mix, when there is more going on behind the lead vocal.

As you may have gathered by now, I am building a multi-dimensional landscape of sound across a stereo picture. During my mixing process, from the 1980s onwards, I have always used my sonic landscape in relation to a picture landscape. To take that further in the mixing stage, I imagine the picture as 3D so that I can analyze the staged layers in a deliberate attempt to separate the instruments for the listener and to help those instruments mold together as one. This should sound to the listener as though the musicians (live or programmed) and singers are in one space—or all on stage—together. I use the word 'picture' deliberately here, because that's how I plan a final mix, like a multi-dimensional landscape painting, with the lead vocal at the front and heart of the picture. I usually have this picture in my head before I even start the production, and I believe this is a fantastic way to plan a commercial pop production. First, you have the vision, which is built around the artist, the direction that they and their manager and label want to go, and then you have the song, either written for or chosen by the artist and the team around them. Then there are various plots, comparisons, influences and directions all chosen by you, the producer(s), in collaboration with the artist, their record label and their manager. That gets thrown into a creative melting pot that you as the producer, engineer and mixer have to deliver, making complete sense and showing that you allowed all of those suggestions to influence the final product. One could compare it to painting by numbers, but it is not quite that strict, as you are constantly reviewing and changing as you go along. In my experience, the most important person to listen to will be the client, whether that is the artist, manager or more frequently the label. The other thing that made sense in my boy band work and this mixing method in the 1990s was that the focus for acts such as East 17, Take That and Boyzone is totally on the boys and their vocals. This is the case whether it's the media (especially radio and television), their manager, the record label or most important of all—the fans. None of these people is initially focusing on the rhythm or the music; if that is all working well behind the vocals, then it is doing its job. This is also the reason we heard so many 1990s boy-band song introductions featuring vocals, often edited from the chorus that would come in within the first minute of the song. It could be said that what I have described above forms my 'Phil Harding Signature Mix Sound', or as Simon Zagorski-Thomas (2014) calls it, "a schematic mental representation".

Step 5. Double-Tracked Lead, Harmony and Backing Vocals

The next step is for me to bring in the rest of the vocals behind the lead vocals—any double- or triple-tracked lead vocals would be 5 dB below the lead vocal. This will give the effect of 'fattening' the lead without losing the character of the main lead that we are likely to have spent hours editing and tuning after the recording. Typically, this works very well in a pop chorus. All of the processing I have described for the main lead vocal would generally go onto the double-track lead vocal except for the crotchet delay. I tend to leave that just for the single-lead track, otherwise it can sound messy if it is on the double as well. Next would be any harmony vocals recorded to the lead vocal, generally a third up or maybe a fifth below. If the harmonies are single-tracked, then they would remain panned center. If they were double-tracked, then I would pan them half left and right, or even tighter. Processing on these would be similar to the lead vocals but with no delay effects. Finally, to complete the vocal stage of the mixing, we move onto the chorus backing vocal blocks, which would often start with double- or quadruple-tracked unisons to the lead vocal in the chorus. This is to add strength and depth, as well as a stereo image with these panned fully left and right. From there, all of the other harmonies in the chorus would be panned from the outside fully or, for instance, half left and right for the midrange harmonies, tight left and right at 10 o'clock and 2 o'clock for the highest harmonies. All of these need to be at least double-tracked once to achieve a true stereo. The processing would be applied on the stereo group fader these vocals are routed to. This saves the computer system DSP by not processing the individual tracks. Typical backing vocal processing would be compression first, set similarly to the lead vocal, equalization, again similar to the lead vocal but less low mid cut and minimal HPF. The vocal reverb would stay the same, though it's worth considering a longer reverb time, four seconds or higher to place the backing vocals farther back from the lead vocals. I would not put the crotchet delay on the backing vocals except for a special, automated effect on one or two words. I send the backing vocals to a small amount of quaver delay overall to give them a different and tighter perspective to the lead vocals. Multiple tests and use of this methodology since the 1990s have proven to me that this is a repeatable formula for all pop and dance mixes. I may wish to vary my iteration of this with more delays and processing for extended and club mixes (especially by more use of the crotchet delay on the backing vocals), but for a radio and video mix, the above techniques almost guarantee an industry standard and accepted sound.

Step 6. Pianos and Main Keyboards

Pianos and keyboards add more musicality and support underneath the new vocal stereo sound spectrum that has been created. In terms of time, steps 1 to 5 could take half a day of your time; so if I have not already done so, I would take a break before step 6. I would advise taking a break every two hours to rest your ears and equally so these days the eyes, which have been constantly staring at computer screens throughout this process. Some people say that our ears are only good for four hours work per day. I am not sure I agree with this, especially if regular breaks are taken. Often I will be happy to stop my day's work after step 5 and come back completely fresh on another day. I would have already prepared my keyboard stereo group fader (auxiliary input track in Pro Tools in my case) and I would bus all the keyboards at this stage to that group fader. Unlike the lead and backing vocal groups, the keyboard group is unlikely to have any processing, auxiliary sends or inserts; it is just there as an overall level control for typical keyboard overdubs such as synthesizer pads, pianos, organ, bells and so on. Most virtual keyboard sounds and hardware keyboard sounds will generally deliver a stereo signal, and my first job would be to make panning decisions. You can, of course, leave everything panned hard left and right, therefore allowing the original patch programmers of those sounds to decide on the stereo image. I prefer to take control of this myself. I will even go as far on a programmed acoustic piano part to split the stereo audio track into individual lefts and rights. I do this so that I can process each side of the piano individually. This raises an important point that applies to drums as well as pianos. You need to make a decision on the performing 'stereo image' of some instruments—are you panning as though you are the performer/player or are you panning as though you are the audience? My preference is the performer's perspective; therefore, you imagine you are the piano player in this case and your left hand, or low-end piano part, is panned hard left and your right hand or mid- to high-end piano part is panned hard right. This puts the listener in the piano seat, a wonderful perspective in my view and particularly effective for the listener on headphones. Therefore, when we get to them, I recommend the same panning perspective on the drums, i.e., put the listener on the drum stool. My typical equalization for a stereo acoustic piano—this is the same whether it is a live piano or a programmed stereo sample. On the left-hand side, I engage a low-pass filter (LPF) down to around 8 kHz or 7 kHz, consider a small boost around 3 kHz to bring out the rasp of the low piano notes. Then, coming down the frequency spectrum, I consider a 2 dB to 6 dB cut in the low mid-frequency (300 Hz to 900 Hz) to get rid of any unwanted 'boxy' sound, as I call it. Finally, I may want to boost some low end to bring out the depth of the piano; I would restrict this from 150 Hz to 200 Hz, as I am saving the frequency ranges below that for the bass (100 Hz) and the kick (50 Hz). I know all this sounds very strict and specific, but it works as a very efficient start point.

For the right hand of the piano, I do almost the opposite to the processing of the left hand. I start with an HPF up to around 150 Hz, perhaps higher. I engage a small low mid-frequency cut between 300 Hz to 900 Hz again, but you could also leave it flat if you prefer. Then I experiment with boosts of around 5 kHz and 10 kHz, with quite large bandwidths to brighten the piano. I find that when you bypass these left and right equalizers and then put them back in for comparison, that width and separation of the stereo perspective of the left- and right-hand piano parts will be enhanced. The other aspect of having split the stereo piano track into mono left and right is that after all of the equalization I have described, you can add a little of the vocal reverb plate to the right-hand side of the piano only; this will help to balance the piano into the track and bring it closer to the vocals. The final thing to consider on the piano is compression, which ideally should be first in the insert chain, so as not to be affected by your equalization choices. Again, I keep the compression simple, starting with a 3:1 ratio and adjusting the threshold to activate the gain reduction on the louder parts.

Step 7. Other Keyboards and Orchestra

Other keyboard parts really have to be treated on their own merits, and ideally one should find a space for them that fits behind either the main piano or pad—or they jump out as a feature of their own. If there were programmed string and orchestra parts, I would deal with them next. Typically, for strings I like to have a longer plate or hall reverb setting, at around four to five seconds in length and ready to go on an auxiliary send. I would have an HPF on the strings up to 150 Hz and definitely no LPF; I prefer a completely flat top end. For brass, programmed or live, I would send them to a small plate or a room setting at around 1.5 seconds in length. Other rhythmic keyboard or sequence parts I leave until I get the drums in, but certainly I would check them at this stage for any need to add compression and equalization, or possibly rhythmic quaver delay to bed them into the track. It's not unusual to have strings and orchestra on pop productions, whether they are real or programmed samples—the mix processing on both are very similar. For the orchestra hall reverb, where I've already explained the settings, I would only send the violins (first and second) to this, plus the harp (often arranged to work with the strings), but a lesser amount on the violas. I would generally keep celli and double basses dry. If there were woodwinds as well, I would send them to the vocal reverb plate so that they are a little tighter than the violins. My ideal string recording setup is described in Howard Massey's excellent book *The Great British Recording Studios* (Massey, 2015: 178). My equalizer recommendations on a live orchestra would be:

a. An HPF on the violins to cut out celli and double bass spillage up to 300 Hz. Then a small boost between 8 kHz to 12 kHz.
b. For the violas, I would use an HPF up to 100 Hz and a small top end boost at 5 kHz.
c. For celli, I would generally leave them flat other than a small boost at 3 kHz if they need it to cut through the balance.
d. The double basses I would keep flat.
e. For the harp, I would consider an HPF up to 200 Hz, a small cut at 300 Hz to 900 Hz and small boost at 8 kHz to 10 kHz.
f. For the woodwinds, a small boost of around 5 kHz to 10 kHz and an HPF up to 100 Hz.

All of this should help the orchestra to blend together and to blend into the track. Notice there has been very little low middle cutting on the orchestra in my loathed 300 Hz to 900 Hz frequencies.

Step 8. Guitars—Acoustic and Electric

I would generally deal with acoustic guitars before electric guitars. If the production were centered on an acoustic singer/songwriter, then the artist's main guitar part would have been my first support instrument of choice while processing the lead vocal. I described in step 3 how I would record an acoustic guitar for a singer/songwriter-orientated production, and I would now record any acoustic guitar part the same way for safety, but in a double- or triple-tracked guitar backing I would only use the microphone that was on the guitar sound-hole (Neumann KM84). The equalization choices for the acoustic guitar on the mix would typically be HPF up to 100 Hz; a 2 dB to 6 dB cut at 300 Hz to 900 Hz, and if it sounds too dull then apply boosts at 5 kHz and 10 kHz. All of this is for the multitracked one microphone acoustic. If you have just one main acoustic and you are using the multi microphone technique and direct injection that I described in step 3, then I would pan the 'body' condenser signal hard or half left and leave it virtually flat other than a 300 Hz to 900 Hz cut, I would even consider a 150 Hz–200 Hz boost or 2 dB to 4 dB to give more depth. I would then pan the sound-hole microphone hard or half right and duplicate the equalization described for the multitrack acoustic but possibly with the HPF up to 200 Hz. The direct injection signal would feed in behind the stereo microphones in the middle, probably kept flat, but at this stage you should check the phasing of the three signals combined and finally consider a tiny bit of the vocal reverb plate on the right-hand signal to help blend the acoustic guitar into the track. Certainly, I would avoid room, ambience or hall reverb.

Electric guitars are so technically varied these days I would again record three signals if I were involved at the recording stage:

a. A dynamic microphone such as the Shure57.
b. A large diaphragm condenser like the Neumann U87 or AKG414 or even a ribbon microphone on the guitar speaker cabinet.
c. A D.I. signal from the guitar for any future re-amping plug-ins to be added at the mix stage.

In the hope that the guitar sounds are well sourced and well played—I do little or nothing at the mix stage. I generally do not touch the low frequencies or lower mid-frequencies, deliberately leaving them in because I am cutting them so much elsewhere. If the guitar sounds at all dull, I only boost around 4.5 kHz, as I am trying to leave 5 kHz and above to the vocals, piano, acoustic guitars and cymbals on the drums. I would only consider any reverb or delays if the guitarist has not used any pedals or guitar amp effects. Thankfully, the technology of guitar effect pedals and the effects on guitar amps is good enough now, I believe, for engineers to trust their quality and low noise ratios. I generally trust the guitarist to deliver the sound that feels right to them on their amplifier to suit the track. If the guitars have been recorded flat and dry, I would add some short plate or ambience reverb between one to two seconds in length and some quaver delay with around 30% feedback for rhythm guitars. Usually I would apply some longer reverb and crotchet delay for solo and lead guitar parts.

Step 9. Bass

Much like the electric guitars, I do very little to the bass. I would hope to have D.I. and amplifier signals and I would use a single large diaphragm condenser microphone on the bass amplifier at the recording stage (Neumann U87). I compress the signals at the first stage of the mix insert chain, being careful to only use 2:1 ratio, as anything above that can destroy the low frequencies. I generally boost both signal paths at 100 Hz (I have been saving this frequency in this 'top-down' mix method exclusively for the bass). Finally, this is where I would try the Roland Dimension D using the same settings and auxiliary send setup as I did on the lead vocal. This will add a stereo perspective to the bass and some more warmth. I also add a small boost around 1 kHz to 2 kHz (the only time I use these frequencies), after the drums are in, if the bass is not cutting through the mix or sounds as if it needs more edge. Regularly now I will also experiment with the Sansamp on the D.I. signal—this is a very useful amplifier simulator plug-in that comes free with all versions of Pro Tools.

Step 10. Drums—Live or Programmed

Finally, we get to the drums, which for the traditional rock, pop and dance genre mix is usually the first step. What I have found strange and yet enlightening in my 40-year career is the similarity in the methods and sometimes even sounds for drums, certainly the processing of them, which can work for all three genres: rock, pop and dance. Generally metal rock drums will need a different specialist approach, but it is astounding that similar compressions, gating, equalization and ambiences will all work fantastically to give drums the same amounts of power required. At this point, I really have only very roughly balanced the various elements from the vocals down to the bass, but the important thing is that I will have looked at everything individually. I will regularly need to mute the vocals to achieve some of the things I have described here, but I will put the vocals back in as I move through each step. Now for the drums, I need to focus mainly on bass, drums and guitars in the balance. It is tough to start gating, compression and equalization accurately on drums when the vocals are still in the mix and prominent. As with a 'bottom-up mix', I would start with the kick but I usually bring the whole kit in roughly and quickly under the current balance to remind myself of what the drums are doing. In my chain of kick inserts, it would be gate first, then compression, then equalization. I prefer to see inserts in that order, as I do not want the equalization to affect my gating and compression settings. I would set up a side chain if I felt that was required. Copying the kick track before you start the processing is a good idea, because the gating and equalization can end up being quite drastic (especially on live drums) and you may wish to balance in the unprocessed original track. The same can be done for the snare track. Typical equalization on the kick would be a 2 dB to 6 dB cut at 300 Hz to 900 Hz and a boost of 2 dB to 6 dB at 50 Hz; if more 'slap' or pedal-beater is required, then boost 2 dB to 6 dB at 3 kHz, but no higher than that as we are saving 4 kHz and upwards for the guitars, keyboards, strings, vocals and cymbals. Typical snare equalization would be a 2 dB to 6 dB cut at 300 Hz to 900 Hz, boosts of 2 dB to 6 dB at 4.5 kHz and 2 dB to 6 dB at 7 kHz or 8 kHz. The kick and snare gating would aim to isolate them from the rest of the drum kit, and typically you would hope to eliminate snare spill on the kick signal and hi-hat spill onto the snare signal, always being careful to allow as much release and 'hold time' to allow the drum decay to breathe before the gate closes. Compression on the kick and snare should be minimal, as they will be likely to drive compression on the overall mix at the end of the session. An individual compressor on the kick should start at a 2:1 ratio for the same reason as mentioned on the bass; higher compression ratios can have the effect of losing low frequencies. A major part of the snare sound will be the choice of room ambience effect, even something like the D-Verb in Pro Tools, set to medium room and below one second in length, can give you a good room ambience for drums and instantly the snare will be more powerful and sound as though it is part of the track. I would also send the tom-toms to the same room ambience effect, then gate the tom-toms to eradicate spill from the rest of the drum kit or, more efficiently, you could edit them to only remain where the tom-toms play throughout the track. Any equalization on tom-toms would be minimal, a moderate boost around 4.5 kHz is all I would do if they are lacking attack or sounding too dull. Overhead cymbals and hi-hats should all have similar equalization to bring out the cymbal sound and to eradicate kick, snare and tom-tom spillage. My equalization for these would be an HPF up to 200 Hz to 300 Hz, a cut of 4 dB to 6 dB at 300 Hz to 900 Hz, a small boost at 5 kHz and 10 kHz. That concludes the drum processing.

An important decision is panning perspective, as was the case with the piano. We need to decide on panning to the 'player perspective': hi-hat left, tom-toms left to right, overheads left to right—or the 'audience perspective': hi-hat right, tom-toms right to left and overheads right to left. I much prefer the player version on drums and piano. If I have some stereo room microphones recorded for the drum session, I make sure that their panning matches the close microphones and consider a similar equalization to the overhead cymbals. I often experiment on the room microphones with some heavier compression at a 6:1 ratio or more and an overdriven threshold setting. You will need to raise the compression output or make up gain, but this can sound very powerful. My personal microphone choices when recording drums are

a.   Kick—Electrovoice RE20
b.   Snare—Shure SM57
c.   Hi-hat—AKG C451

   d.   Tom-toms—Beyer 101s
   e.   Overheads—AKG C451s
   f.   Room microphones—Neumann U87s
   g.   Ride cymbal AKG C451 on the bell of the ride cymbal and route that
      to its own track to enable control of this during guitar solos and so on

Step 11. Final Stereo Balancing and Tweaks

Everything has now been individually processed and each step has dealt
with a group of overdubs or a single instrument at a time. After the drum
step, the main lead vocals should still be prominent in the balance when
I bring them back in. Now is the time to achieve the overall mix balance, and
if I have been listening on large studio monitors or quite loudly on nearfield
monitors—one often has to when processing individual sounds, I would
consider taking a break. I believe it our duty as creative technicians to make
sure that the overall mix is playable without ear damage at high volume. This
is because many listeners, who we hope are excited by the music we create,
want to listen loud. But at this stage of the mix you should come down to a
quiet or medium volume on your nearfield speakers to achieve the final stereo balance. In an ideal world, I would approach step 11 on a fresh day—or
at least after an extended break. I find that these final tweaks can take up to
half a day, maybe checking on headphones, flipping between speakers and
volumes and possibly running a test mix wav file or CD to listen to in different environments, such as your home hi-fi system or car stereo. Also, at this
final step, I would send the lead vocals and bass to the Roland Dimension
D, as mentioned earlier, to help the lead vocal stand out from the track; the
Dimension D has an effect of bringing the vocal forward. While we do not
want the bass to be as loud as the lead vocal, the Dimension D effect brings
it forward and gives a stereo perspective that is generally a nice touch.

Step 12. Master Bus Compression

Overall limiting or compression, subtly, on the stereo mix master or bus
is the last thing to look at. A low compression ratio of only 2:1 or 3:1 is a
good starting point, and the idea of limiting or compression here is only to
subtly control any peaks in the overall mix (not to boost the level—that is
a different process). We should always bounce or export an unmastered 24
bit wav file for future mastering. For a listening copy to go to the clients and
artists, I would now insert a digital compressor/maximizer such as 'Maxim'
in Pro Tools. This can raise the overall mix level to commercial levels, which
means that should the clients or artists compare the mix to another commercial release, the mix will be close to that kind of volume and peak level. See
the Maxim screenshot in Figure 4.1 for one of my typical settings.
    As a first mix, I would generally expect clients and artists to come back
with comments and suggestions that will require a revisit to the saved multitrack file and session. For this common industry reason and practice,

Figure 4.1   Typical Maxim settings, inserted on the stereo Master Fader in Pro Tools

I have generally dispensed with using automation on commercial pop tracks that are mixed entirely in the DAW. If I have achieved the whole mix 'in the box' with no outboard hardware, recalling the mix is fast and easy. I hope that the requested adjustments will also be fast and easy, whether I agree with them or not. If, for instance, I have automated the lead vocal on the initial mix, it becomes very awkward to make level adjustments on other instruments because everything would still need to work with the potentially rigidly automated lead vocal levels and rides. This is quite hard to explain but will become obvious with continued practice. The way around this is to go through the lead vocal track, zooming in on the wave display to help you, and then go through the song and make the required adjustments, up or down, by using the audio 'gain rendering'. I know this sounds a little strange and even unprofessional, but by my continued use of this task it became second nature, and by applying mix adjustments of 4 dB to 6 dB or higher (common in pop), it became easier and I found myself no longer battling automation left over from the first mix. The only things I tend to automate now are extra vocal delay-send boosts (to emphasize one or two words) or quick equalization changes to help any vocal 'pops' that have slipped through on the recording.

Conclusions and Final Notes

The idea with this method of mixing is that one is concentrated on the song from the minute of starting the mix, and although I have described the process from the vocals down, it would also work for instrumentals and other non-vocal-orientated tracks where you would start from your lead instrument or 'theme'. Fauconnier and Turner (2002) talk about "conceptual blending" as an unconscious human activity, and that's exactly what I seem to do in my mixing approach. Short ambience on the drums are blended with medium- to long-plate reverb on the vocals, then those vocal reverbs are blended with my crotchet mono delay as highlighted in step 4. These are good examples that describe the types of technology we engage while mixing a record using today's technology. Another common industry practice in 2016 that I have not mentioned is the dropbox file sharing system. This is useful when either engaging others to perform a mix for you or also if you have been hired for a mix. Exchanging multitrack files via dropbox has become common practice, and I recommend signing up for an account to enable participation in this.

Technical Appendix: Export and Import Procedure to a Mix Template File

My current (2016) pop production technology process is to start compositions and productions in Logic Pro. At the point of a completed production, I then export all of the individual tracks, both MIDI and audio, that are still open in the final session file, as wav audio files 24 bit / 44.1 kHz (or 48 kHz) or higher as the project permits. Before doing this, for efficiency, I would go through all of the tracks left to right (L-R) on the mix or edit window screen and add a number from 01 onwards to all of the track names. This means that when you or someone else imports them, all of the tracks will line up L-R in numerical order on any typical music software mix window. This is a fast and efficient process in Logic Pro. Highlight all of the open tracks from bar 1 beat 1 (this is standard industry practice) to the end of the song. From the software pull-down menu select File > Export > All Tracks as Audio Files and be sure to bypass Effect Plug-ins and do not include Volume/Pan Automation. See the screenshot in Figure 4.2. This will now be exported as raw audio data, ready to be imported into Pro Tools or any music software of your choice. It would also be useful to export one MIDI file from Logic Pro (I would suggest a pad or piano) to retain the song arrangement markers and tempo; this saves a lot of time in Pro Tools.

I would suggest, before importing this into a Pro Tools session, that some time is spent creating a blank (no audio or MIDI data) Pro Tools mix template file (see the screenshots in Figures 4.3 and 4.4). This is a useful time-saver. It is a tiny file, and each time you are ready to start a new mix, you create a new folder, copy your mix template Pro Tools file into it, launch Pro Tools from that file, rename the song in a 'save as' from the file menu option and start by importing the MIDI track from Logic Pro. Position and highlight the track to the left of the auxiliary groups (refer to the template screenshot in

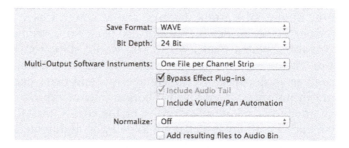

Figure 4.2   Logic Pro: export 'All Tracks as Audio Files' menu choice recommendation

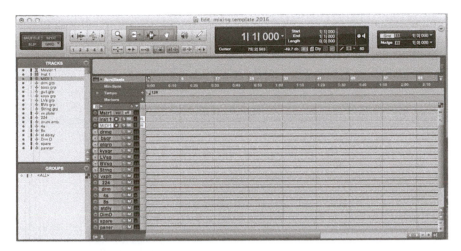

Figure 4.3　Pro Tools Mix Template edit window

Figure 4.4　Pro Tools Mix Template mix window

Figure 4.4) and then import all of the audio tracks, also at bar 1 beat 1. The final task before starting the 12-step mixing program is to position all of the auxiliary 'group VCAs' to the right of the cluster of audio tracks you want them to control. Route those audio tracks, via buses, to the corresponding input buses of the VCA groups, for instance, all drums and percussion can go to bus 11–12 (we have saved buses 1–10 for auxiliary effects sends). See the Pro Tools mix template screenshots in Figures 4.3 and 4.4 to see how this final process should look.

Now refer back to step 1.

Notes

1 Please note that throughout these steps I refer to the 300 Hz to 900 Hz frequencies as disliked or loathed by myself. I have never found an occasion when it is useful to boost this frequency range. I find it to be 'muddy and boxy' when boosted and will generally recommend either cutting these frequencies or leaving them flat.
2 There is a technical appendix at the end of this chapter that describes my audio and MIDI file export procedure from Logic Pro to Pro Tools. You may want to read that before embarking on step 1.

Bibliography

Cauty, J. and Drummond, B. (1988). *The Manual (How to Have a Number One the Easy Way)*. London: KLF Publications/Ellipsis.
Fauconnier, G. and Turner, M. (2002). *The Way We Think (Conceptual Blending and the Mind's Hidden Complexities)*. New York: Basic Books.
Harding, P. (2010). *PWL from the Factory Floor*. London: Cherry Red Books.
Massey, H. (2015). *The Great British Recording Studios*. San Francisco, CA: Hal Leonard Books.
Moylan, W. (2015). *Understanding and Crafting the Mix*. Oxford: Focal Press.
Zagorski-Thomas, S. (2014). *The Musicology of Record Production*. Cambridge: Cambridge University Press.

Discography

Tracks mixed 'Top Down'

'Stay Another Day' by East 17. Five weeks at #1 in the UK singles chart from late December 1994 into January 1995. Cat # LONCD 354.
'Words' by Boyzone. #1 in the UK singles chart October 1996. Cat # 575537–2.
'Go on and Tell Him' by Cliff Richard & The Temptations from #10 UK album 'Soulicious', 2011. Cat # 50999 0 88152 2 4.

Tracks mixed 'Bottom Up'

'You Spin Me Round' by Dead or Alive. #1 in the UK singles chart March 1985. Cat # TX 4861
'Hand on Your Heart' by Kylie Minogue. #1 in the UK singles chart April 1989. Cat # PWCD 35
'She Wants to Dance With Me' by Rick Astley. #6 in the UK singles chart September 1988. Cat # PD 42910.

5

Mixing in the Box

Justin Paterson

Introduction

In the context of this chapter, 'mixing music' should be understood as the adjustment of some aspect of recorded audio to enhance the aesthetic of the end-listener experience. Although such mediation has its roots in the control of 'live' audio at the point of capture, in this text 'mixing' will only refer to retrospective adjustments of multitrack audio sources.

Mixing is a crucial stage in the production of most popular music. In the broadest sense, mixing has a number of functions: to balance the various musical parts in terms of their volume and spectral content, to correct technical anomalies that have crept into the recording process, and to offer a creative platform upon which to enhance the basic tracks.[1]

For many years, mixing was executed using only physical hardware equipment, gradually being augmented by software. However, in more recent times computer technology has developed to the point where high-quality audio playback and mediation can happen entirely within a computer. Such an approach is often referred to as mixing 'in the box' (ITB). Signal path complicates this labeling, since an audio interface might typically output multiple channels into a hardware mixing console for summing, further processing and to act as effect sends to hardware devices. For the purposes of this chapter, unless otherwise stated, it will be assumed that the interface only outputs a digitally pre-summed analog stereo audio signal. Also, whereas many of the issues discussed might equally pertain to routing a digital audio workstation (DAW) through a console, here they will be viewed only in their native desktop setting.

Traditionally, hardware-based mixing entailed running multitrack audio through a mixing console that was typically able to apply insert processing (from external or sometimes internal processors), equalization, panning, gain adjustment and grouping of tracks together via buses. It could also send signals to external processors and return their output. These processors might typically offer time-based effects, dynamics processing or more exotic forms of timbral manipulation. The console would ultimately combine all the audio and via the console's 'mix bus' offer a single stereo output. This might be captured[2] to represent a finished mix of the music. Such consoles and processors could be analog, digital or a hybrid of the two.

In order to realize the above process, the equipment needed to be large enough to offer 'finger-sized' parametric control over it, as well as control over associated feedback mechanisms such as the values of said parameters, metering and displays. The physical environment needed to be large enough to accommodate all of this equipment.[3] The hardware equipment would typically have sonic qualities that were characteristic of its circuit design, manufacturer or even age. Thus, mixing in the hardware studio required a quite specific workflow, one that involved physical movement around the environment (with the corresponding change in perception of the sound) to access the various controls, operations that involved both hands, physical routing of signals, taking a visual overview of the settings of multiple parameters, and functional workflow guided by an idiosyncratic order of priorities that suited the specific situation.

In contrast, an ITB approach will be centered on one or more computer displays, thus precipitating a static position in the room, hopefully in an idealized acoustic position. All available parameters must be accessed via the displays, so only a finite number can be seen at any one time, and further, adjustment of them will be through an established human–computer interface (HCI) device such as a mouse or trackpad, although a hardware control surface might augment this. Operation of the former will be largely one-handed.

This chapter will consider the ITB paradigm, and identify & discuss a number of salient factors associated with its workflow. The treatment will commence with the overlap of ITB and other creative processes, and develop to consider the integration of editing. It will then discuss interface implications and the influence of the toolset, and move on to consider the clustering of mediators. Finally, it will close with some conjecture for the future. It is intended not as a comprehensive treatment of the subject, since there are many other areas specific to mixing ITB that are worthy of investigation, but rather it might serve as an introduction to a number of the key concepts.

Overlaps

While mixing per se has always been a profoundly creative act, mixing ITB has amplified this potential and expanded it to overlap with other major aspects of the creative process: composition, arrangement, pre-production, editing, production and mastering. The DAW has presented the integrated environment that allows these aspects to be incorporated (at least in part) into the mixing stage, or indeed for the mix to start to coalesce before its technical inception. Moorefield (2010) presented his argument for the producer as composer, underpinned by considering both the bidirectional interface of illusion and reality, as well as auteurism in music production. While composition and arrangement could be guided from a DAW arrange page with the mix itself performed on hardware, it is the integration of mixing into the computer domain of ITB that facilitates hybridization of the creative process. "It's not surprising that far from removing 'sacred cow auteurs', modern technology has simply shifted the metaphor from exceptional accomplishment on paper by 'composers' to exceptional accomplishment on hard disk by 'producers'" (Moorefield, 2010: 111).

Although Moorefield refers to the producer in this context, mixing ITB allows the metaphor to be extended from producer to mix engineer. Ever more often, the two roles tend to overlap, and it is ITB that offers the increasingly common scenario where the producer and the mix engineer are one and the same. Assuming they are not, consider two workflows: the producer might elect to print an effect that will then be subjected to the mix process, or the mix engineer might apply an equivalent effect of his or her own volition. While either mode is valid, the former could be said to offer the ultimate auteurism to the producer, but of course it can also bring its own disadvantages to bear. The production aesthetic becomes entirely constrained by the subjectivity, stylistic preferences and limitations of the individual; as such, auto-editorial control becomes an increasingly relevant skill. The latter offers its own considerable advantages by bringing the objectivity and specialist skills of a professional mix engineer to build upon the creative foundation laid by a producer,[4] but it is the mode of producer/mixer working in the integrated environment of ITB that offers a greater level of dynamism and the ultimate flexibility. For many (at all levels), mixing ITB has also now facilitated a broadening from Moorefield's auteur to atelier—in which the neo-craftspeople of music production might compete with the established orthodoxy or simply enthuse in their newly found capabilities. The apogee of their aspiration is ever more accessible.

Traditionally, the mix engineer would work with a number of relatively fixed performances, optimizing these spatially, spectrally and dynamically. Digital editing brought an enormous fourth dimension to mixing—temporal adjustment. Preceding this and building on the earlier tape-based work from musique concrète, John Cage (1952) introduced the notion of (what is now often referred to as) hyper-editing[5] in his piece, *Williams Mix*, which featured hundreds of random tape edits of short fragments of sound. Although aleatoric, in principle this deviation from the assumed linearity of music in Western traditions was to have a profound impact on future studio approaches. Such detailed editing might be termed 'micro editing'. On the other hand, 'macro editing' might refer to editing longer sections of music. One example of this is how, in jazz, Rudy Van Gelder pioneered tape splicing as a creative tool in the 1950s (Skea, 2015) to concatenate sections of ensemble performances into an idealized composite rendition of a given tune.

The advent of non-linear recording and editing made both micro and macro modes of adjustment more accessible. Micro editing came to be used for phase-aligning multi-microphone recordings such as drum kits—time-slipping the different channels in order that they most precisely align with a key element such as the snare-drum-microphone recording.

This technique could greatly enhance the transients and low-frequency coherence of the target channel when all were played back together while also reducing comb filtering, and such maneuvers gained popularity both at a production stage and also when mixing ITB. Originally, such adjustments were done visually by aligning the transients of the waveform visualizations[6]—a metaphor of the clapperboard in movie synchronization. Later, academic research produced algorithms to automate such adjustments, for instance by Paterson (2007) in the time domain, and then more effectively in the frequency domain by Clifford and Reiss (2010), operating in real time. Sound Radix (2010) released the first commercial plug-in to automate such actions, namely, 'Auto-Align'.

Micro editing was first done by hand, but later evolved to offer audio quantization. Such processes could be performed by a number of algorithms that may or may not attempt to preserve transients or perform time stretching. Both producers and mix engineers found that tightening performances to align the transients (regardless of any alignment to a grid) could have a profound influence on the mix. Such temporal adjustment could impact on the timbre as mentioned above, but of course also offer profound opportunities to adjust the groove of the music.[7] Although subjective, groove is an enormous part of any contemporary track and is tightly bound to both EQ and compression requirements, and options. Consider, for example, adjusting the transients[8] of both a kick drum and bass part that are loose, but 'feeling good', as opposed to if the transients are tightly aligned; the processing requirements could be quite different. Although matters of groove should really be addressed at an earlier stage of workflow under direction of the producer, this type of operation might be best executed holistically via ITB techniques by the producer-mixer, whichever 'stage' that happens to fall in.

Macro editing would traditionally have been associated with composition or arrangement, but there are many types of overlap. For example, if the mix engineer had finalized a precisely honed section of a performance, or even an entire chorus, it might be appropriate to copy it to other musically identical sections of the tune. Similarly, if at the arrangement stage it was decided to print an echo effect in order to harmonize the pitches, then that could be viewed as mixing. Mixing ITB has allowed (even encouraged) the overlap of many such aspects. DAW-based workflow has also led pre-production to integrate with mixing ITB—it is common for musical elements from a demo to be retained, and perhaps polished towards the final mix.

Although the term was first coined when referring to radio (Orban, 1979), one technical aspect of the more recent Loudness War[9] has been the prevalence of software mastering tools, especially limiters, but also entire mastering-orientated effect chains, all of which are regularly available when working ITB. Although there are many possible actors, from artist to label, that might contribute to the drive towards louder tracks, producers and mix engineers are sometimes tempted to push the overall track level into a limiter on the mix bus. By such mechanisms, mastering too has sometimes come to overlap with mixing ITB. Mastering engineer

John Webber (2013) suggests that if the intention is ultimately to make the master very loud, then the mix engineer preempts this by monitoring with a limiter on the mix bus, but removing it before printing. Webber recommends that the limiting be applied only at the end of the actual mastering.

The Interface

One problem that can be encountered when mixing ITB is the lack of proprioceptive control. When using a traditional hardware interface, the user can typically operate a fader or knob without looking at it, using a combination of proprioception and aural feedback to exert control. When using a mouse, trackpad or touchscreen, parametric control is much less intuitive, and it can be difficult to control the rate of parametric change or even remember the correct mode of interaction with a given plug-in (e.g., rotary versus linear action)—each of these presents a momentary impediment to the primary aural intention. Further, typically only one parameter can be changed at a time,[10] which is often less than optimal when, for instance, applying EQ. Although a dedicated hardware DAW-control-surface (or to some extent, one of the increasingly popular tablet-based remote-control apps, which are of course multi-touch) can mitigate such awkwardness, such systems can introduce further complications such as orientation when using a subset of soft controls that frequently change assignment.

Of DAWs, Mycroft et al. (2015: 687) found that

> Under complex navigation, as often found in large multi-track audio mixes, user orientation becomes a key issue and providing well-designed global views of the data is an important criterion for successfully navigating the information space,

and that

> Reducing the need to navigate the interface to find visual information can significantly improve the user's abilities to hear concurrent audio changes to the programme material.

This provides a logical rationale for the intuitive preference for multiple displays when working ITB, and helpfully sets a path towards further consideration of the influence of the interface upon workflow.

It is commonly understood that the multimodal layers of information provided by the DAW extend the operator's sonic perception beyond typical hardware metering (which is often purely gain based). This is exemplified, for example, by ITB representations of pitch and timing. There is, however, a trade-off. Ihde (2013) discusses the broader phenomenology of instrumentation, and while acknowledging that instruments might embody human experience and therefore extend our senses, he emphasizes that the use of such tools is non-neutral. The typical ITB interface attempts to seduce the user with its feature-set and encourages certain behavior. Of course, the hardware studio also does this, but its configuration and functionality might be considered to be relatively standardized. Thus, the ITB workflow (induced by a particular DAW) and its effect are likely to be *less* neutral, and the tool exerts more influence over the musical artifact. An example of this might be the spectrogram. Spectrograms can be incredibly useful in providing additional information about a sound source, particularly when overlaid on an EQ curve, but these can once again influence our listening and perception. The more assertive the graphical user interfaces (GUI), the more likely it is to provide a compelling influence on our aural judgement.

Johnson (1999) affirmed that as the complexity of an interface design increases, the user experience becomes more of the interface and less of its effect. Of the user being able to view only a subset of the available parameters, Mycroft and Paterson (2011) stated:

> There is a balance between a visualization that aids, reinforces and expands our innate aural abilities, and one in which the instrumentation translates all other aspects into visible results.

There might appear to be a useful trajectory forming for faster, more ergonomic and intuitive mixing ITB in the future. The latest GUI additions to Apple Logic Pro X[11] appear to reflect such a conscious move towards finger-sized simplicity, although more likely driven by the iPad 'Logic

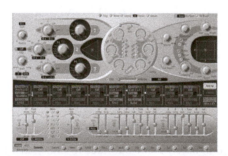

Figure 5.1   Top: the older ES2 Synthesizer from Apple Logic Pro; Bottom: the newer Clav plug-in. Both images are to scale, revealing the difference in GUI style
Images captured from Apple Logic Pro X

Remote' app, or perhaps a future move towards a multi-touch desktop paradigm. As ever, the trade-off is against access to a sufficient number of parameters and mitigating excessive navigation within the workspace.

Further, Tano et al. (2012) found that complex GUIs in creative-support software tended to impede short-term memory (STM), and beyond the 3D-sketching paradigm that they were investigating, the findings might be easily translated, especially when aural analysis is required simultaneously. Mixing ITB epitomizes a situation for susceptibility to such sensory overload. Mycroft et al. (2013) found that the typical GUI of a DAW could tend to overload the user and impede aural perception, and in particular, they found that scrolling displays had a detrimental effect on critical listening reaction time, which along with STM is a key component in the appraisal of a mix. Such phenomena will also translate to the use of hardware control surfaces with soft controls, which might typically present banks of channel strips that can be dynamically assigned to a subset of the total number, and they also apply to the tablet-based control apps. It is therefore possible that the ITB engineer's efficiency might be compromised by immersion in the desktop and its extensions—again, through their experience *of* the interface. Of course, those mixing on a console will still likely be using a GUI visualization of the musical arrangement for orientation, but this will be largely confined to a single view and make lesser demands on the retention of visual information from out-of-sight windows. They also have the advantage of a constant panoramic view of many parameters across the console and effects racks.

Out of sight is not necessarily a disadvantage in itself. As an example, when applying EQ with a console, although not always the case, it is possible for the physical size of the fingers to obscure the view of the parametric values. The user would therefore tend to rely upon aural feedback to judge when the setting is appropriate and get used to working that way by default. With a software interface, the temptation exists to look at numerical values and make logic-based judgments on what is appropriate, and these can conflict—consciously or subconsciously—with the aural feedback.

When the user is conscious of any such conflicts, the dichotomy has to be considered, possibly with the aid of experimentation with parameter values to 'calibrate' perception, or possibly making a decision on which source of information to align to. Again, these processes can take immediacy away from the job at hand. In the case of a subconscious conflict, a danger exists (particularly for the inexperienced mixer) to choose a reassuring number that aurally might be less than optimal. There is also a tendency to obsess over tiny details and minuscule adjustments, perhaps becoming preoccupied with numerical parameter readings or graphical feedback. While precise control is necessary, responding to number ranges close to or even below the resolution of perception will only consume time without yielding significant sonic benefits.

'Gear envy' is common with regard to outboard hardware and is perhaps particularly common towards those who own revered (and expensive) vintage equipment. Despite this, hardware-based mix engineers tend to have a tolerance and camaraderie around the use of a console, of whichever model. Although, again speaking more generally, Ihde (2013: 40) states, "it is also possible to 'objectify' the instrument such that it is understood to be animated, to have its own 'ghost', and thus be reified". While this observation could be applied to hardware, software users are often (fiercely) patriotic and frequently relish championing their preferred system to users of others. It is perhaps symptomatic of an established familiarity, yet fragility of mastery over a given highly complex system, and the naiveté to the feature set of other apparently arcane systems that give rise to this partisanship. As such, with regard to mixing ITB, Ihde's (2013) objectification could be considered beyond reification and towards deification. While console users frequently sit at different models in different studios ('latest' plug-ins aside), it is rare for the ITB disciple to change deity.

Tools

Acquisition of new software is part of daily life in the world of music production and mixing. It is often exciting to first launch a new tool and explore its capabilities, very likely before consulting its manual. The naiveté that one might hold when first approaching such a tool can lead to immediate gratification, and often with the application of the tool to a greater extreme than the experienced user might employ. Thus, it is a raw creative process free from preconceptions, quite akin to Hans-Georg Gadamer's (2004) being "lost in play",[12] which is an approach to recognition and interpretation, as related to musical improvisation and beyond, first suggested by Ramshaw (2005). Once experience with the tool accrues, it is likely that preferred modes of working will emerge, very often resulting in a rather more subtle application. While such a pattern equally applies to more general software processes, it is the frequency with which this situation arises in the mixing ITB environment (particularly through plug-ins) that causes it to exert more impact on this workflow.

Once such a tool proliferates and it is heard in a number of different contexts with different interpretations, that initial excitement might typically wane. Should the tool have a powerful unique identity, it may become meme, perhaps even to the point of cliché. A good example of this was the definitive early 'extreme' application of Antares Auto-Tune in Cher's (1998) *Believe*. Such an application falls into the category of creative abuse (Keep, 2005), and it is notable that the producers Mark Taylor and Brian Rawling published an exposé that originally attempted to hide their new tool (Sillitoe, 1999) in order to maintain the mystique of their technique. The application of Auto-Tune (and its equivalents) is currently all but ubiquitous in pop music, both as an effect and when applied more transparently, perhaps as the tool was originally intended. Perhaps the example of Auto-Tune is one that is simultaneously both meme and cliché.

As in the above example, in the mode of ITB, innovation is often driven by technology, in that periodically it is emergent technologies that facilitate new modes of mixing[13] and corresponding new sounds. Further, these increments can be distilled into a smaller number of seminal moments or indeed into a much greater number of smaller gradations, which may or may not be harder to quantify. Regardless, acceptance of the technology driver aligns with Rothwell's (1994) 'technology push' model of linear innovation, but this calls into question the ability of those who mix in the box to genuinely innovate. It is common for early adopters of relevant technologies to be the first to disseminate the application and therefore to be associated with such innovation as practitioners. As such, there is a case to accept this work as innovative simply because it lacks precedent. Further, Von Hippel's (1986) user-innovation model is accommodating of such uptake and application, and thus qualifies the mixing ITB pioneer as 'innovator'. Further, in terms of cultural impact, would the layperson attribute the sound of Auto-Tune to Cher or Antares?[14]

On the action of von Hippelian innovation reflexively influencing successive iterations of 'technology push', Zagorski-Thomas (2014: 149) remarks that "this, in turn, encourages further creative practice in the domain of new product design, and the cycle continues". The caveat, however, is context—here, related to genre. In a given genre, extant mixes must be devoid of a particular sound or musical gesture, and then feature it, and it is only the context which allows this transition that facilitates innovation. Innovation might therefore be regarded as the application of a technique that hitherto had not existed in that context, whether the seed for this technique came from a novel artifact or simply a different context. It is the context that is in fact the nexus of innovation when mixing ITB.

One issue worth discussion is that of the extent to which the tools are engaged with when engineers work ITB.

> As the tool-set becomes larger with the addition of DSP-based processes, a larger number of parameters may be considered and refined. The new tools, though shiny and exiting, may not be as sturdy as the old well-worn ones. A combination of both old and new may be able to refine the job still further. The skill of a modern producer to truly extend production values requires an awareness of how to use all tools to produce best effect. The understanding gleaned from being in a situation where a tool had to be mastered as it was the only one in the box is an important asset when presented with a much wider choice.
>
> (Phillips, 2012)

The motivation to go beyond the intended functionality of a given piece of equipment is guided by need, available options, experience and curiosity. As Phillips points out, having only a single tool forces the user to develop more flexibility with it. In the (analog) hardware era, a greater proportion of users tended to covet their finite range of equipment, which naturally led to experimentation, and even subversion, in an effort to gain the maximum range of functionality. As a given user gained more equipment of increasing complexity, they would typically have less time to engage with the full range of the available feature sets. The user might not read the accompanying manuals comprehensively, and overall they became less likely to develop and extend lateral functionality.

Of course, the professional user might still find time for all of this, but such a pattern might impact more on the amateur. Such a paradigm is not just a function of the physicality of the hardware tool, since a parameter-light vintage compressor will impact upon such situations in a different way than would a complex Eventide harmonizer.

Perhaps it is the presence of the embedded software operating system that makes the difference in this example. Such operating systems became increasingly ubiquitous in the sampler/workstation era,[15] vastly increasing potential functionality, but also making its intended boundaries more nebulous. The prevalence of user-installed software on a computer greatly multiplied the complexities that were typically dealt with, multiplied again by the transition from MIDI sequencer to DAW plus third-party systems, and ever more time was spent reading manuals and solving problems just to stay abreast of expected operation. As the palette of functionality continued to multiply exponentially, the depth of engagement with the tools diminished further, a simple function of available time versus the need to complete certain tasks. To some, this was amplified as both the intrinsic and financial value of software decreased.

Regardless, the contemporary ITB engineer needs to deal with an ever-larger toolset. While many neophiliacs might use this to pursue a sonic avant-garde, pragmatism and focus must remain. As Skea (2015: 1) observed of Rudy Van Gelder, who worked long before the software age, "the quality of Van Gelder's output rests not necessarily on technical innovation but on determination to master successive waves of state-of-the-art technology available to him and a legendary degree of perfectionism".

The use of preset parameter sets can make the array of tools much more accessible, although often at the expense of precise function or even appropriateness. Expert users might tend to shy away from presets in equipment with which they are very comfortable, yet still employ them in the interests of pragmatism with less familiar systems (Paterson, 2011). The sheer number of such presets is increasingly difficult to navigate, yet the range of their functionality can be increasingly flexible and exotic. Manufacturers are starting to respond with more intelligent preset options that simplify user engagement or standardize parameter sets across a range of equipment. In part, this is an attempt to keep (less involved) users abreast of the equipment, but it also introduces a danger of homogeneity and 'user irrelevance' when working ITB. Naturally, there are still a great many practitioners who actively pursue the novel, and there are ever-larger opportunities to implement this, but as with Van Gelder, adaptability is the key in the fast-paced fluxive arena.

mixMacros

In his quest to model compositional style by computer, David Cope (2004) defines "signatures" as musical motifs that occur in more than one work by a given composer. These are patterns that are "composites of melody, harmony, and rhythm" (Cope, 2004: 109), over a small number of beats. Such signatures provide a useful metaphor for many mix processes, but especially when mixing ITB. In the world of hardware, Michael Brauer's trademarked 'Brauerize' multi-bus compression technique (Tingen, 2008) is a notable example. Brauer assigns groups of instruments to one of a number of buses, each of which features a chain of compressors and equalizers. The console faders function pre-compressor chain, and everything is mixed through one of the chains. Although the specific configurations can be variable depending on the situation, the effect of this process produces a replicable sonic signature, derived from what are effectively templates. Each of these might here be coined in a more general sense as a 'mixMacro'.

While Brauer's large collection of hardware facilitates this, it is beyond the reach of many. In the software world, however, not just extensive compressor chains, but mixMacros of automation curves, complex signal routing configurations, groove templates, all manner of plug-in combinations, and more, are easily configurable in sets, and importantly replicable via copy/paste or embedding in DAW template sessions. The deployment of mixMacros allows the ITB engineer to form a custom tool kit from which to draw, to form signatures of sonic mediation. Counter to such convenience, such an approach can create an interesting tension for those who wish to avoid 'preset' sounds in the name of bespoke treatment or integrity. The nub is perhaps dependent on whether a manufacturer's effect chain is used in the mixMacro, or whether the engineer develops a wholly bespoke and considered one, perhaps even over many years, as was the case with Brauer. Either way, if the pre-configured mixMacro is consciously allowed to influence the artistic direction, it gains a degree of autonomy, something that Zagorski-Thomas (2014) describes as its residing within the actor-network theory (ANT) of social theory. The mixMacro is an actor concatenating with Ihde's (2013) non-neutral instruments.

So, in mixing ITB, mixMacros are autonomous objects that might be assembled in a near-infinite order, in an actually infinite number of contexts (tunes), yet each always contributes its own unique 'signature'. This allows an extension to a further metaphor from visual arts, specifically the actual implementation of these components could be regarded as parallel to collage or montage. As Cutler (2004: 145) notes, "with so many precedents in the world of the visual arts . . . it does seem surprising that it took so long for there to be similar developments in the world of music".

Conversely, Theberge (1997: 206) states, "the artistic practices of collage, assemblage and montage used in popular music virtually destroy the organic integrity of 'the work'". He is speaking specifically of songs,[16] and clearly does not hold that aspect of them in high regard. Of course, the semantics of "organic integrity" might be analyzed; surely, such a statement must relate to the singer-songwriter or the paper-based composer. It is interesting that in the intervening years since that was written, both modes of creator have increasingly turned to computer assistance to empower their craft. Crucially, however, if in the context of mixing ITB integrity is placed within these metaphors from the outset,[17] then they cannot be viewed as invasive to a self-aware art form, especially when composition might be in tandem with production and mixing ITB. The mixMacros might be autonomous, they might be actors, but they are still just tools.

Within a given DAW setup, it is relatively easy to perceive the full scope and boundaries of the mediation in an ITB session. This is reified in the saved file of that session, independently of the musical performances to which it applies. The session could be thought of as containing an aggregate of mixMacros, a die that could in theory be applied to different tunes (time-dependent components such as automation, apart) to which customizations were also added. This is analogous to successive tunes by a rock band of fixed instrumentation running from tape through the same console setup, but goes beyond simply replicating such a configuration in a DAW. When required, custom session templates and cross-session importing of settings facilitate this type of workflow when ITB. Of musical analysis, Butterfield (2002: 327) discusses the concepts of 'autonomy' and 'musical objects', and quoting McClary states, "Works rely for their meaning not an abstract, formal or structural coherence, but on 'codes of social signification such as affective vocabularies and narrative schemata'" (McClary, 1993, as quoted in Butterfield 2002: 328).

This model could be translated either to individual mixMacros or indeed aggregates, and perhaps if augmented could be used to formalize specific approaches to ITB.

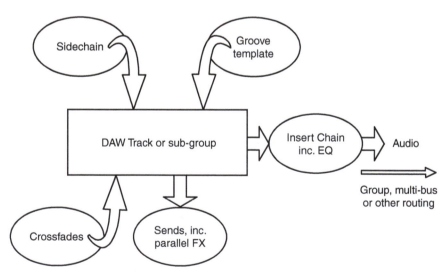

Figure 5.2 A schematic of a mixMacro for one or a group of tracks, illustrating some of the factors that define its 'signature'

Recall

One of the biggest advantages to mixing ITB is total recall, and as such, it deserves specific mention here. Opening a DAW session re-creates the exact state that it was in when saved, whereas almost any analog part of the signal path would have to be manually restored. Manual restoration can virtually never be totally exact (to phase accuracy). More importantly, it is the time taken to re-patch the hardware and set all the parameters that makes the difference. Mix engineer Mark 'Spike' Stent attests:

> In addition, producers, record companies, and artists are used now to the fact that they can call you, even two months after your mix, and request a change, and you just bring up the Session and five or 10 minutes later the change is made. So mixing in the box is about time and being flexible, and of course it also saves on the budget.
>
> (Tingen, 2010)

However, there are also a number of reasons why mixing ITB can be more time consuming than legacy hardware approaches, at least during the primary mix. When performing a given operation such as compression, the mixer must decide whether to opt for a favorite plug-in or perhaps try out a number of options based on prior recommendations or curiosity. Typically the number of plug-ins available to the mixer will be an order of magnitude greater than those in a hardware studio, and many will offer unusual modes of operation that require considerable experimentation and/or reading of the manual. As previously mentioned, it is easy to become fixated on micro editing, not just of audio, but also automation and numerical parameters—such operations often require highly repetitive and multi-stage workflow. Despite this, an increasing number of professional mix engineers are turning towards an ITB approach in the interests of pragmatism.

Future

The current paradigm of mixing ITB is almost entirely based around the metaphor of the traditional mixing console. New modes of mixing are now being proposed. There are innovative designs, such as that of Cartwright et al. (2014), which presented a 2D map around which a cursor could be moved to influence the EQ and gain settings of every channel simultaneously on a holistic like or dislike basis. Automated mixing is being developed as alluded to in chapter 15 by Joshua Reiss, and this is likely to realize many procedural tasks for the professional, and perhaps one day precipitate user irrelevance with regards to technical operations, allowing the user to focus on artistic and strategic choices.

Butterfield's (2002) model is yet to be fully adapted and specified to define specific ITB approaches, but perhaps if combined with Moylan's (2007) mix descriptors, then a taxonomy of such approaches could be formed for the digital era. This could be augmented with Zagorski-Thomas's (2014) sociological approaches to align with the emergent musicology of record production.

It is inevitable that further new and surprising tool sets will evolve and apply artificial intelligence to many of the processes, extending Moorefield's (2010: xiii) "reality of illusion". Perhaps the trajectory of illusion can be illustrated even more clearly if parallels to other art forms are made. Whereas once, audiences were amazed by ghostly projections of actors onto pieces of glass on the theatre stage, we currently have markerless facial motion-tracking (Faceshift, 2015) in cinema and games, and this technology is still in its nascency. In this context, the physical characteristics of an actor are now irrelevant, and further, the very performance can be retrospectively adjusted. Intel RealSense (2015) is poised to bring similar 3D sensing and beyond to the consumer, with many developers yet to identify potential markets. It takes little imagination to translate this into an equivalent form of sonic control, as is already being demonstrated by the augmented performance capabilities of Zoundio (2015), where lesser-skilled performers can deliver virtuosic performances, yet still control the artistic direction. A Guitar Hero (Activision, 2016) metaphor for real instruments might yet become the norm in the future, with long-honed traditional dexterity as obsolete as touch typing in the current voice-recognition era.

In addition to new 2D mixing-interface paradigms, 3D gesture specifically for musical control is already at hand with the Leap Motion controller and the Geco MIDI app, and the emergence of augmented reality for the DAW is likely to extend this, making the action of mixing ITB ever more intuitive and powerful. At the time of writing, Bullock (2015) is developing algorithms for two-handed sonic manipulation in free space, supported by a 3D real-time visualization. Such approaches might develop into a new mode of HCI when mixing ITB and be applied in numerous contexts. Retina tracking and augmented-reality workflows are also likely to emerge commercially in the near future, and all the time, brainwave-based control systems become ever more feasible.

As component modelling develops to the point where software is ubiquitously accepted as being genuinely capable of accurate emulation of consoles and hardware outboard, workflow will be the most compelling demarcation, although this too will evolve in parallel to the growing demographic that feels independent of a console/outboard-based workflow. Mixing ITB will become ever more powerful, autonomous and independent of its legacy, allowing more of us to reach the current zenith more rapidly. If reaching an ideal takes less time, then reaching a number of equivalent and musically interesting ideals in the same time might become possible. The emergence of new dynamic or interactive playback formats that require multiple mixes will be empowered by this. This mode might go beyond the zenith, and for a given tune offer multiple new points on the celestial sphere of music.

Notes

1 It is particularly pertinent in a chapter on mixing in the box to emphasize that the content of a 'track' might be either recorded audio or a MIDI sequence.
2 For example, recorded to ½" magnetic tape, digital audio tape (DAT) or CD. Such mixes were often recorded, and artists would decide which they preferred for the final release by retrospectively listening to each.
3 This text will not concern itself with room acoustics, which of course influence the size and shape of any mixing environment.
4 Who will generally create a rough mix that includes printed effects for guidance.
5 A term derived from intensive MIDI editing in the 1990s, employed by artists such as Aphex Twin.
6 Sometimes, this alignment was done with a bespoke hand clap before the music commenced. It is perhaps interesting to speculate that if the drummer were to clap in a natural position, one foot above the snare, then the synchronization would be susceptible to approximately 2 ms error; 1 ms farther from the (assumed close) snare drum microphone and 1 ms closer to the overhead microphones.
7 Of course, this is actually the primary function of quantization.
8 For instance, with compression attack and hi-mid EQ.
9 As discussed, for instance by Vickers (2010).
10 Although this is currently being challenged by emergent multi-touch systems such as the Slate Raven (Slate Pro Audio, n.d.).
11 Logic Pro has long sported screen-space efficient GUIs that packed many small fields into a limited area, e.g., the ES2 synthesizer.
12 First published in 1960.
13 As well as, of course, the greater field of music production.
14 Cher, most likely, since they will not have heard of a niche company. Perhaps the scholar might prefer that the debate on such a musical gesture be between Homer Dudley and Sparky the Magic Piano.
15 Which commenced with the launch of the Fairlight CMI in 1979.
16 It would be unfair to presume his holding any objection to mixing as collage, but it serves as a useful foil for this metaphor.
17 It must be so, given that the DAW tool set is ubiquitous across genres.

Bibliography

Activision. (2016). Guitar Hero Live The Game | Official Site of Guitar Hero. *Activision.com*. Available at: https://www.guitarhero.com/uk/en/game [Accessed: 18 April 2016].

Bullock, J.B. (2015). Transforming Transformation: 3D Models for Interactive Sound Design. *Gateway to Research*. Available at: http://gtr.rcuk.ac.uk/project/23DDE2F7-3184-4D94-97C6-85B50B09602C [Accessed: 6 January 2016].

Butterfield, M. (2002). The Musical Object Revisited. *Music Analysis* 21 (3): 327–380. Available at: http://www.jstor.org/stable/3840795 [Accessed: 14 March 2013].

Cage, J. (1952). Cage: The 25-Year Retrospective Concert of the Music of John Cage by Various Artists. *iTunes*. Available at: https://itunes.apple.com/gb/album/cage-25-year-retrospective/id521720437 [Accessed: 24 June 2014].

Cartwright, M., Pardo, B. and Reiss, J.D. (2014). 'MIXPLORATION: Rethinking the Audio Mixer Interface.' In *19th International Conference on Intelligent User Interfaces*, Haifa, Israel, 365–370.

Cher. (1998). *Believe*. Available at: https://itunes.apple.com/gb/album/believe/id73273758 [Accessed: 20 October 2015].

Clifford, A. and Reiss, J. (2010). 'Calculating Time Delays of Multiple Active Sources in Live Sound.' *Audio Engineering Society*. Available at: http://www.aes.org/e-lib/online/browse.cfm?elib=15580 [Accessed: 7 December 2015].

Cope, D. (2004). *Virtual Music: Computer Synthesis of Musical Style*. Cambridge, MA: MIT Press.

Cutler, C. (2004). 'Plunderphonia.' In *Audio Culture: Readings in Modern Music*, eds. C. Cox and D. Warner. New York City: Continuum International Publishing Group Ltd., pp. 138–156.

Faceshift. (2015). *Faceshift*. Available at: http://www.faceshift.com/ [Accessed: 14 July 2015].

Gadamer. H.-G. (2004). *Truth and Method*, New edition. London; New York: Continuum.

Ihde, D. (2013). *Technics and Praxis: A Philosophy of Technology: Volume 130*, 1979 edition. New York: Springer.

Intel. (2015). Intel® RealSenseTM Technology. *Intel*. Available at: http://www.intel.com/content/www/uk/en/architecture-and-technology/realsense-overview.html [Accessed: 15 April 2016].

Johnson, S.A. (1999). *Interface Culture: How New Technology Transforms the Way We Create and Communicate*, New edition. New York: Basic Books.

Keep, A. (2005). 'Does Creative Abuse Drive Developments in Record Production?' In *First Conference of the Art of Record Production*, London, UK. Available at: http://www.artof recordproduction.com/index.php/arp-conferences/arp-2005/17-arp-conference-archive/arp-2005/72-keep-2005 [Accessed: 7 May 2014].

Moorefield, V. (2010). *The Producer as Composer*, First edition. Cambridge, MA: MIT Press.

Moylan, W. (2007). *Understanding and Crafting the Mix: The Art of Recording*, Second edition. Burlington, MA: Focal Press.

Mycroft, J. and Paterson, J.L. (2011). 'Activity Flow in Music Equalization: The Cognitive and Creative Implications of Interface Design.' In *Proceedings of the 130th Audio Engineering Society Convention*. Available at: http://www.aes.org/e-lib/browse.cfm?elib=16568 [Accessed: 19 March 2013].

Mycroft, J., Reiss, J.D. and Stockman, T. (2013). 'The Influence of Graphical User Interface Design on Critical Listening Skills.' In *Proceedings of the Sound and Music Computing*, Stockholm, Sweden, 146–151.

Mycroft, J., Stockman, T. and Reiss, J.D. (2015). 'Audio Mixing Displays: The Influence of Overviews on Information Search and Critical Listening.' In *Proceeding of the 11th International Symposium on CMMR*, Plymouth, UK, 682–688. Available at: http://cmr.soc.plymouth.ac.uk/cmmr2015/downloads.html [Accessed: 26 November 2015].

Orban, R. (1979). FM Broadcast Quality. *Stereo Review*.

Paterson, J.L. (2007). Phase Experiments in Multi-Microphone Recordings: A Practical Exploration. *Journal on the Art of Record Production* 1. Available at: http://arpjournal.com/1803/phase-experiments-in-multi-microphone-recordings-a-practical-exploration/ (Accessed: 19 March 2013).

Paterson, J.L. (2011). 'The Preset Is Dead; Long Live the Preset.' In *Proceedings of the 130th Audio Engineering Society Convention*, London, UK. Available at: http://www.aes.org/e-lib/browse.cfm?elib=16569 [Accessed: 19 March 2013].

Phillips, M. (2012). Interview with Martyn Phillips.

Ramshaw, P. (2005). 'Lost in Music: Understanding the Hermeneutic Process in Creative Musical Composition.' In Vilnius, Lithuania. Available at: https://www.academia.edu/172365/Lost_in_Music_understanding_the_hermeneutic_process_in_creative_musical_composition [Accessed: 5 January 2016].

Rothwell, R. (1994). Towards the Fifth-Generation Innovation Process. *International Marketing Review* 11 (1): 7–31. Available at: http://www.emeraldinsight.com/doi/abs/10.1108/02651339410057491 [Accessed: 7 December 2015].

Sillitoe, S. (1999). *Recording Cher's 'Believe'*. Available at: https://www.soundonsound.com/sos/feb99/articles/tracks661.htm [Accessed: 20 October 2015].

Skea, D. (2015). Rudy Van Gelder in Hackensack: Defining the Jazz Sound in the 1950s | Jazz Studies Online. *Jazz Studies Online*. Journal. Available at: http://webcache.google usercontent.com/search?q=cache:xWTfR-HmMo4J:jazzstudiesonline.org/resource/rudy-van-gelder-hackensack-defining-jazz-sound-1950s+&cd=1&hl=en&ct=clnk&gl=uk [Accessed: 8 July 2015].

Slate Pro Audio. (n.d.). *RAVEN MTX: Slate Pro Audio*. Available at: http://www.slateproaudio.com/products/raven-mtx/ [Accessed: 13 December 2013].

Sound Radix. (2010). *Sound Radix*. Available at: http://www.soundradix.com/ [Accessed: 7 December 2015].

Tano, S., Yamamoto, S., Dzulkhiflee, M., Ichino, J., Hashiyama, T. and Iwata, M. (2012). 'Three Design Principles Learned through Developing a Series of 3D Sketch Systems: "Memory Capacity", "Cognitive Mode", and "Life-size and Operability"'. In *2012 IEEE International Conference on Systems, Man, and Cybernetics (SMC)*, 880–887.

Theberge, P. (1997). *Any Sound You Can Imagine: Making Music/Consuming Technology*. Hanover and London: Wesleyan University Press.

Tingen, P. (2008). Secrets of the Mix Engineers: Michael Brauer. *Sound on Sound*. Available at: http://www.soundonsound.com/sos/nov08/articles/itbrauer.htm [Accessed: 2 July 2014].

Tingen, P. (2010). Secrets of the Mix Engineers: Mark 'Spike' Stent. *Sound on Sound*. Available at: http://www.soundonsound.com/sos/feb10/articles/it_0210.htm [Accessed: 14 December 2015].

Vickers, E. (2010). 'The Loudness War: Background, Speculation, and Recommendations.' In *Audio Engineering Society Convention* 129. Available at: http://www.aes.org/e-lib/browse.cfm?elib=15598.

von Hippel, E. (1986). Lead Users: A Source of Novel Product Concepts. *Management Science* 32 (7): 791–805. Available at: http://mansci.journal.informs.org/content/32/7/791.abstract [Accessed: 26 August 2011].

Webber, J. Interviewed by J.L. Paterson (2013).

Zagorski-Thomas, S. (2014). *The Musicology of Record Production*. Cambridge University Press. Available at: https://www-dawsonera-com.ezproxy.uwl.ac.uk/readonline/9781316014035 [Accessed: 23 April 2015].

Zoundio. (2015). Zoundio. *Zoundio*. Available at: http://www.zoundio.net/zoundiomain/ [Accessed: 14 July 2015].

6

Audio Editing In/and Mixing

Alastair Sims with Jay Hodgson

Audio editing has played an important role in record production since razor blade first touched tape, as it were. That said, modern computer-based modalities comprise an entirely new and crucial genus of that aesthetic species. So crucial has this new genus become to record production in general, in fact, that few working recordists would blink if I suggested to them that modern record production is defined or characterized almost entirely by it. Indeed, production styles are no longer marked by the amount of audio editing they encompass, as they once were, simply because every style of professional music production now entails the same high degree of editing. Those who know what to listen for can hear audio editing permeate every musical nook-and-cranny of modern record production, regardless of genre. And yet, despite the tremendous growth in research on record production in the last few decades, very little research focuses directly on this crucial new musical competency.

What follows is meant to address two of the lacunae I identify above. It provides a broad description of audio editing as a musical competency in and of itself, and it elucidates its position within—and with regards to—modern record production at large. An overview of common editing techniques is then provided, and musical examples supplement the discussion to help readers hear those techniques in action. These audio examples can be downloaded at http://ww.hepworth-hodgson.com, and it is strongly recommended that readers audition each example precisely where indicated in the text.

The manner of explanation modeled below would best be considered a methodological hybrid, mixing ethnographic interview techniques with traditional musicological analysis. It is my firm contention that, to truly understand modern professional production practices, a methodological broadening is required that allows practitioners themselves to speak in academic contexts. And I believe the hybrid below models one among many possible such 'broadenings'. I chose a series of questions, in a particular sequence that I felt addressed the lacunae comprising the primary subject of this chapter. And I sought the expertise of Alastair Sims, one of Canada's most successful audio editing engineers currently at work, to provide those answers.[1] Sims and I spoke for a few hours, a transcript of our conversation was made and edited, and then Sims was given the transcript to approve and edit further however he saw fit. I identified areas where I thought listening examples might help concretize some of the concepts and techniques Sims discussed, and Sims provided them in turn. The Download and Listen tracks are listed later in the chapter and are available on www.routledge.com/9781138218734 and www.hodgsonhepworth.com. It is ultimately our hope that, in doing all this, we have provided analysts with a useful toolbox that will help them hear audio editing permeate the modern recorded soundscape completely. In turn, the musical role that audio editing plays in modern record production should clearly emerge to the analytic fore.

Editing In/and Record Production

Digital editing is a relatively new—and, thus, relatively unknown—process. How would you define what you do, in the broadest possible sense?

The easiest way to explain what I do is to say that I make musicians sound 'right' for the track. In other words, I make recorded material sound 'on time' and 'in tune'. For instance, one of the simplest types of editing, if a musician flubbed a note, I might replace it with a note from a different take, or with a note from later within the same take, so the performance sounds 'right'. The three main aspects of editing are timing (rhythm), tuning (pitch) and timbre.

When you're handed a session file from the control room, what judgment calls do the producer(s) and engineer(s) assume you will make without needing to tell you? Are you usually given explicit instructions each time you're given a file, or are there certain tasks that one can assume the audio editor should do regardless of project particulars?

When I'm working with a producer, at the beginning of the project we'll discuss the editing needs for the whole project in general terms. For instance, whether I'll be simply cleaning the tracks or completely quantizing and tuning the song(s). After this point, I generally know what needs to happen to every file I'm given. I'll get a session file which is already comped, so it's just one 'playlist' I'm working on ('playlist' is the name given to a single composite track in Pro Tools). At this point, my options are limited anyway, so there's really no need to give me special instructions. I'm not expected to choose different takes, to create a comped track, for instance. It's more about cleaning up a performance or track. For example, a take may have the attitude or feel that the producer and band want to capture, but the musician played out of time and stomped their foot loudly during the recording. Happy with the performance, keeping in mind that they know what I can 'fix' in editing, the producer will then send the track to me. They know that when they get the track back, the performance will sound 'tighter' and the undesirable sounds will be gone.

How long does editing usually take?

That depends on the instructions I receive at the beginning of the project, and of course on the material I'm given. If the production only requires me to clean up noise and fix blatant mistakes, for example, one bad drum fill in a song, it might only take thirty minutes for an entire song. However, if the producer wants a fully quantized and tuned performance, it can take substantially longer. Completely quantizing an easy guitar part—perhaps a guitar strumming whole notes, for example—I can be done in about fifteen to thirty minutes. Eighth-note or sixteenth-note 'power chord' guitar or bass parts (think punk rock) can take a lot longer to edit, upwards of a day to fully quantize. Drums generally take half of a day to a full day if there are some difficult edits to make.

Is it safe to say that nearly every Top 40 song these days is subject to editing in some form or another?

Yes, there is almost guaranteed to be editing on all recordings at some level. Famous musicians can generally perform quite well—they're famous for a reason. But there will still be some editing on their songs, even with virtuoso musicians. Almost every track, on every record, has some editing, whether it be as small as comping together takes or cleaning noise or as large as quantizing and tuning an entire song. This may sound jaded, but that's just the nature of the beast nowadays.

How did you first discover audio editing?

I was referred to an editing position in a production team by a former employer and mentor, so I discovered the role while receiving 'job skills training', as it were. Normally, the career path for someone who does editing has a standard route, very similar to other positions in the recording industry (assistant engineer, studio intern). You're hired on as an unpaid assistant, similar to an intern. The people employing you expect you to show up, say nothing and don't mess anything up. It may sound harsh, but they need to see if they can trust you before they give you any responsibility. They want to know that you won't say stupid things around clients or share sensitive information, because as an editor you will be listening to and possibly fixing very intimate aspects of a recording. Then you get given a task and, if you do a good job, you get given another, and so on, until you are an indispensable member of the production team.

When I started working with Gavin Brown in Toronto, there were two other editors already working there. He, the engineer and the other two editors made it clear that they did not want any input from me. They just wanted me to sit and observe their work, to learn what they do and how they do it. The first assignment they gave me was editing bass for a songwriting demo from Gavin [Brown]. I don't know if it ever got sent to the artist, but it was essentially a test to see if I could run Pro Tools and do some basic editing in the DAW. Looking back, if I did that first task now it would only take me about thirty minutes to edit. But back then it took me about three days! I was nervous and so I questioned everything I did. I'd think I was finished, and the other editors would come and listen to my work, and find more and more mistakes. It was brutal!

What types of technical concerns do editing engineers worry about?

One of the most important aspects in editing is to be able to find the exact beginning of a sound. That might sound simple, but it's very much an art in and of itself. With guitar, for example, the beginning of a guitar tone is not necessarily when the pick first scrapes the strings on a guitar, even though you'd intuitively think it was. For the performer, yes, that is where the note begins. But for the editing engineer, we are concerned with where the tone, and pitch, and sustain information builds to a sufficient point that it 'sounds like' a note at mix level. So you have to be able to hear like an editing engineer before you can even begin to edit. Finding the beginning of notes, and knowing how to 'smooth' your edits so they aren't audible, correcting pitch are the main technical concerns. Another facet of being an editor that never really gets discussed is that you are generally considered the computer guy. Whether that means you are also the Pro Tools op (operator) for the session or just the guy that they come to when the session keeps crashing or won't allow a certain function to operate properly, you'll be the guy they'll call. So having in-depth knowledge of the software and programs being used as well as of computers themselves is invaluable.

Can you elaborate a bit? Maybe provide some concrete examples of types of edits you might make on a track?

Let's focus on timing first, taking a musical performance and adjusting the rhythm of individual notes or longer passages (see Figure 6.1). If a musician played a note before the beat, and I move it back onto the beat, there will be

Figure 6.1 Quantizing audio, 1

Figure 6.2 Quantizing audio, 2

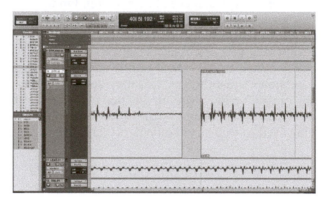

Figure 6.3 Quantizing audio, 3

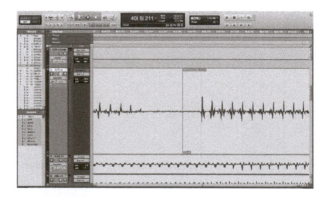

Figure 6.4 Quantizing audio, 4

silence left over between where the cut was made and where the note was moved to (see Figure 6.2), as well as clicks at the cut points (see Figure 6.3).

Normally, when working in a clip/region-based DAW, you pull the beginning of the second clip back to meet the end of the first (see Figure 6.4), and place a crossfade there, roughly five to eight milliseconds'

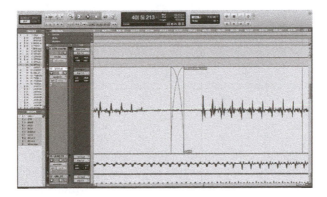

Figure 6.5   Quantizing audio, 5

Figure 6.6   Quantizing audio, 6

Figure 6.7   Quantizing audio, 7

Figure 6.8   Quantizing audio, 8

duration, or more, as a starting point (see Figures 6.5 and 6.6). The shorter the crossfade the better, as a general rule. Occasionally, this process will cause some audio to repeat itself (see Figure 6.7) which can be addressed by adjusting the length of the crossfade (see Figure 6.8). Sometimes, though, the fade needs to happen earlier into the note.

Failing these options, you might see if the same note is played somewhere else in the song, and use it in place of the problem note. You might also use time compression or expansion to smooth edits, digitally lengthening a note to fill the gap. You might say there are three primary ways to approach editing timing. Each one of these approaches—(i) using fades, (ii) replacing notes and (iii) time compression and expansion—are essential techniques or tools for us and can be applied whether you are quantizing to a grid or nudging notes manually.

Pitch correction and tuning is another aspect of editing, usually a little more automated than timing. You would generally use a plug-in (Melodyne, Auto-Tune) to achieve the edited result. Just like timing, pitch correction can be applied to the whole track or simply a few off notes depending on the desired aesthetic for the song.

Musicians are sometimes very suspicious of editing,
especially if they have very little experience with modern
recording workflows. How do you address such concerns?

I've never encountered someone who said, "No! You can't auto-tune me!" The majority of people that I work with, whether professional or amateurs, tend to say, "Oh, you can do that? That's awesome! That saves time . . . That makes me sound better . . ."

Why do practice rooms exist on music campuses? Why do musicians spend hours and hours practicing? It's because they want to be perfect, or as close to perfect as they can get, when they perform. Drummers practice to a metronome, to learn to play on time. String players tune their instruments, and practice their intonation, for example, so that they sound perfectly in tune, and they spend lifetimes perfecting their tone. So why would anyone get frustrated when I add one more small layer of production to help them achieve the vision of their sound? I edit with the aesthetic aims of the recording at large in mind, after all. It's not like I change pitches around on performers to what I like instead, or needlessly quantize an already amazing piece of music or take of a song. Editing engineers do their very best to 'serve the performance', the same way producers do their very best to 'serve the song'.

In the past, musicians were required to record in extended takes with minor overdubs. How has editing changed this, in your experience?

The process of recording is still very much like this. The majority of material is captured in long takes, with small overdubs throughout if needed. The difference is that musicians don't need to feel pressure from this anymore. Faced with a difficult passage or solo that they can't get in one take, there are many ways to capture it. Be it using multiple takes and comping them together, or punching in a number of times, or maybe even punching in every single note or chord. It doesn't matter now how you do it now, because with the use of editing in DAWs, it has not only made it fast and easy but sound transparent.

What would you say is the main incentive for hiring an editing engineer for a recording session?

There are many, but two come immediately to mind. The main reason to hire an editor has to do with budgetary concerns. A band can rent out a studio for two months and get near-perfect takes through grueling punching and repetition, or they can spend two weeks recording, get the tracks 85% 'correct', and hire someone to edit the remaining 15%. The latter is a far more efficient process, compared to the way recording sessions were done before. And it's only getting more efficient. During the early days of editing and comping, you would record all of your material and then spend a couple of weeks editing it into shape after. Now we do editing and tracking at more or less the same time. You can have entirely finished songs ready to be mixed now, by the time you've torn down and are leaving the studio. That said, though, you can't rely on editing as a crutch during tracking. What you record has to be as good as possible, and as close to the final product as you can get without editing, so artist morale stays high and ideas keep flowing in sessions.

Editors also fulfill an aesthetic function, however. Popular music now has a particular sound to it that comes from electronic music (drum machines, sequencers), and people expect records to sound that way (whether they realize it or not). Katy Perry, for instance, would not benefit from a 'live off the floor' early-1970s Black Sabbath production mentality, right? In fact, a majority of listeners likely interpret that sort of editing (or lack thereof) as 'bad' production. So editing is really a production tool now, used to achieve that 'shiny pop perfection' sound that people come to expect from radio hits.

How important is the equipment you use for editing?

I have a hierarchy that I visualize when trying to explain how important certain aspects are to the recording process, and it moves from source to destination.

Song > Player > Instrument > Microphone > Equipment > Processing (Editing)

Figure 6.9   Order of importance

Gear *is* important. But it is by no means top of my list. Everything starts with the song, the player and the instruments you record. If any one step in the chain is faulty, it cannot be fixed by the next tool in the chain. A great song, or an amazing player, can make up for a bad microphone, for example, but a bad song cannot be fixed by a good microphone or preamp or editing for that matter. So you have to follow the chain, and get each step 'right' before moving on to the next one.

You have worked with some of the most successful recording acts in the world. What have you learned from working with them?

Most of the successful people I work with are uncompromising about quality. This suits me just fine because, in editing, the point is to make everything sound 'perfect'. So I learned from those artists to *never* compromise. Ask yourself, "Is it perfect?" If it isn't, you have to make it perfect. And you should have this attitude at every stage of production, even when you're recording a demo. In fact, I once sent a demo to someone who was financing a project to show them the progress, and while it was still very much a songwriting demo, they sent it around to interested management, promoters, record labels and so on. So I was happy that I'd done my very best to make it sound as good as possible at that early stage in the process. Not focusing on the fact it was a demo, simply focusing on making the best I could regardless if it was the first time it was captured.

When editing, you can crossfade, find a note elsewhere in the song and replace it. You can use time compression and expansion and pitch-correction software. You still even have the option to re-record the part! You should be able to find a way to make a recording 'right', it may just take a little longer than you hope, but punching in every single chord can sound great, and is done surprisingly often. You can apply this ethos to songwriting, engineering . . . *anything*. If you're not getting the sound you want, then you change it. It's that simple.

You mentioned earlier that the main aspects of editing are timing, tuning but also timbre. Can you explain this?

The timbre or tone aspect of editing is an interesting one; this is where you start adding to the attitude and performance more. Taking part of the performance that has more attitude or more of a quality that you want and moving it to other parts of the song. For example in a guitar take, the first chorus, the guitar player was really laying into the guitar, therefore hitting the amp harder giving a more distorted, edgy sound. In the second chorus, though, he was relaxing and playing smoother, perhaps improving the timing and tuning of the chorus but taking away the energy. Now take the edited (timed and tuned) first chorus and paste it to the other choruses so they maintain the tone and energy of the first and you've just edited the tone or timbre of a track.

Editing In/and Mixing

Is editing now a required tool in the modern mix engineer's arsenal? Do mix engineers need to know how to 'tune' and 'time' tracks the same way tracking engineers know how to use, say, a compressor?

I would say yes, editing is definitely a required tool in the modern mix engineer's arsenal, especially now that mixing is so often done via the Internet. You can get sent a session file from across the world, and anything could be wrong with it. And you need to send back something that is great. So, as a mixer, you're the last line of defense before the talent and audience hears the track; it's on you to make it right. If there's a fill that's out of time, or one word with tuning issues on it, then you need to fix it. There are certainly times when you're told that they have edited everything, and it is as good as they can get. Being able to take it that much further if there are still tuning issues or timing issues is incredibly important. So editing can certainly impact and be a part of mixing. It doesn't matter if you, as the mixer, are going back and tuning and timing yourself, or hiring an assistant to help you with the editing. As long as the impact of editing on a finished mix is known, that's the most important part.

This all said, I wouldn't say that editing is *expected* in mixing. The mixer's job is to balance instruments and tracks in a song as well as to shape the overall tone to fit the final vision of the project. This makes them one of the last stages of 'quality control', so it does often happen that mistakes are found and need to be rectified by the mixer. Often the new balance they are creating will reveal or boost a flaw in the track, which is there because the engineer/producer/artist was working in a poor listening environment.

How does editing ramify later, during mixing?

Some things are hard to mix when they're unedited. When you have a bass and kick drum that don't line up, it can be a hassle to get them to work together in a way that is usable at mix level. Pulling them together, 'timing' them, so they align to grid, makes my job way easier when I'm mixing, because doing this makes the track sound 'tighter' and adds more punch to the low end in general.

Do you find that editing is such an integral and expected part of tracking now that, even before you get session files to mix, the material is already mostly edited?

The way much of music is written and performed now via samplers, sequencers, MIDI and loops means that a lot of mix elements are already 'edited' even before they're flown into the arrange window. You shape the tone and quantize a MIDI part, and it's done. What this means, though, is that when you don't have something that's edited like that MIDI part—like a live vocal or guitar—it could sound 'off' next to that MIDI track, and you'll need to address that with the kind of editing I do. One of the best places to hear this is on a metal/hardcore guitar track. When the notes end is almost as important as when the note starts in this genre, so having a very edited and clean guitar track it vital. ('Physical Education' by Animals as Leaders or 'Lost in the Static' by After the Burial) You get a really rhythmic sound on those tracks, more than you do with rhythm guitars in a lot of other genres. And, thanks to editing, those guitar parts are almost robotically precise! This is a crucial element of the genre, in fact. If you don't edit the guitar parts, the track sounds 'wrong' to interested listeners. And you just wouldn't have been able to achieve this level of precision in mixing before. You would have had to play around with gates, and expanders, and do a lot of punching in. Humans just can't achieve this level of precision on their own. So, yeah, editing changes not only how tracking happens but also how songwriting, production and mixing happens, insofar as you can conceive and achieve a wide variety of sounds that you couldn't before digital editing became commonplace.

So is editing post-production or production, then?

Editing was first viewed as a post-production tool, separate from all other processes. Now editing has expanded into every part of the production process. It can be used as a tool that works in the background, which you don't even hear but you still know is there. Like compressors which used to be used to manage only level and dynamic contour, now they're used to produce truly creative sounds, like the famous drum sound in 'When the Levee Breaks' by Led Zeppelin or side-chain pumping you hear in electronic music. The same transition has been (and is) happening with editing—take Auto-Tune, for example. Engineers used to use Auto-Tune primarily to fix a note or two when it was first available. Then Cher came along, with her song 'Believe', and all of a sudden Auto-Tune becomes a texture in and of itself. The later an example of editing playing a role is you'd have to call 'production' rather than 'post-production'.

How do you know how much you're expected to edit,
then, when you've only been hired to mix a track? That is,
how much editing is tacitly expected when handed the mix
brief?

I would say the best way to know what kind of editing is required would be listening to the reference tracks sent to you by the artist or producer. They have a vision in mind and will usually talk about other songs or bands they're looking to sound like. Listening to those tracks you should be able to get an idea for what you'll need to do to achieve that sound. The tough part is when they have a very different productions style (i.e., no editing and very loose) from the reference tracks they've passed along. When this happens, you have to figure out if the budget as well as the schedule will allow for editing. If they sent me a session file, and a few hundred dollars, and said, "Mix it", then I might fix one or two pitches on the vocals, or move a drum fill around, but that's about it. If there's more money in the budget, as well as time in the schedule, then I would sit down and edit the drums, put in drum samples, edit the bass and guitars, tune the vocals, time the vocals, and then start mixing. All the editing would be done before mixing.

Hearing Audio Editing

Do you hear editing when you listen to a track?

Yes, of course.

Do you hear it because the editing introduces certain
'telltale' sonic artifacts?

For the most part, yes. Vocals are the main element of a pop song, so they are very forward in the mix and it gives you the chance to hear any editing on them fairly well. So hearing tuning or timing on lead vocals is pretty common. Some things you might hear aside from extreme tuning (think T-Pain or Cher) on vocals are additional overtones or stuttering of consonance if a vocal track is timed using certain digital tools like ElasticAudio in Pro Tools, FlexTime in Logic or the audio plug-in Vocalign. This can have a huge impact on the vocal performance and tone, especially when fully timing, or quantizing, the vocals. If you take the time to chop and move a vocal part around by hand and re-fade the tracks, however, it can be hard to hear what I'm talking about. One method is not better than the other, you just get different results and they require a different amount of time and effort.

Can you describe some more of these artifacts? For example, I know that if the attack time setting on your auto-tuner is dialed into a setting that's too fast for the part, and the vocalist has a pronounced vibrato, the result will be what I can only describe as a robotic 'warbling'.

You can get 'the warble' when your auto-tuner snaps between two different notes very rapidly, for sure. The other thing you get with tuning is, if you tune the note too far from its original pitch, the track will start to sound weird and move out of what sounds like a normal singing range. The singer will either sound like a chipmunk or like they are trying too hard to sound 'manly'.

On the timing side of things, you'd probably have the easiest time hearing artifacts from the Vocalign plug-in. That processor aligns two different takes, very useful when you want to have doubled vocals that are very tight. When you have hard consonants like *kuh* or *guh*, though they are lower in volume compared to many of the other elements in speech and singing, they appear as quite sharp transients in the waveform. Sometimes Vocalign will misinterpret the sonic nature of those transients and will double or triple them, resulting in an obvious 'flamming' in the vocals.

Download and Listen
to tracks 6.5, 6.6, and 6.7

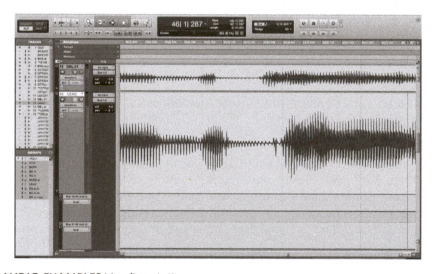

AUDIO EXAMPLES Vocalign stutter

Similarly, in Pro Tools, you might use something like ElasticAudio or X-Form to time a vocal track. In Logic, you'd probably use the FlexTime algorithm. What you get from these editing tools, when they're pushed, is a weird 'carrier' frequency underneath the recorded parts. There's a strange midrange frequency, or tone, that they add to everything. It makes singers sound like a duck or a frog, I'd say; they add a distinct resonance to vocals that wasn't there before, and it changes the way those vocals sound. This artifact arises most obviously when the edited track is stretched to extremes and there's not enough recorded audio to support a time stretch. When you're stretching audio to the point that you're exceeding the limits of what was actually recorded, the plug-ins need to add in frequencies or repeat audio to 'fill-in' the silences that result, and you get that 'ducky' or 'froggy' resonant sound.

Download and Listen
to tracks 6.1 and 6.2

Download and Listen
to tracks 6.3 and 6.4

Have you heard 'ducky' or 'froggy' vocals on commercial releases?

Yes, definitely. Most Top 40 tracks are obviously edited nowadays—editing isn't something that engineers feel they need to shamefully conceal any-more—so you're going to hear evidence of editing everywhere.

You can hear some X-Form-like processing in the song 'The Heart Wants What It Wants' by Selena Gomez. Right off the top you can hear it on the lead vocal.

We've covered the sound of editing on vocal parts. Can you describe similar 'telltale' sounds of editing that you might hear on other instruments?

On all the other instruments, the most typical way to edit is using the 'chopping and fading' method, that is, by cutting audio, fading each cut and aligning them all to grid or some broader timing scheme. You can also use ElasticAudio and FlexTime on those instruments, if you want, but I would generally not advise it, because the chopping and fading method is more transparent. You don't get the same sort of artifacting I just told you about when you chop and fade. Though again, it is more time consuming.

In terms of artifacts from the chop and fade method, they will vary, but always be based around the fade and the phase relationship of the two clips or regions being faded together. There are three ways to 'hear the fade', as it were.

1. A large gap that has been smoothed
2. An in or out of phase fade
3. A shortened sustain

The first, a filled gap in the audio, is the most common. When a piece of audio is played out of time, cutting at the beginning of the note or the transient of a hit and quantizing those pieces of audio is the most common practice. As a consequence of the quantization process, though, two pieces of audio might be moved apart from one another, leading to a gap in the audio. The simplest way to remedy this gap is to pull back the beginning of the second note or hit until the gap is filled. Because you are pulling out the beginning of the second note, you will be repeating a section that fills the gap twice. Once in the first note and again in the part of the second clip used to fill the gap. This will cause an audible effect of doubling or stuttering that part of the audio file. To solve this repeat artifact, you simply extend the length of the fade back into the first note (extending it towards the second note would mean the transient or beginning of the second note would be repeated, which only increases how audible the edit is), blending the two notes together making it one smooth, longer note.

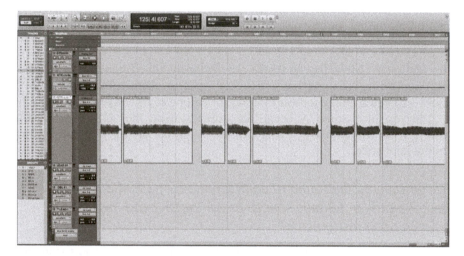

AUDIO EXAMPLES Electric guitar stutter, no fades

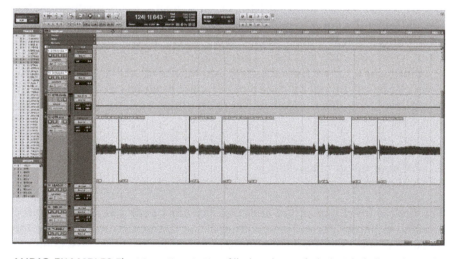

AUDIO EXAMPLES Electric guitar stutter, filled and crossfaded with fades adjusted

The second type of 'fade artifact' you can hear is a fade that is either in phase or out of phase. Imagine again two notes that have been cut and quantized and they move very little or exactly the length of half or one of the wavelengths of the note being played. In the instance that the two are moved only a very small amount when you make a fade, particularly a longer fade, you will hear comb filtering. The best way to avoid comb filtering is by making the fade as short as possible. If the audio clips are moved over by either one full wavelength or half a wavelength of the note played, you will have a fade that is either perfectly in phase (one wavelength), or out of phase (half a wavelength). This will lead to a quick volume increase in the case of in-phase audio or a volume decrease in the case of out-of-phase audio. To avoid these changes in volume, you need to change the type of fade you're using. There are two basic shapes of fades, equal gain and equal power. Equal gain is a linear fade, while equal power is logarithmic. To avoid a volume increase in the case of in-phase audio, you would use an equal gain fade, and to avoid a volume decrease you would use an equal power fade. While the chances are small that the two audio files will phase match perfectly in an additive or subtractive way, it will happen from time to time. I suggest starting always with an equal power fade that is short, around five to ten milliseconds, and changing them as needed.

Download and Listen
to tracks 6.8 through 6.14

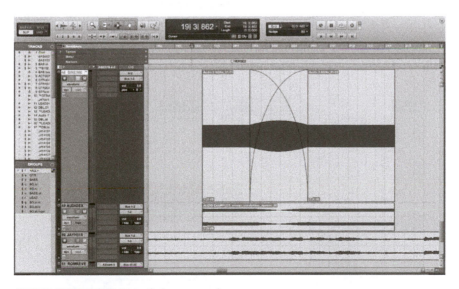

AUDIO EXAMPLES Out of phase, equal power

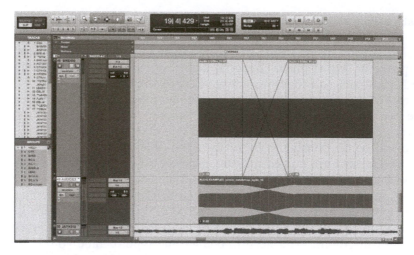

AUDIO EXAMPLES Out of phase, equal gain

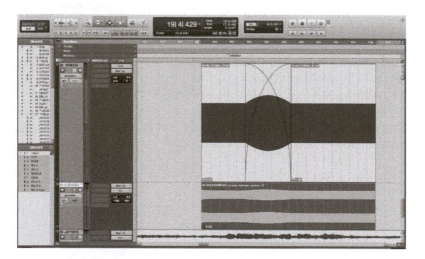

AUDIO EXAMPLES In phase, equal power

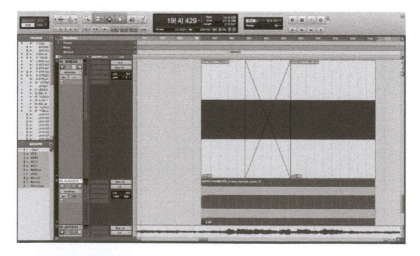

AUDIO EXAMPLES In phase, equal gain

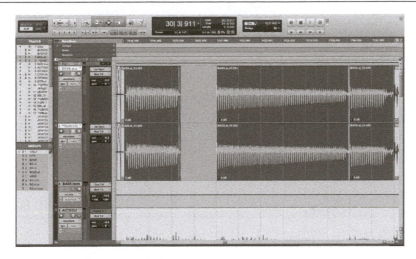

AUDIO EXAMPLES Bass, original

The third type of 'fade artifact' is when the sustain of a note is shortened because you've moved two notes closer together. Imagine a piano with the sustain pedal held down, allowing for notes to ring out under one another. If we were to then cut up passages played on this piano and move them closer together then fade them, you would have a quick decrease in volume between the sustain of the first note to the now-moved forward second note, which has a fixed amount of sustain bleeding over from the first note. This causes an interesting effect on the source being edited; it makes for a disjointed and jumpy sound. To fix this type of edit you adjust the length of fade, making the fade longer, even to the point of it being most of the length of the first note—the jump between different sustain levels is smoothed out across the fade, making for a natural-sounding note transition.

Download and Listen
to tracks 6.15, 6.16
and 6.17

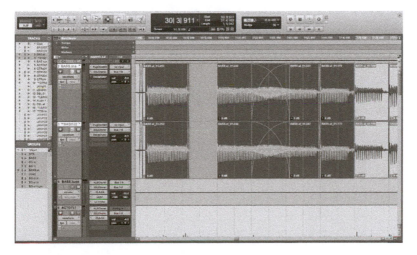

AUDIO EXAMPLES Bass, out of phase, equal power

AUDIO EXAMPLES Acoustic, original

AUDIO EXAMPLES Acoustic, lumpy

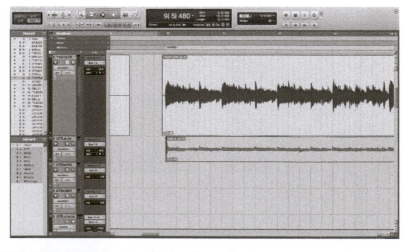

AUDIO EXAMPLES Acoustic, edited and smoothed

While editing, these three types of fade happen all over the place and all together. You might have an out-of-phase fade that has been pushed closer together, so you're also hearing a jump in sustain volume. This is just what happens. You have to spend time adjusting the size, shape and location of the fade. If that doesn't work, find the note later in the take; if the note is only played once, you can try adjusting the length of the note using time expansion or compression tools, or if the problem is really that bad even record another pass of the song.

What are some terms that editing engineers use? Are there any that you hear a lot?

The hard thing with [terminology] when describing music is that the words are descriptive, not objective, similar to trying to describe the taste of food. You might say something like "That sounds froggy" or "That sounds ducky" because of certain types of vocal editing. Or perhaps when you get those short fades and sustains getting cut off, like I just described, you'd say "It sounds choppy" or "It sounds lumpy". These are terms I use often when describing editing, because that's what they sound like to me. A few other terms I use frequently and throughout this article are tuning (using pitch corrections), timing (quantizing audio), chopping or cutting (the act of separating the notes in a performance to be quantized), and a big one I didn't mention are 'triggers'. Triggers are used in drum editing; think of them as placeholders for where the exact beginning of the drum hit is. I use these to help time the drums properly, as well as properly put in drum samples for mixing.

What do you think the future holds for audio editing?

Editing has been commonplace in music production for years, and has started to be used even artistically within music now. I see editing becoming even more integrated into the production workflow. It is used for its unique and different sounds, to save more time and money in the studio, and to help artists further their craft and allow them to create new sounds and better their performances.

Note

1 Sims has worked as an editor on recent releases from Rush, The Tragically Hip, Barenaked Ladies, Three Days Grace, and many more household names.

7

Pre-Production in Mixing

Mixing in Pre-Production

Dylan Lauzon

Delineating roles and rules in music creation isn't something I often run into in the field, working professionally in the music industry. While certain tasks are saved for certain phases of the recording process, there is rarely as wide a gulf between phases like production and pre-production as is often portrayed in music academia. Jay Hodgson discusses it further:

> From a practical perspective, such divisions will always be artificial. Each time recordists select a particular microphone to record a particular sound source, for instance, they filter the frequency content of that sound source in particular ways; in so doing, they equalize and mix their records, even at this very early stage. Recording practice is an entirely holistic procedure, after all. Tracking, signal processions, mixing and mastering cannot be separated—not in practice, at least. They are easily excised in theory, though, because each procedure is tailored to produce a different result. During the tracking phase, for example, recordists capture raw audio singles which they later massage into final form using a variety of signal processing, mixing and mastering techniques. During the signal processing phase, recordists filter and refine the "raw audio signals" they collect during tracking; and, moreover, we will see that many kinds of signal processing are done during tracking. Mixing is done to spatially organize the component tracks in a multi-track production into well-proportioned shapes, and during mastering recordists apply a finishing layer of audio to varnish their mixes, to ensure they sound at their optimal best on a variety of playback machines and in a variety of different formats.

(Hodgson, 2010: xii)

Mixing isn't a process that begins entirely in the formal mixing phase of the recording process. In practice, mixing begins in pre-production—when the first note is recorded. To reiterate Hodgson, every decision throughout the creation process has a profound effect on the mixing process. To further explore this phenomenon, I intend to thoroughly explore my own personal pre-production process and attempt to describe how mix decisions are occurring throughout. Further compounding the confusion is modern recording technology. Recordists at all levels now have access to sample libraries recorded in rooms like Abbey Road and digital emulations of synthesizers and guitar amplifiers that were historically inaccessible to most users. These technologies allow us to add pre-mixed content to a project at any stage of creation. The sheer quality and complexity of these tools blur lines between the creative phases of recording even further. While I will explore mix moves in pre-production, I am not positing that mixing isn't a discrete process unto itself within the mix project, but rather that mix moves occur throughout pre-production and into record production proper. What follows is my personal definition and understanding of pre-production and a thorough outline of the process taken by my writing team throughout the development of a song.

What Is Pre-Production?

Pre-production, in my experience, has become an overused buzzword among producers, songwriters and audio engineers. One often hears it as a meaningless platitude used to encourage hard work, or to scare a band or songwriter into working hard. For example, "If you guys don't do your pre-production, this record won't be half as good as it could be!" or "Do your pre-pro! We're going to have a great time if we're prepared!"

What is it really, though? It can broadly be described as the fixing of certain musical elements of a song into place prior to entering the actual studio production process—in other words—writing the song and preparing the parts before recording. Colloquially, however, it often just means that the band needs to get its act together prior to recording. Rehearse all your parts so that they don't eat up precious studio time and money, write the song, come up with some production ideas, work out arrangements and organize all of the administrative details of the whole recording process, like lining up musicians, accommodations, etc. While often minimized, I believe the pre-production stage of song development is arguably the most important part of the whole process, as it encompasses both the inception and the execution of the musical idea.

An issue with this chapter lies in defining terms and roles within the recording process. Where is the line between songwriting and pre-production, between pre-production and production, between production and mixing? The answer is beyond the scope of this paper, but I believe it can be argued that musicians are making production and mix decisions at every phase of the creation process. For example, microphone selection on some level is intrinsically a mix decision in that it's an irreversible sonic decision that profoundly affects how an element will behave in the final mix. Even decisions as basic as the key in which the song is played have far-reaching effects on the final production and mix of a song. The most obvious example is the difference between a comfortably sung midrange vocal and a belted vocal in the upper reaches of a vocalist's range. A belted vocal will have a major impact on how the whole mix and production is presented to the end listener. While reading, you might note that what I consider pre-production you may consider production, or that what I consider pre-production you may even consider mixing. The hard line that I draw while creating and producing material is drawn at the entrance of a commercial studio. A track enters the production stage, in my professional opinion, when it is vetted beyond the project studio stage and enters its final stages in a commercial studio, or in the case of smaller acts, becomes a formal song. That is, when the song is formally recorded for release, rather than just a bedroom demo.

Pre-production can be a very personal process, and it doesn't necessarily have any rules. Some producers, like Max Martin in the pop world or Devin Townsend in the rock and metal world, take pre-production to the extreme. Tracks enter the studio almost 90% complete, really only coming in for mix, final production tweaks only apparent on a tuned system, and the recording of instruments that can't be easily captured in a project environment like drums, strings or horns. Other producers, with whom I have personally worked, prefer to depend on 'electricity' in the studio, so to speak. That is, to minimize pre-production, simply often write in the studio and rely on luck and skill to get the song done on the fly rather than do extensive preparations beforehand.

For the purposes of this chapter, I will outline the process I take in developing a track with a limited budget for proper recording and mixing. The approach that I'll outline is quite similar to that of Max Martin's and Devin Townsend's processes: it takes the song as close to completion as possible in the pre-production and project studio stages. The process focuses on creating tones at the project stage that translate into a mix-ready, and often premixed, product. It also develops a DAW (digital audio workstation) session to the point where it can easily be transferred to a more professional listening environment and taken to mixing with minimal changes and overdubbing.

I hope to prove that the mix process doesn't have to be treated or conceptualized as a self-contained stage of the music creation process. Steps can, and need to, be taken throughout the entire development of a song in order to expedite and simplify the production and mix processes.

The Process

The purpose of this section is not to provide an instruction manual on how to write or produce a track, but rather to describe my process and how the final mix factors into the process—writing a mix, if you like. While at moments it will seem rather repetitive or basic, the mix implications and reasons for steps are intended to be the focal point, rather than the steps themselves. It should also be noted that this entire process typically takes place in the home or project studio and encompasses the writing process as well as the recording process. Writing is taking place as the parts are being recorded.

i) Idea Generation

It's beyond the scope of this chapter to discuss the creative process and how ideas are generated from scratch, but we can discuss how best to capture this stage. Everyone has their own method, but at this stage simple and transparent is better. It's about capturing the content, rather than capturing tones and creating productions. The simplest in most cases is to use the dictaphone built into most modern smart phones. The purpose here is to capture the idea for later use in the pre-production studio.

ii) Session Setup

Though difficult at this early stage of recording, it's still important to consider mix decisions when deciding on tones for this stage of recording. One shouldn't hesitate to make mix-relevant choices while recording the initial tracks, or initial ideas. When tracking acoustic guitar or vocals, especially for more modern music, it can be common practice to compress and equalize heavily going in. It's also easy to be persuaded by the emptiness of the track to capture a bass-heavy basic tone on acoustic guitar or piano, despite the fact that at mix stage the instrument would probably be aggressively high passed, removing excessive bass from the track. If the decision is made at this stage to capture the tone in a mix-ready state, we've already taken the first step toward mixing!

iii) Developing the Bulk of the Song

In order for this process to be successful, one must focus on creative momentum rather than perfection at each stage of the process. I do posit that it's important to make mix decisions as one goes forward, but that in no way means that one should labor over these concepts. Make decisions with confidence and move on. Mistakes do occasionally happen that will make certain tracks worthless, but the gains made via creative momentum more than make up for the losses, in my experience.

Writing and capturing the bulk of the ideas is a hugely variable part of pre-production. Everyone can and should have their own process for developing the production of the track. It even varies genre to genre for me personally. It's a necessity, however, to reiterate that for our purposes the pre-production process is aimed at a mix-ready or even mixed process.

The foundation of most modern popular music is a powerful 'backbeat'. There are many drum and percussion libraries available commercially that are close to mix ready out of the box. With intelligent application of compression and processing, it's simple to achieve a commercially viable product in a project studio setting. Also, as a side note: the importance of humanization (randomization of velocity and time in MIDI) is often overstated. Popular songs are often fully edited to be 100% tempo accurate anyhow, so in many cases, especially in electronic music, programmed drums are effective without using the humanizing functions of DAWs.

This ties in closely with the process used by Canadian metal and progressive producer Devin Townsend of Strapping Young Lad and Steve Vai and the WildHearts fame. In December 2015, ToonTrack (creators of the Drumkit from Hell plugin) commissioned Townsend to create an entire track from scratch live, forcing him to slam together the pre-production,

production and mixing stages of creation. The following video provides a unique insight into this modern form of pre-production.

Townsend describes his process for developing drums for his track at the beginning of development that often get used in full on the final version of the record.

> I pick presets, and I really make sure not to over analyze or second guess myself. I often just pick loops that sound good in the context of things, and move on not changing anything. It's more about momentum than it is about having the perfect or ideal part.
>
> https://www.youtube.com/watch?v=2Bxzr9n_HK8

The insight here is that the focal point needs to be on the development of the song rather than the tones or actual part played often. Mix issues can be linked to structural and musical problems within the song as often as they can be linked to actual technical and engineering issues.

The common next step is to develop the bulk of the musical elements of a track; this often includes bass, guitar, synthesizers, banjo, kazoo. Really, anything other than vocals and drums. As always, a huge amount of variety of processes exists at this part of the pre-production process. As in the session set-up step, it's very important here to capture a tone that's easily reproducible and can be used as an element in a mix. That's not to say to shy away from unique or interesting tones, but rather make sure to document your processes, so that decisions can be recreated.

As an aside, this is a good moment to begin developing a collection of production moves if it's something you don't have as a producer. A collection of presets can really facilitate flow and momentum in songwriting and idea creation. That is, have sounds prepared ahead of time for particular scenarios; a Mellotron pad patch for dramatic moments, a clean, dirty and distorted guitar tone, a solid funky bass tone.

It's important to make mix decisions as these ideas are developed. Committing equalization curves and compression settings not only allow one to begin creating mix-ready tones early in the process, it sits elements in their proper places, encouraging the inclusion of additional elements that wouldn't become relevant until the production and mixing phases of recording.

While tracking in pre-production, it's also important to track to the same extent one would in studio. Specifically, things like guitar doubling are important. Mix decisions based on a mono guitar will be very different from those based on a wide and full stereo guitar, as will mix decisions based on a simple single performance versus a powerful quad-tracked heavy performance. Sometimes, underdeveloped elements can lead to an overpopulation of elements that otherwise don't need to be in a mix.

In the video linked above, you see Devin Townsend taking these steps in creating his own production. Each step taken is with the final mix in mind. As he captures even the least reliable instruments, he's always considering the mix. For example, during the video he is seen recording an acoustic guitar that has less than desirable intonation. Though the guitar could never be used as a focal point in the performance, he realizes at this early stage its value in the final mix—specifically, in this case, as a rhythmic element rather than a harmonic one. Without a mix-forward perspective during this stage of pre-production, Townsend may have completely foregone the acoustic guitar altogether, resulting in increased workload in studio or a complete lack of the element in the final mix.

iv) Post-Tracking Song Analysis

One of the biggest advantages of using this pre-production method is the efficacy with which you can analyze and critique a song prior to taking it to the studio. It can be difficult to accurately judge whether the tempo or key of a song is ideal at the dictaphone stage of the pre-production method. In fact, for me, the song is almost universally ten or fifteen beats per minute too slow, due to the quiet nature of recording it, and a tone or semitone too high for the vocalist, since it's often easier to sing high in a quiet falsetto than it is to belt out the note.

Once the rhythmic elements and a rough vocal are in place, problems with the basic structure of a song become apparent. An intro that may have seemed excessive with just one instrument may become an interesting musical element with some auxiliary instrumentation. A tempo that may have seemed perfect may come off as dragging once all the rhythmic elements are present. Many small problems become highly apparent at this stage of recording, which is immensely valuable for any project heading into studio.

When the mix is taken into account at all stages of the pre-production process, creative moves often become much more obvious. Where a guitar seems perfect without considering mix, we can often realize it needs synth support, or when a simple acoustic and vocal arrangement can make a song seem passable, a full arrangement may highlight boring or repetitive moments. This all adds up to a savings in valuable studio time and creative energy.

v) Vocal Capture

The message here isn't to obtain the best equipment possible to capture the vocal, but rather to make decisions. High pass where necessary, compress and equalize to balance with the tracks captured in previous stages. The tracking process shouldn't be unlike one that would be taken in studio: capture it part by part or line by line if necessary, and create a near-perfect composite of all the takes. It's even viable to automate the vocal as one would during the mix stage. A vocal that is too full and too loud can make necessary backing vocals and harmonies less obvious, while a mix-ready vocal necessitates backing vocals at key moments.

Following lead vocal, backing vocals are fully fleshed out, almost beyond true necessity at mix level. While I'll dig into this more deeply in the next phase of the process, the excess harmonies will be muted in the editing portion of the process. The simplest method is to do blocks of harmony for each part—thirds, perfect fifths, low and high octaves. This gives clarity as to what more interesting and complex harmonies are necessary.

vi) Edit, Rough Mix and Hard Mix

The final phase of my process is to fully clean and edit this session. When I say editing, I mean editing in full. That is, cut up and quantize all components of the song (if such methods are right for the song), remove any and all dead space with unwanted noise, and fully tune and comp all vocal performances. Some would argue that doing this prior to recording is an exercise in futility, since often the vocal is redone anyway, as is much of the editing, and I would have defended that ideal two or three years ago. With the methods described in this chapter, however, editing becomes a very valuable component of the creative process. It allows us to take song analysis to a level over and above that described in part *iv.* New flaws become apparent as editing is completed, especially as room is being made by removing noise and making interaction tighter between tracks. Often one realizes that more tracks are necessary before proceeding to the mix stage after the editing stage.

Once editing is complete, it's time to do a rough mix—something that sounds as close to what you'd hear on the radio as possible without spending an excessive amount of time. This is going to be a process more focused on sonics than on the musical placement of elements. In a typical session, this would include adding basic reverb and delay sends and generally making every element of the song presentable.

At this point, the tracks would be sounding clean, but are likely to be dense and busy, since across the process we focused on keeping creative momentum in prior steps. To correct this, the next step is something I call 'hard mixing'. It's the process of selectively muting swathes of audio in the interest of improving the song. In some sections, you'll find yourself removing guitar or vocal doubles, and sometimes you'll even drop down to just vocal and drums. This is a very aggressive form of mixing within the pre-production process. It may lack the technical finesse of traditional mixing, but the musical effects are profound. Hard mixing is important for guiding the final mix of the song, especially in terms of dynamic range and opening up space for production elements like delay trails and cavernous reverbs.

vii) Pre-Studio, Pre-Mix and Final Processes

At this point, the song is nearly ready for the studio. Administrative tasks such as preparing the inputs and outputs for the specific recording studio, preparing sessions for the studio's DAW, and general simplification and tidying of sessions needs to happen, but are outside the remit of this chapter.

When this pre-production process is complete, the song should be nearly complete. Many of the tones should be heard in the final process, and the time in a commercial studio could be cut down to a bare minimum.

The goals of the studio should be to capture tones that are impossible in a project's setting stage, i.e., instrumentation that require immense input lists like drums, or perhaps a more exciting vocal tone using rare and vintage microphones. Even choirs, strings or horns if they are right for the song are a great thing to capture in a commercial studio.

As outlined in the introduction, this process truly does gray the line between traditional notions of production and pre-production. In many cases the writing process never stops, even when the pre-production process is arguably over and the song has entered the studio. Modernization and availability of recording equipment has truly blurred the lines between all roles within the recording process. It's as difficult to pin down where engineering begins and production ends as it is to pin down where pre-production ends and mixing begins.

As you step into a studio or a mix environment following your pre-production sessions, you really do come to realize how valuable it can be to make conscious mix decisions in pre-production. Lacking components become much more apparent, while conversely overproduction is easily visible, saving valuable resources and, more importantly, time and creative energy.

Academically, this needs to be explored further. I believe that it is apparent that pre-production is, today, an integral part of mixing as a whole. To put it simply, one needs material to mix in the first place, and pre-production provides some of that material, thus informing the mix as a whole. Even though pre-production processes are individual and varied in nature, they will always affect the mix in that sense. In closing, I hope that outlining my process has successfully illuminated my connections that stretch between pre-production and mixing.

Bibliography

Hodgson, Jay. (2010). *Understanding Records: A Field Guide to Recording Practice*. New York: Bloomsbury.

8

Between the Speakers

Discussions on Mixing

Dean Nelson

The recording community of today has the most superior equipment, studios and processing power. Almost everyone now has access to pieces that were once only accessible to those who were lucky and talented enough to work in the great rooms around the world such as Abbey Road, Olympic, United Western or Sunset Sound, to name just a few. Anyone can load a $20k Fairchild 660 to every single channel of the DAW. While I believe the digital emulations are somewhat 'apples and oranges' compared to their analog counterparts, I do somehow prefer the consistency and reliability of digital plugins.

So, I ask myself, are we then creating, recording and mixing history's most epic records? We should be, and I think we are for the most part! However, I believe other mixers can become sidetracked by the latest piece of gear, plugin or update. They sometimes appear to be some holy carrot hung before us. It's so tempting to endlessly research all the facts and figures, including the gear used on so many great records, and for me the result is that I'm buried in a referential foundation. I too am a victim of these distractions and they do frustrate me. I believe in a healthy embrace of the past but with a strong ear towards the future. Depending on the project, I value a blend of past and present. I'd rather be forging into the future and not wasting energy.

There is no question about it, we have a bounty of resources, and I believe in knowing recording history. However, that only gets one a hair closer to completing a mix from beginning to end. Once we place ourselves between the speakers, the ultimate goal is to make the speakers move and elicit an emotional response. That's where I want to be, 'between speakers'. Ideally, the mix should be a journey that wraps around each ear, through the eardrum to the inner ear and thus setting off emotional neurons. I'd like to share a few tips to make this journey from speaker to ear happen with speed and efficiency, and find the ceiling while embracing the liberation of limitation in the areas of tones and textures. I'll share some of my mixing staples. I'll explain my fondness for the woolliness of my blue sweater, i.e., the classic Neve with loads of iron-wrapped transformers, and how I establish a bold creative point of view while serving the song and production.

Each of the aspects that will be examined include both the technical and artistic side of mixing. Additionally, while keeping the practical in mind, areas such as time and budgetary constraints in the realm of prepping, working from presets and templates from the beginning to finishing a mix will also be considered. The digital world provides the ability to work with amazing speed and efficiency, thus allowing for stellar records to be made for a budget that would have been considered a demo budget twenty years ago. I began in the industry at the end of the 1990s/2000, during what I would say was a sea change of the dominant analog recordings to the digital world, with Pro Tools being the leader in DAWs. This allowed me a great vantage point of the two worlds of analog and digital. There are many positive workflow and efficiency aspects to DAWs, but I do encounter how easy it is to overdo it due to the unlimited ability of recalls, which can jeopardize the end result by overprocessing sonically. I've learned the beauty of chance and happy accidents provided by the analog world when mixing. I have found further delaying decisions can be counterproductive. The mixer oftentimes can wait for the next day and the mood has changed, and a vicious cycle is entered of unlimited recalls and a never-ending mix.

The way I've learned to circumvent infinite possibilities is with limitations. Limitations for me are a way to direct and harness my full potential of creativity. In the coming pages, I will explore the virtues of narrowing down the playing field in the area of workflow, which includes session preparation of routing and layout, to the beginning and end of the mix. While I'm not suggesting that new plugins shouldn't be tried, I do find boundaries are a helpful place to start. Just because we can eat food from across the globe doesn't mean we can throw together a mixture of ingredients on a plate with a hodgepodge of spices. As a mixer, you're trying to obtain a balance of flavors. My spice cabinet is a blend of tubes, transformers and transistors. A way to manage these choices and to reach for the appropriate sound is to know to which group the EQ or compressor belongs. As I stated, there is no shortage of gear and equipment choices today, which means that knowing how and where to start can be daunting. As with cooking, you become handy with certain methods, for example, sautéing, frying and developing a competency and dexterity with certain knives: paring, boning and chef's knives. When mixing, we can use a compressor for controlling dynamics and elevating and embellishing the feel, but we can also establish the mood based on the type of compressor: tube, transformer or transistor. Having these go-to staples creates a familiar landscape for you to work within. These are the pigments of your palette, your familiar tones, and the compass to use throughout your mix.

Before I start with pre-mixing details, let me give my thoughts on what I think of the role of a mixer. First, I definitely never elevate my role as a mixer above the producer, band or, most importantly, the song. My job is to serve the song and the production. I believe I'm hired for the outsider perspective and as a second set of ears. I believe it can become even more of a challenge if you also recorded the record. Sometimes it becomes apparent that you can't see the forest through the trees. The goal is taking what's there and highlighting and magnifying all of the beautiful moments. At times, it can be the last 25% which may be felt and not heard. That's what I believe mixing to be in a nutshell.

I remember, when I was just starting out, how intrigued I was by how people made their way into the mixing bowl, so I share my brief route. My path came by way of a recording school called the Ontario Institute of Audio Recording Technology in London, Ontario, Canada. After graduation, I got an offer via a past grad at the school to be a tech at a studio in Burbank, California. After a week-long cross-country drive from North Carolina, I met with the studio owner and we quickly surmised that the tech position was not for me. I was terrible at soldering. However, the current assistant was looking to move on, which meant that there was the chance at assisting, but I'd have to start out as a runner, learn the room and get thrown into a couple sessions to see if I sank or swam. I definitely swam, although the gracefulness can be questioned. The studio was Ocean Recording, and when I started in June 2001, they had recently installed and refurbished two old 80 series Neves. They essentially wired the two together for a higher channel count. But, while the place was amazing for tracking, there was no automation, which meant not too much mixing. They did eventually install Flying Faders sometime after I left. The great thing about starting off in a tracking studio with a healthy amount of classic mics, outboard gear and vintage Neves is that it allowed me to have a strong reference point of understanding the most sonically bare state that a record could be. Enter ear training 101. I think the biggest instruments I got to know were the drums in their raw state. The industry was starting to feel the early effects of file sharing and Napster, thus the loss of record sales. In turn, the budgets were being cut, which meant the majority of bands were coming in to track drums or beds and then moved on to cheaper studios for overdubs. So, I got to hear some of the finest drummers and kits (vintage and modern) in Los Angeles, as well as top-notch engineers and their various approaches to miking. This was amazing ear training for me, as drums typically play a large part in the mix foundation. After a couple years, I felt I had a pretty good grasp on the foundational side of tracking, for example, the placement of the musicians in the room and mic and compressor choices, as well as basic EQ'ing approaches. After bands finished up their sessions, I would often visit them when they moved onto other studios for mixing. What I heard at the mixing point seemed miles beyond the tracking stage. In the early 2000s, mixing in the box was still in its infancy, mainly due to processing power and, more specific to Pro Tools, the lack of delay compensation. Mixing as you go had not yet reached its stride. I decided to move on from Ocean and ended up at a studio called Chalice, where there was a decent amount of mixing going on. The cool thing about Chalice was the mixture of various genres of music from rock, urban, RnB and hip hop. It was great to see the different approaches based on genre. I think the reason they catered to the variety was that they had an SSL 9000J and a Neve 88R. My time there was a bit short-lived, as Chalice was just getting going, and when the bookings slowed down around Christmas, the owner freaked out and fired most of the staff. Eventually, I ended up at Ocean Way, which has now been changed back to the original name of United Recording Studios. I was assigned to Studio A where Jack Joseph Puig had been in residence for over eight years. I started out as the standard assistant and by the time I left, I was working more in the role as engineer. When I was on the job hunt, my goal was to be in a place where I could experience more mixing and possibly a mentor-type role.

Lastly, before moving on, I'd like to mention the importance of the undervalued but incredibly beneficial role of mentoring. Mentoring provides the environment for some consistency, i.e., with the console, gear and room. I believe when one is developing their ear, having constants such as the same mixer's method, gear and room allows the ear to be more perceptive to changes in a particular area. Similarly, when one is A/B-ing mics, you keep the position and distance, along with the mic pre the same then substitute out the mic. Once you have a mic you like, then you move on to various pre's and then compressors. Essentially, you change one element at a time. In the assistant role with Jack, there was a controlled environment of the studio, console and gear, and the variables were the incoming mixes by various bands/producers and engineers. What I've realized is that seeing the record-making process from various vantage points has provided particular valuable insights. From the tracking side of things, I was able to see all of the mic choices, drum placement and instruments in the tracking room. As a result, I felt I had a strong understanding of how a well-tracked record should sound upon arrival to the final mixing stage. When I get a session to mix now, for example, I can rate the quality of the song from a mixing standpoint. I analyze it and determine if I have some fixing or polishing to do, or a combination of both. When you are at the mixing stage, one gets the vantage point of hearing all types of tracking and production jobs/styles from solid to moderate to subpar. Once you completely move out on your own, you don't get as much of that special window into all of the different tracking and production styles.

After five years with Jack, I moved on to work with Beck as his house engineer and mixer for about two years. As with Jack, the consistency of working with the same person was a fruitful environment for my cultivation as a mixer. Eventually, though, you move on and try to stand on your own, and that's where I'm at now.

Beginning of the Mix

It's no secret that there is an inherent conflict between art and commerce, and it is within that conflict that a challenging divide in record making falls. A mixer's goal is to excel and execute all that is needed creatively, yet the end goal is also to produce an album that sells. Anyone can trim extraneous time spent on a mix so that there will be money saved for you and the band. One area that can burn time away unnecessarily is getting started. For example, when you receive a session and can't start immediately, it can be a creative damper. In my time with Jack, I learned the importance of sitting down and going from zero to sixty mph in a couple of seconds. In other words, hitting the spacebar and unmuting and bypassing some channels and plug-ins to see the tonal and sonic possibilities of the mix come to life. This ability comes from one of the benefits of the DAW world. Starting a mix again (prep work of routing, choosing plugins, figuring out naming, etc.) can take from thirty minutes to a couple hours. That extra time, however, adds up over the course of a couple weeks, months or years. And one must ask, "who's paying for that and how it will it affect the ultimate course of the record?" Therefore, I offer some insight on how to expedite and streamline this process to allow you to dive into what you really want to be doing, which is the creative side of mixing.

During the first hour of any project, I like to find my bearings by starting to address points noted: the main areas of timing and tuning to the specifics of tones along with any bad recording anomalies. These undesired artifacts can be clipping, distortion, edits without fades or improper fades. If there is an excess of timing or tuning issues, I'll get in touch with the band/producer. If it's a small thing, I'll make the edit and try to accommodate and work around the issue. It's ideal for a mixer to receive sessions in which all tracks are cleaned up with completed edits, fades applied and consolidated clips, and that playlists are cleaned out unless there is need for an alternate take. Markers should be located in the correct spot, and all tracks should be labeled with the name of the instrument. The comments section is a great spot for information such as the mic, pre and part that the instrument is used in. I can't stress that enough. Somebody's name is not an instrument, for example. Lastly, there should not be any unused audio tracks, and if any virtual instrument tracks have been used those should be printed as audio tracks. Sessions sent having the playlist and clip list cleaned out helps with file size in that you are not sending any unused parts, which allows for faster uploading and downloading time. Plus, it allows for lower chances of confusion in regards to any unnecessary clips or tracks which may be questioned for the intended use in the mix.

Other important considerations are needed when beginning a mix. If you have the luxury, decide whether you will mix completely in the box or use a console/summing mixer. My preference is a hybrid of the two: part in the box and part on the desk. This allows some variation in the tone as well as the tactile and ergonomic variety of not looking at a screen. This hybrid approach is how I mixed Buck 65's last record, *Neverlove*. I discuss this record more at the end of the chapter. One reason that I like to mix on a desk, and without total recall, is that when the day or session is finished, the mixes were either right or not. I like that challenge—it forces me to make decisions in the moment. Buck's *Neverlove* was mixed on a desk without recall, as was what I mixed for Beck. There's a certain amount of fear that pushes me in that type of situation by knowing that I'll never have that time again and I can't just go home and open the session to make changes. Thus, the safety net of recall creates sterility and mediocrity. There are times, however, when I don't have the luxury of mixing on a desk and will use an old Yamaha summing mixer from the 1970s. It is a simple six-channel mixer with mic/line selection and three band EQs. Usually, I'll send the drums and bass to channels 1 and 2, the remaining music to channels 3 and 4 and all the vocals to channels 5 and 6. This is a basic type of stem processing;

it glues the major sections together. Next, I may use a compressor on the output before I print back into Pro Tools. My go-to bus compressors are the SSL, SmartC2 and the Neve 33609. With this setup, I like to have a solid set of analog-to-digital convertors going back in, because if the convertors are average, it seems to cancel out the benefits of summing through a mixer altogether. The D/A A/D conversion process with poor electronics is far more degrading than the benefits of analog summing.

Over time, I've created mix template sessions, which are sessions only populated with various auxiliary tracks that contain plug-in chains on the inserts. The use of template sessions is another concept that I picked up and developed from my time with Jack. These not only save time but also accelerate the mixing pace. These include dynamic manipulation: parallel compressors, gates for kicks, snares and bass; harmonic variance/tone control: saturation to distortion; width: stereo image exaggeration; as well as pitch, reverbs and delays. If starting the first song of a full record or series of songs, I may import some auxiliary channels or these chains directly from the template sessions to the tracks. When I import them, they are already named and a bus is selected; thus, I may only have to add a send to the channel. This allows me to simply unmute several possibilities in a short amount of time and gain vast perspectives quickly, as a result. This approach prevents me from having to change settings and go through a bunch of pages of the plugin. Example compressor options in these template sessions offer a variety of colors and ADSR shaping. The tones are largely determined by their electronic design and range from various tubes, transformers and transistors. These options will have various settings for attack and release time along with ratios for the ability to increase attack for punchiness or create a longer release for sustain, thus helping to bring out the body of a snare. I'll use high ratios to limiting (10–20–1000 to 1) for putting a part front and center. In the reverb world, they vary in types again based on personality and tone: from classic metallic and shimmering plates and chambers to the odd irregularities of a spring. I continue these options with delays (tape based to digital), harmonic effects (tape and tube emulation to guitar amp simulators) and stereo image effects. I really love the versatility of the Soundtoys EchoBoy.

Another time-saving strategy when moving between songs is an approach I learned during my days working in a largely analog world. The goal is to not to zero out the console, but to leave the outboard gear patched in. Generally, a band or artist records an album in chunks or in one fell swoop of tracking with a similar setup of instruments and mics, thus creating the possibility for a very consistent sound for the record. Chances are that you can leave all of the outboard gear patched in and, similarly with DAWs, route or import all the of same plugin chains on top of the corresponding tracks of the next song. Then you can simply toggle between the insert points being engaged and out to hear EQ or dynamic changes. These are presets in a sense, ones that you've created, and more than a single effect but chains of processing. Now, you may have to do some slight adjustments but you are not beginning from scratch, which can save half an hour or so because it eliminates the need to redo all the routing and plug-in selection. In the analog world, this patching can eat up hours, along with valuable time in DAWs as well. In Pro Tools, you can do this by setting up a mixing template or using import session data to copy inserts from your previous session to the current session's inserts of the respective channels. Chances are the EQ and dynamic adjustment should be close to what you attained on the previous song. When you import from a template session, you bring in a few choices per key insert points on the main instrument tracks. The result is easily toggling the bypass in and out to offer a variety of sonic possibilities. Add to this a couple back buses scenarios and you can go from zero to sixty in a few minutes. This allows you to save time and to progress creatively in the mix exponentially. For every record I mix, after I'm done with the mix I typically will make a template session or export key plugin chains to these mix template sessions. To clarify, I'll have several back bused (parallel) chains for main elements: kick, snare, overheads, toms, bass, guitars, vocals, etc. Then, there a variety of reverbs and delays. This works in line with the concept of using presets of plugins as a starting point. All of these chains usually have some tonal commonalities: for example, an EMI compressor to an EMI EQ. As I stated at the beginning of this section, the way I have these organized falls basically into two categories: colorful to transparent. This is usually determined by whether they have an origin ranging from tube to transistor. Other specs considered are discrete, transformer coupled or not, and the implementation on an integrated circuit. Possibly, the more metal involved equals more color. Knowing these specs can help to eliminate the guesswork when trying to capture the desired tonal shape and color. With mixing, you are trying to capture a moment in time, and that requires the ability to work with efficiency and effectiveness. One should work to know which pencil, crayon or knife to grab while it does take time to experiment. This will help in your speed and the ability to execute the sonic landscape that you or the artist is searching for in the mix. Along with the color aspect of the plug-in, I also consider how it allows me to detail and control the ADSR of a signal. Broadly speaking, I decide whether I need to retain the attack, add attack, truncate the sustain or exaggerate the body. I may also consider whether I want to knock down and smooth out the transients or add some punch. Knowing if the plugin is based on a tube or transistor will offer insight into how fast the compressor will respond. Generally speaking, the beauty of the transistors over time are their speed of capturing transients, thus allowing for more attack. The downside to them is the apparent loss of color. Let's check out a couple of examples. A typical chain for the kick could be a couple of different mics: in, out and sub. I may give each a touch of individual EQ'ing and then sum down to an auxiliary track for global control. For the most part, the strategy for this setup is as follows: the inside kick captures the attack/beater for some nice punch (the 120 Hz area) or sub depending on the mics used, the goal of the outside mic on the resonant head is to complement the inside mic with sub or punch, and the sub-mic like the converted NS-10 or the Yamaha sub mic is just that sub, 20 Hz–50 Hz. What you have in this scenario of the low-end world is the kick in segments with some overlaps, along with what I call nesting qualities like 20 Hz–40 Hz from the sub mic, 40 Hz–100 Hz and then 100 Hz and above. This is very close to how the kick was treated on Buck's 'Danger and Play'. This is a broad breakdown, but provides one with many options for a kick sound. What to expect from the mics is an important consideration. The mics are funneled down this aux, which allows you to carve further to taste with respect to the song. In regards to the kick chain, I may use some kick back busing options of compressors broken down in this way: one for punch, one for sub and one for attack. That's it. I essentially imagine the ADSR chain for each signal, which enables me to break it down and separate it. Basically, this is what you can do with multiband compressors, but for my simple mind I like to break it down into these planes or snapshots of time. Wrapping up using templates, a pitfall may be that it pushes you to familiar territory, so if the band, artist or song is requiring something fresh, don't go that route.

Now that we've talked a bit about the prepping process, let's move on to the beginning stages of the mix.

As I mentioned earlier, my goal as a mixer is to serve the song. So, where does one start after the prep stage? My strategy is to break down the mix into two broad parts: production and engineering. Some things I'm listening for from a production standpoint are the arrangement, the lead and ancillary parts, instrumentation, fills and transitions (handshakes), the execution of the performances in the areas of timing and tuning, phrasing and the dramatic arc of song: the use of tension and release, how it ebbs and flows, where the climax may be, the melodic and harmonic structure and overall feel and mood. From a technical standpoint, I'm listening for clarity, definition, the overall balance from low to high and possible holes or frequencies that are missing or the buildup of frequencies. I consider whether the overall balance is muddy or too bright. Other sonic considerations are tone and timbre of the instruments: is there an overall warm and fuzzy feel, or does it feel clinical and sterile? Also, I determine if there has been any processing in the area of compression—too much, too little? In the area of ambience, I consider if the mix is overall too dry, or is it more of an established ambient setting? Is it too wet or too dry? Are there too many effects, etc.? I'll listen to the whole song from top to bottom in these respective areas along with soloing certain elements to see how they sound on their own to see if there are any anomalies or artifacts. Some things can sound awful soloed, but in the mix is what matters with the relative relationships. As you listen, take notes, not just mentally, but actually write them down and make a checklist. Have a keen ear to the lyrics and note key words or phrases that should have automation pushes (rides), as well as effects, delays or reverb tails. What I hope to obtain by this approach is a sense of any major concerns, pros and cons, whether I need to cover up a poorly played part that cannot be fixed with editing or try to provide clarity for the vocals. If the song sounds like it's in a good spot regarding production and engineering, I typically just take the approach of buffing, highlighting all the beautiful moments and building off of what's already there. Mixing is this process of relationships and relativity. If you adjust one element, you should consider the effect on the surrounding parts, not just what is happening in the immediate vicinity (section), but down the road in the following sections of the song. Analysis to me is what's working within the mix and what seems unrealized or what's not working to its fullest potential. It all goes back to the song. As a mixer, I believe you toggle the hat of a producer and engineer. On the production side, I ask, "What does the song need?" Sometimes I believe that songs, while technically complete, may feel like something is missing, and that is something you can provide during mixing. You may have to fly a part around, mute something that is arranging, or rely on some bells and whistles of 'tricks', weird delays, reverbs or effects. It can be like a sleight-of-hand card trick. Mixing is guiding the listener through an aural landscape of a sonic story. Some artists seem surprised or caught off-guard by this approach when they realize that a mix is not just simply a process in sonics. It very much has to do with a holistic understanding of the song and why it might then be necessary to mute a part or fly something around. I can operate from a strictly sonic/engineering angle and work with EQ, compression, panning, etc., but to me it's the full picture from a production and engineering standpoint.

Starting: Where the Artist Left Off or From Scratch

There are two places to start the mix: (1) where the band/artist left off or (2) completely from scratch with the faders set to zero and without any plugins. I believe that picking up where the artist left off, which could possibly mean leaving all the plugins inserted and routing intact that was used during the tracking and recording process, allows one to hear the mix from the exact place that the tracking process was left off. Mixing from this point of view allows for continuity in tone and shape. In this approach, bypass plug-ins to see how the treatment is working. Does the EQ'ing feel accurate? Do the compression times and ratios feel good? How about muting effect returns? Ask yourself if this treatment is adding to or taking away from the song. Continuing in this way, you can have the client print a rough mix at the end of the tracking process before they pass it on to be mixed. Then, begin by zeroing out the faders and throwing all the plugins into bypass. Next, build the mix back up based on the rough mix and reinstate the plugins as you go. The result may be that you ultimately tweak parameters to your liking and substitute your own preferred plugins. In a sense, this can be a hybrid of how the mix came in with your own magic wand waving on top. Beginning the mix from scratch is another possibility for starting a mix. In this process, you would not refer to any rough mix and rebalance from zeroed-out faders. This allows for the most creative freedom from a mixing standpoint. After deciding which of the above ways to take, there are then a couple of places to start with the song itself, instrumentally speaking. The two main ways I begin a mix are either with the foundational—low-end world of drums and bass—to the lead melodic element, which in most cases are the vocals. See Phil Harding's chapter in this book (chapter 4) for an expanded discussion on this.

I tend to think of mixing as starting from a broad 10,000-foot view and moving towards a very minute microscopic view. It's like beginning with broad brushstrokes followed by touch-ups. This can be a cyclical process: starting broad and moving into fine details and then zooming out to broad again. I mix in phases, building up the mix with equalization and compression in stages. I try to be mindful to not build the drums and bass up too much in size and lose space to fit other main elements, such as guitars, keys, and vocals in. It's a gradual process of using broad sweeps on the EQs, along with first cutting any undesirable desirable frequency. I'm a big fan of the importance of low mids. To me that area is the 'gut' of the mix with some muscle. This area is tricky from an equalization standpoint, because it is intertwined with the mud range of 250 Hz. A common thing I hear is the cutting of too much low mids with a bell and high-pass filtering too far on instruments like vocals, acoustic guitar and electric guitar. The result starts to lean towards more of a smiley-face mix, which is not my style. In my arsenal or palette, I like to have my staple EQ and compressors and the fun one-trick ponies. My current picks for EQs fluctuate between the Neve 1073, API 550a and the REQ by Waves. The REQ has a bit of color, so it is not totally transparent. A really cool thing about the REQ is it makes use of the resonant shelf design that Pultecs incorporate. The one limitation I find, however, is the Q, which sometimes doesn't get as narrow as I'd like. In that case, the QEQ can be a better choice. My first move in the chain, if needed, is subtractive EQ to clean up any harsh, muddy or excessive buildup of frequencies. Next, for any additive EQ along with some tone color, I'll go to the 1073 or 550. A common question that arises is what should sit on the extreme very bottom of the mix. The most common options are bass, eclectic or synth, or kick on the bottom. One other way to frame it is to have the two elements work as one unified thump. In the early stages of the mix, I do not mind starting with extra fat on the bone. I build it up as large as possible, somewhat grand within reason to the tune, like the Technicolor version. For example, there's only so far you can take an acoustic guitar and voice song. The framing of the song can be larger than life, a grand holiday meal, or a bit more realistic, like your average daily honest family meal. Therefore, I build it up as big as possible and then scale it back. For me it's very similar to tracking. I'll start with seeing what kind of extension I can get on the top and bottom of the frequency spectrum. The analogy I like to think of is sculpting with a blank piece of marble and chipping away. For example, in the low end I like to imagine the song as having more of a defined sub/round shape bottom around the area of 60 Hz–80 Hz. The other shape would be one that is punchier, around the area of 120 Hz–130 Hz. These two shapes could be used in one song, though in different sections. One way this can be determined is by tempo—slower tempos have more time between the beats to let the sustain ring out and more up-tempo pieces typically need a kick with shorter decay. Other factors in defining the low end, of course, is the key of the song and the fundamental frequencies involved. For different shapes of the kick and bass elements, check out 'Grey/Afro', 'Santorini' and 'Danger and Play' from Buck 65. After addressing the kick, I'll look at the snare and overheads. My typical approach for overheads is a solid composite picture of the entire drum kit. However, this is entirely reliant upon the tracking engineer's approach, because one belief of the purpose of the overheads is more for the cymbals and some snare. Of course, the kick might not be so represented, but there can be some frequencies present in the area of 80 Hz and above, so I just won't throw a high-pass filter on up to 200 Hz. This type of equalizing ties in to the overall approach to seeing how far or large I can build up the song. I may dial a touch out of the mids in the overheads where the snare is prominent on this initial balance of the drums. Next, I'll go through the remaining drum mics and on to the bass. Once I get the rhythm section feeling good, I'll throw in the vocals or the other supporting element and work from there.

Tone and Texture

Since I was a kid, one of the consistent draws to record making has been my obsession with tone. Of course, I did not know that then, but as much as the song stuck with me there were also the tones and textures. A large part of my tonal education came from my time spent at Ocean Way with JJP and the chief tech Bruce Marien. I knew the instrument choices played a large role in the sound, but it wasn't until I got into recording that I learned that the other part of the equation were the mics, pre-amps, compressors, tape machines and the studios themselves. Words I use to describe tone are warm, woolly, fuzzy, dull, dusty, woody, vibrant, shrill, metallic, glassy and crystal clear, just to name a few. From an educational and financial standpoint, an awesome thing about working with virtual plugins is the ability to a get a gist of the classic pieces. The cons are, however, the access to so many immediately. I often struggle with narrowing down the max choices. It can be difficult to match the piece of equipment to that specific sound in your head. In a very broad way, I think of tone as ranging from really transparent (very true to how an instrument sounds before going through the signal chain) to extremely colorful and vibrant. In order to capture the spectrum of equalizers and compressors, a very general way is to narrow down tone control choices into the fields of tubes and transistors, along with the inclusion of transformers or op-amps to the current use of ICs. Again, this is a very general division in order to begin narrowing down gear/plugin choices. What you are getting from these different types of electronics is the ability to shape the ADSR plus apply a certain amount of color or not. Transistors provide better transient response, as well as less color with the removal of transformers. For example, with Neve 1073 you get transformers with a healthy amount of iron on the input and output, resulting in some nice warmth and thickness and with a transformerless GML a very true and clear sound. On mixes, I occasionally try to challenge myself to sticking to one model for EQ and compression. Depending upon the sound and style of the band, with this approach you get great tonal consistency. An analogy I like to make uses the medium of visual arts. In drawing, let's say you choose to stick with only pencils, charcoal or pastels. You may find that you can still do all the dimensional detail, but because there is only one medium being used by default, there is tonal continuity. All the records that have come out of historic studios like Olympic with Helios, Abbey Road with EMI, Motown with Quad 8/Electrodyne and Trident with Trident have a particular sound, a distinctive sonic fingerprint. I'm guessing this is because there were specific and unique desks in each of those rooms. In these classic studios, they designed and built their own consoles and gear. And if they did buy stock equipment, chances are that the studio techs modified it in some way that aligned with their aesthetic. Hot rodding of sorts. One great example would be the Altec 436 mod EMI, which resulted in the RS124. (By the way, there is no other compressor that sounds and responds like the RS124.) Beck had two Altec 436s with a version of the EMI mod, and those can be heard on the overhead for 'Ramona', 'Threshold' and 'Summertime' of the Scott Pilgrim soundtrack. All these studios, for better or worse, have sonic fingerprints. Having worked at Ocean Way, I can hear the fingerprint of the chambers on records from Frank Sinatra to Beck.

On Buck 65's last record, the songs I worked on were done in Studio C at Revolution Recording in Toronto. The desk at that time was a custom Ward Beck. Sonically, it was clean and punchy, and while it had a superb balance top to bottom EQ-wise, it did not have the ultra-round bottom of a Neve, leaving out the 'darkness' on top. One very appealing factor of working in that room and on that desk is the fact that it's one of a kind. No one's going to go to the software inserts and load that plug-in. It only exists at that studio (or it used to—I'm elsewhere now). Maybe I'm selfish, but for me this is really cool and ties to my values of record making. Please see the Buck 65 examples.

A concept based on this method of tonal continuity is to limit my specific model choices to the same pre/line amp, EQs or compressors on all the needed elements of the record. For example, the Waves NLS (Non-Linear Summer) has three models of consoles, a Neve, EMI and SSL. When starting a mix, I may put a Waves NLS on every channel. Or, for a bit of a more modern angle, I may have all the drums routed through the Neve setting and the guitars routed through EMI and across the stereo bus the SSL. A little mixture of flavors distributed to the instruments, which tend to be flattered by the inherent electronics design of each model. A classic combo is that of tracking through a Neve and mixing on an SSL. Or, as with the Record Club Skip Spence, INXS and Yanni, those were tracked on the custom API at Sunset Sound and mixed back at Beck's studio on the Neve 5315. One can start by tracking all of the warm and woolly color of the Neve and then try to whip out the punchiness and aggression by mixing through an SSL desk or, at the very least, using an SSL compressor strapped across the stereo bus. A ballpark setup for bus compression for me is a low ratio of 2:1 or 4:1 with about 3 dB–4 dB of gain reduction. Attack and release are dependent upon the tempo. A helpful starting point is to start off with the slowest attack and fastest release. Start increasing the attack and once the transients of the snare, for example, disappear, back off. I will insert the stereo bus compressor fairly early on in the mix and get some rough settings. Adjust the settings on the loudest section of the song. If you set the stereo compressor on a quiet section, like a verse, when a louder section, like a chorus, comes in, the compressor will jump on it and suffocate the mix. The gain reduction will be too much. Keep in mind throughout the mix to frequently check the bus compressor to see that it's not working too hard, killing needed transients or pumping and breathing erratically. All the examples listed have an SSL comp on the 2 mix.

I haven't really hit on tube pieces yet—they're all about the harmonics! In the world of classic tubes, I think of the Pultec EQP 1A, LA-2A or the Fairchild. These are pieces I'll insert onto just about anything, certainly any element that I feel is lacking what I would call 'vibe'. For example, acoustic guitar, vocal, piano and overheads really shine going through one of the above. Richness is a common word to describe the sound. I'm not even looking to do much in the area of compression, but just to add some harmonic color. The Pultec has an amazing roundness to the low end and a super smooth top end. The unique feature of the Pultec is the use of a resonant shelf, which has the ability to cut and boost the same frequency. I used the PuigTec on the Kick for Buck's 'Superhero in My Heart'. I like to use it on the stereo bus with a little bump on the top and bottom. I was lucky enough to be working for Jack when he did the PuigTec and Puig-Child plug-ins with Waves. Jack sent his Fairchild to the Waves in Tel-Aviv and once they did all the electronic measurements and coded up the first beta, they sent us a copy. When we first opened it up, it responded like his 670, but it was missing that 'x' factor. There is a beautiful open and silky sound in the high end that comes from just inserting the unit into the signal path without even pushing it to compression. So they sent us a version, essentially with the hood open with a variety of parameters to tweak. Eventually the 'shimmer'/x-factor sound got dialed into Jack's liking—a fascinating experience. I guess when you have fourteen transformers and twenty tubes, something epic should occur. In the end, I think the plugs have the essence of the analog counterpart.

So far I've been referring to the use of tone control from the standard use point of view, meaning 'the technical and user manual correct way'. What happens when these units were used in a slightly 'wrong' way is what's fascinating to me. I grew up in the 1980s and '90s on hip hop, punk rock and skateboarding, and none of those things was about following the user manual. Mixing for me gets really interesting when you start pushing gear past the comfort zone, like hitting the input hotter and bringing down the output. This is when you can really start to hear the true personality of some gear. Pushing the sonic comfort zone goes hand-in-hand with my ethos of finding the ceiling. Working with Beck on a mix once, he said, "sometimes you just have to throw orange paint on it". I believe that was his way of saying the mix was too safe, boring or lame and it needed something extreme and beyond textbook logic. One route to explore in this way is the ability to overdrive various stages of gear from the aforementioned tubes, transistors and transformers. Thinking from opposite parts of the spectrum from mild to extreme, a very mild palette change is tape saturations or hitting the inputs of a mic/line amp that is transformer based, thus overdriving the input transformer for more coziness and warmth. But if you want something a bit more raw and ratty, overdriving the input of the op-amp-based APIs is honing in on more edge. Many plug-ins have the ability to do some type of overdrive. The guitars on the Scott Pilgrim tunes 'Summertime' and 'Threshold' are good examples in regards to overdriving op-amps. The Waves NLS have the drive parameter, which comes from the ability to overdrive the line input of consoles coming from tape or Pro Tools. Typically, overdriving the input requires you then to drop the output as not to clip the bus or DA convertors. Throughout the recording process to the end of mixing, there are several stages and places to attain distortion and saturation. Some of my go-to plug-ins for tone control to extreme distortion come from Massey Tape Head, Waves Krammer Tape, SoundToys Decapitator, Radiator, Devil Loc and iZotope's Trash. All of these have some nice mild settings to very extreme ones. There's probably not one instrument to which I haven't tried to add some sort of saturation or distortion. The classic for me would be SansAmp, which I still use quite often on drums, bass, guitar, keys and vocals. The king of SansAmp, hands down, is Tchad Blake. His recordings and mixes, to me, are always a benchmark and have unique tonal qualities.

Moving on from the sheer edge of saturation and distortion in the vein of finding the ceiling, I hold strong to the belief that you can't inch your way to it! From compression to the use of delay and reverbs, sometimes you have to have some extremes of wet and dry to develop emotion. In the world of compression, I approach dynamic control broadly, from knocking down some transients and tightening the overall feel to extreme exaggerating of the attack or sustain. My signal chain for exaggerated compression is usually placed on a parallel auxiliary. I find it easier to disguise exaggerated compression on a parallel channel. Hit the compressor with a high ratio and substantial gain reduction, hear it work, and then peel back the gain reduction and ratio a touch. With EQ'ing, don't look at the knob and turn until you hear it jump. There's no magic number, so don't be surprised if you are adding 5 dB–12 dB in some places. One thing that took me some time to figure out was how what's on paper or in your head can be completely different from what happens when the elements come together out of the speakers. Theory versus reality. In the ambient world of reverb, here's a way to push for the ceiling. For example, I was mixing a folky tune with pedal steel playing lead, and the main rhythm was electric guitar. It was mid-tempo, and I added reverb to the pedal steel. Although I think that it already had some reverb from pedals and I had added even more, the two instruments still sounded too close together. So, I just basically turned off or turned down about 80% of the direct sound of the pedal steel, turned up the wet reverb signal and added a slight bit of top end. This adjustment opened up the top end and added astonishing depth and space for both the slide and electric. An effective approach to find the ceiling of reverb use is to think of it in terms of explicitly wet or dry. As opposed to adding a blend of wet and dry sound in parallel, take the processing and add it directly to the instrument fully wet. Once you find the ceiling of what's too much, you can just reel it back. Keep in mind, however, that sometimes you only need to push one element to the ceiling. If you push every element, you lose the use of juxtaposition. It all comes back to the point of arrangement: aim for a glaring effect of reverb or delay or a crunch of distortion to get the attention of the listener and continue guiding the ear through the mix.

My approach to developing moods that have been established in the production shall be discussed here. For example, if the song is leaning towards the aggressive side, I might add distortion and extreme compression on the drums, snare, bass or vocals. If there is something mysterious or unresolved, add some small pitch fluctuations in a reverb or delay. I try to play off the moods of the song, whether they are happy, sad, ironic, pensive, melancholic, dark or light. These are basically the same adjectives one uses to describe the moods established by keys and chords of the song. As I have already mentioned possible options for tonal development, another aspect to consider is depth of space. In my head, I imagine something like the textures and the painting technique of impasto, similar to impressionistic paintings where you can see the ridges, valleys and blurred edges of the brushstrokes. A way to develop depth and size is with the use of high and low pass filters along with delays and reverbs. I start by trying to establish the ambient setting of the piece, hopefully with any natural ambience from the recording. For example, I'll bring up the room mics, and use compression on the drums or any other element that I can extract some natural ambience out of. I start by considering close and far, foreground or long shots in composing the dimensional depth aspect. I like to see how close I can pull elements forward in the mix. One way to do this is by adding top end with a shelf, somewhere around 6 kHz and above along with a high pass filter. To push elements towards the background, I'll use a low pass filter and roll off the top end. Some basic settings I use to divide up my delays are 10 ms–30 ms for building up size, 30 ms–250 ms for trailing highlights of slap, 250 ms and above for the clear intention of delay. My general view of reverbs are to develop the size and timbre of the ambient space. I'll set up a near (less than 1 s, nonlinear reverbs like the AMS-RMX), mid (1 ms–4 ms) and far (4 ms and above) to divide up the distance from foreground to background. I'm a big fan of plates and chambers. I find myself using the Rverb by Waves and McDSP's Revolver quite a bit, along with Dverb by Avid. It has it has a nice grittiness to it; it works great on tambourines. A mix doesn't have to be this nice linear dissension into the horizon, though. For dynamics and boldness you could have a couple of elements in the foreground and just throw a glockenspiel, pads or tambourine way in the back drenched in reverb. Treating vocals with a type of backlighting effect is an interesting approach to developing mood. Think of the contrast of a tree with the sun setting or rising behind it. The overall light balance is dusk or dawn. For me, this fits into the mood of the song based on the lyrics or the key: minor/major or something bittersweet with the use of sevenths or sus chords. The tree from the front perspective is dark. It's mostly a silhouette, while the sky could have some hues of pink, red, orange and blue. By brightening the reverb up with a shelf, around 7 kHz, allows the mix to open and shimmer and illuminates the vocals. To further contrast the bright reverb, the vocals could have a bit of the top end rolled off. Another thought is to take the reverb signal and add a high shelf.

One last way to avoid predictability and to establish ambiguity or unresolved type of moods in songs is with asymmetrical panning. When listening to music, it's obviously happening over time, so we can thus include it as a dimension that we are mixing in. For example, in a verse, the right side could be totally empty, or possibly include some stereo overheads, toms or hat. But the song only has one lead/rhythm guitar on the left, while in the chorus the doubled guitar comes in on the right or a keyboard part. That empty space creates some tension, and the resolve is the balance in left and right attained as the doubled part comes in at the chorus. For the most, part of what is going to musically happen in a chorus is to provide some sort of harmonic resolve. I separate panning into two categories: left, center and right (LCR) and internal. The idea of only using LCR has a bit to do with the old Delcon Console that was in Studio B at Ocean Way. It did not have variable panning, so you were forced to only use hard left, right or both. I typically start off by panning LCR, so elements like piano and overheads recorded in stereo will be panned hard left and right. The actual width is determined by the stereo miking configuration. XYs will have a better center image and spaced pairs will have better width. Sometimes I'll pan the piano hard left and center, as a piano would be stage left in a mental image. One distinction you can make for the image or landscape you are trying to create is to set it up as a band would be on stage at a performance. Or, make it more free in the sense that things are floating in this ethereal ether of space. For example, if you have a singer/songwriter with very simple instrumentation of vocals, acoustic guitar and percussion, you could pan the acoustic guitar hard right, separating it from the vocals, thus creating distance and space. This distance could, possibly, be interpreted by the listener as a loss of intimacy. On the flip side, in line with how you would see a live performance, the guitar should be panned down the center. For me, separation between elements creates more of a surreal, bold, dream-like feeling, and keeping things panned closer together can be more associated with seeing a live performance. In keeping with the LCR concept, each portion of left, center and right can be separated into foreground (near), midground and background. I like to reserve the center part of the spectrum for very strong mono elements, such as vocals, bass, kick and snare. On the left and right, things like a rhythm guitar playing an eighth-note part can sit with a Rhodes playing chords. One thing to consider when panning is the rhythm being played. Be cautious of panning similar rhythmic parts on top of each other, because if they are not dead nuts in the pocket together it can sound messy. If there is a shaker playing eighth notes and the hi-hat is doing the same, I would pan them off each other, situating one on the left and one on the right. Internal panning for me is the space between center and hard left or right. I'm hesitant to just use every empty space, because I think it's a bit like cheating in the sense of not EQ'ing properly. Also, things can get messy with every spot filled. Watch the clutter. Sometimes when you give the ear connection points from center all the way out to the outer edges of the stereo spectrum, I believe you lose some of the width illusion, because the ear has this straight line to connect. When there's information in the center and the next bit exists on the perimeter of the spectrum, you can have a better illusion of expansiveness and width. Keep in mind that there will be information in that space if there were stereo mic setups used. I try and keep that space for any part that needs to jump or needs to have a little of the spotlight. For example, toms, hats or a little lead keyboard part.

In the end, listen, listen, listen (if you are spending most of your time twisting knobs you probably should take some time to listen and walk around the mix/console). Sometimes I'll just pace around the desk for thirty to forty-five minutes just listening between the speakers in the next room. Leave room for chance and error, use limitations to drive creativity. Have your staples that you trust and know to move efficiently throughout the mix process and grab the listener's ears by finding the ceiling colored 'orange'!

My Time Spent with Buck and Beck

As a whole, Beck and Buck 65 allowed for lots of creative experimentation and exploration. I would say that is a common thread between Buck and Beck. They both have such an incredible sonic intuition and desire for the sonic tapestry to align with the story of the song. I spent about two years with Beck as his in-house engineer and mixer. At this time, he was spending more time at home with his family and taking a bit of a breather from touring. Up to that point, for twenty-plus years he had been immersed in the cycle of making a record and touring, album after album. I was hired on shortly after he had just finished up the touring cycle for the *Modern Guilt* record and was nearing the end of producing the Charlotte Gainsbourg record *IRM*. Most of our time together was actually spent working on other artists' records: Charlotte Gainsbourg, Jamie Lidell, Stephen Malkmus, Bat For Lashes and Thurston Moore. We also did some film and TV projects along with the Record Club series, where I was lucky enough to have a first go at serious mixing. The Record Club series was amazing because Beck and various friends would go into the studio and cover a record in a single day. We would then mix and release a song/video every week for each song off of the record. The mixing setup was hybrid, Pro Tools HD through a Neve 5315 w/ 33114 pre/EQs, which is class A/B discrete. It's a nice desk, but compared to the 1073's, it's just on the 'ok' side of things. The first tune I mixed was 'Grey/Afro' of the Skip Spence record *OAR*. I think my approach was to further develop the effects Jamie Lidell was doing live. He essentially had a live dub setup, which he was tweaking with various delay, reverb and effect pedals. I remember using some GRM plugins to tweak the vocals quite a bit. For the some of the tunes that Beck did for the movie *Scott Pilgrim*, the approach was pretty garagey/speaker ripping, mainly coming from the guitar tones. This concept and sonic angle was developed during the recording phase. From what I recall, a portion of those tunes were tracked to half-inch 8-track. The mix approach was to take what came in from the tracking side and further that vision. For this, I pushed the midrange of the guitars fairly extreme, 10 dB+ in the 2–3kHz range. It's a delicate balance because you don't want the mix to be fatiguing while still remaining both forceful and edgy. I recall Beck pushing me further and further with the tones and thinking "really, I can EQ the guitars that far?!"

In regards to Buck 65, we met just before I moved to Canada from LA. Once I got to Canada, we worked on a track called 'Dolores', and one thing I remember doing that seemed adventurous was recording a marimba through a Leslie cabinet. It was really spooky sounding and surely caught Buck's ear, as he next asked me to produce/record/mix some of the tunes off the next record. One tune that comes to mind in particular is 'Baby Blanket', a really dark and sad piece. This is one that was developed from the recording process. At Revolution Studios in Toronto, they have a really cool pump organ that we used; one thing we did was to make a loop of the mechanical noise of the organ, which provides this really creaky and dark vibe. In mixing this one, we transferred the multitracks to an 8-track 2-inch Studer. In this process, I summed the multitracks down to those 8 tracks and then printed them back into Pro Tools for some final mixing. Going to tape helped further the warmth, depth and vibe of the final track.

At the of the day, having the opportunity to spend time growing and evolving with such exceptional artists—and to have their trust with creative exploration—is both a huge honor and education. Throughout my career, including my time with Jack Joseph, there are moments of extreme tutelage and wood shopping. With Jack, I spent months simply listening and watching, and slowly the trust formed and I moved into engineering with more responsibilities. But it wasn't until my time with Beck that I was placed directly into the hands of the artist and left on my own to mix. You must become the hands to the artist's vision yet be guided by your own intuition and solid foundation.

Discography

Buck 65 *Neverlove* 2–548156
Beck and Bat For Lashes "Let's Get Lost" / *Eclipse Original Motion Picture Soundtrack* 523836-2
Scott Pilgrim 0343–2
Beck: Record Club
Skip Spence 'Grey/Afro' https://vimeo.com/9638358
INXS 'Guns in the Sky' https://vimeo.com/10245433
'Never Tear Us Apart' https://vimeo.com/11991409
'New Sensation' https://vimeo.com/10634950
'Calling All Nations' https://vimeo.com/12662870
Yanni 'Santorini' https://vimeo.com/12996440
'Keys to Imagination' https://vimeo.com/13184967

9

Mixing for Markets[1]

Alex Krotz with Jay Hodgson

Mixing is often discussed—and now taught—as though it were a singular and straightforwardly linear procedure, with a single definitive output, namely, 'the mix', which is then sent off for client approval and, eventually, for mastering. But mixing is a far more complicated procedure than this linear narrative suggests. Mixes are often made, and remade, and remade again, before they ever make market, and this is done to suit a host of often competing interests. Moreover, to account for the specific musical market(s) for which a record is intended, mix engineers will often alter their approaches in a number of significant ways. A kick drum on a heavy metal record sounds very different from the kick drum on a jazz record, for instance; and with pop, rock, EDM and folk records, the vocals sit in very different places in their respective mixes. Indeed, a galaxy of unique sonic details marks a mix for one market over another, and every mix engineer adjusts their technique in light of those details. Few sources provide concrete information about the adjustments they make, however, beyond simply suggesting that they do so to suit the demands of the various markets they address by their work.

What follows is based on a simple premise: that understanding the different approaches a mix engineer takes to the same sonic material, but intended for different markets, will broaden our understanding of modern mixing per se. Moreover, it is my contention that only a successful working mix engineer can provide us with the sort of concrete and authentic detail that is needed to address this concern. As such, I contacted Alex Chuck Krotz, an engineer who works at Toronto's celebrated Noble Street Studios. Krotz's credits include work on tracks by some of music's most successful acts at present, including Drake, Three Days Grace, Walk Off The Earth, Billy Talent, The Trews, Mother Mother and Shawn Mendes, to name only a few. I interviewed him on his creative practice as a mix engineer, his thoughts about mixing in general, and about some of the different things he might do to mix the same sonic material, but if it were intended for different markets. He was provided with the transcript, which he edited, and he supplied me with numerous audio examples to help concretize some of the points he makes along the way. These examples can be accessed by following the appropriate links at http://www.hepworthhodgson.com, and they should be auditioned where mentioned in the text. In doing this, I have done my very best to let the artist speak directly to an academic readership about his craft. It is my sincere hope that what follows—which I would call a methodological hybrid of ethnographic interview work and straightforward musicological analysis—models a viable method for academics to draw successful artists directly into academic discussions about their work.

How would you define 'mixing' for non-specialists?

Mixing is balancing all elements in a track, and giving each track its own place. At the same time, it's guiding listeners through a song, dictating what they should hear, precisely when you want them to hear it, to help bring out the emotion of the song.

How is mixing different from other tasks in record production?

Mixing is different in that it is a lot more technical than many other tasks. When you mix, you are trying to tease out the emotions in a track. The producer makes decisions like, "Let's play this chord here", and "Let's arrange the song this way". The editors edit it, and the mastering engineer masters it, all to ensure the production sounds good as a final product. It is the mixer, though, who takes all of the elements the producers and engineers and musicians throw into the stew and makes it musically and technically coherent and pleasing. It's the mix engineer's job to make a track 'sound good', both from a technical and aesthetic standpoint. And you can't make a record without mixing, of course. So I'd say it's one of the most essential tasks in record production.

How do you learn to do it?

Engineers learn to mix by 'playing'. You hear something, and you think, "How can I change that? How can I make it better? How can I make it different?" You say, "What can I do with these sounds so they make sense as a mix and as a song? And what tools do I need to make that happen?" Then you just 'play' with the tracks until you get them to sound more or less how you want. That's the only way to learn, in my opinion—by playing, and making a lot of mistakes, and doing a lot of things wrong, and learning from it all. At the same time, you should be watching people who are good at mixing and learning from what they do. When you watch people work, you might hear some things you like and some things you don't like. So you're watching and you think, "I kind of like this, and I don't like that", and then you can adapt what you see to your own way of mixing things. In the end, if I'm doing what I should be while I'm watching the pros work, which is 'playing around' with some mixes on my own, I'm going to learn to mix tracks in my own unique way, even if I'm basing much of what I do on the things I see and hear the pros do. And that's what makes every mixer different. It's how every mix engineer eventually finds their own voice.

In your opinion, do any particular talents suggest that an engineer might be good at mixing?

I don't know if there's any particular talents that might indicate someone would be good at mixing. That said, you have to *want* to do it. You have to want to mix to get good at it. And that's not as common as you might think. Mixing can be super hard, and it's often long and drawn out as a process. Unnecessarily so, in fact, given the wrong client! People might initially think they'd like to be a mix engineer, but after their first few sessions, or after they deal with their first difficult client, they're not so interested in becoming mix engineers anymore! You have to want to be a mix engineer despite how grueling it is.

Is mixing usually considered its own unique phase of record production? Or do most producers 'mix as you go'?

It's a bit of both, to be honest. The way that I work, and the way of working I've been around most in the last few years, is 'mixing as you go'. It's a very popular way of working because you can keep throwing things at the canvas—you can retain an experimental approach to arrangement, without having to fully commit to anything—knowing that, if you change your mind later, you can always just delete whatever you added to the arrangement the day before. That said, you still have to make the production sound good, you have to make it sound like something that gets the musicians inspired to do more to it, so you mix as you go.

In a lot of ways, you actually have no choice but to mix as you go. If you record a guitar too high a level, for example, and you don't mix it into the track a bit before moving on to the next track, then you won't hear anything but that blaring electric guitar when you listen. And so much of production is listening back and making decisions. So you'll have to mix that guitar a bit, make it fit into the production, before you can move on. Then the way you mix it may inspire you to go back and record the part a little differently, or do different things here and there. But that's part of the production process. Things are a lot more open now, even if you know precisely where you want to get to going into a project.

Another thing I'd add is that part of production in general is to figure out where there are holes in the frequency spectrum. You have to 'mix as you go' to hear that sort of thing.

Of course, this all said, once you're done with the engineering phase, and all that 'mixing as you go', then there is still the proper, standard 'mixing' phase. In this case, you know you're not adding any more elements. So you're more interested in making all the elements you've got fit together and make emotional sense in relation to the song. Sometimes that's done with the producer in the room, dictating what the mix engineer should do. Other times, the producer just hands off the session file to a mix engineer, with maybe a few words about where they'd like the production to go, and leave everything else up to the mix engineer. In that case, the mix engineer is expected to do a bit of subtractive arranging. They might mute some tracks, move other tracks around, maybe add some delays and phase effects here and there. They could say, "it would be cool with delays here", and they get creative with it on their end as another stage of the production. At that point, it's another phase, though. Generally the mixer is provided with a 'rough mix' that the producer is happy with so he knows the overall sound of the track, such as distorted vocals in the verses then big clean reverb vocals on the chorus. This is guide for the mixer, but doesn't mean he has to or will do the same things. He can take some liberties but at least has a starting point.

When mixing is its own phase, there's usually an expectation that producers and musicians will be able to make 'mix notes' or request 'revisions'. What is, or has, this revision process been like for you?

The revision process is funny. Sometimes you get one note for revision, and sometimes you get a hundred. It always varies. It just depends on who's on the project, and how good you are at capturing whatever it is they said they wanted when they handed off the session file, and how much your creative tastes match up with what they like.

Generally, there's no set person, or agency in the process, who gives mix engineers their notes. It really depends on the project. Sometimes the producer has a bunch of notes. Sometimes it's the band who has the notes. And sometimes it's the management team that's got a million notes for you. So many people hear the mixes you submit, and often a lot of different people will get a say in approving mixes, so there's lots of opportunity for people to disagree with the decisions you make while you're mixing. But usually, over time, as you get to work with people on a more regular basis, the revision process becomes less and less drawn out—your clients will learn to trust you a bit more with every mix they approve. And if you're willing to listen to their notes, and really honestly try to deliver the mix exactly as they want it, and you don't force your vision of the song on them, they will trust you even more in the end!

In my case, I usually mix a track and then the producer listens. They'll have some notes before anyone else involved in the project ever hears the mix. When I get their notes, I say, "Great!" It's their project, after all, so what do I care if they think, say, the snare should be a dB up or down, and maybe the drums could be a bit louder in the chorus? I just do the revisions and send the track back to them. They usually approve, or maybe have another note or two, but then the mix goes off to the band for their approval. The band will have some notes, usually, which are generally not as useful as the producer's notes. Usually band members just want to hear more of themselves—that's the honest truth! You might get a note from the guitar player, saying something like, "I want my guitar louder." Then the keyboard player chimes in, saying, "I want my keyboard louder." And then the drummer says "I can't hear my kick!" The revision process becomes a bit of balancing act, at that point. Part of being a mixer is dealing with clients in these situations; there's a client management aspect to the process.

In fact, here is where the mix engineer might actually 'overrule' a mix note. But you have to be really careful how you do it. You have to overrule musicians, often, because it's just not practical to raise the volume or profile of tracks any more than you've already done, but they just don't understand that. Sometimes, and this is being honest, the way to 'overrule' a mix note is to tell a little lie, saying, "Yep! I did that. Great ears!" And you literally did nothing to the mix, but all of a sudden they like it better. This makes me wonder if, sometimes, mix notes aren't just a way for musicians to feel like they're staying involved in the process, because a lot of musicians can feel like the project is 'out of their hands' once it leaves the tracking stage.

Also, if there are just too many revisions to deal with, you have to ask, "Was I the right guy to mix this? Maybe I shouldn't be the guy to mix. Maybe you should try another engineer." That does happen quite a bit, actually. Engineers have to say, "Maybe I'm not the right person for this", a fair bit. It doesn't happen on lower budget projects as much, because most projects blow their budget on the bare minimum. But it happens on projects with larger budgets as well. It's not always the mixer who says that either. On other occasions, the producer or band realize after a few rounds of mix revisions that "this guy just isn't getting our vision", and they kindly say thank you and move on to another mixer.

In your experience, how many revisions will a track typically undergo before it's finished?

That really depends on the production. A project I just did, for example, was recorded in one day. And we did a 'rough mix' as we were recording it. At the end of the day, we had a rough mix, and everything was edited. Then we did a 'mix session' on a later day. The band then had a couple notes—they wanted to try harmonies in a couple different spots, mute the guitar here and there, and that was it—and then the track was off for mastering. But that's an exceptionally quick turnaround. And it's rare that a project gets done like that without having some corners cut as a result of everything being so rushed.

On the flip side, I'm working on a project right now where I've submitted four different mixes, each of which was supposed to be a final mix. Two other engineers mixed it, before I got the gig, plus there were 'rough mixes' made along the way before it even reached mix stage. And now it's being sent off to another mix engineer, who will produce separate mixes for the single releases. Once we get those mixes back, I'll have to remix the rest of the album to match balances with them! These tracks have been mixed so often, now, that you just know there has to be a *very* sizeable budget involved! And if you can afford to do it, why wouldn't you? Why not labor over your tracks until everyone involved feels like they sound as good as possible? Once a track's been released, especially when a band's operating at the very upper echelon of the industry, you can't just turn around and say, "Thanks for buying our album but we've decided we don't like the mixes anymore, so we're going to have the album remixed and then we'll re-release it!" Once it's released, you have to live with it. And your reputation depends on what you release. So I understand why producers and bands are very, very careful about what they put out there, and why the mix process is sometimes very, very drawn out as a result.

The Mixing Process

How do you usually start a mix?

It depends on the song, really. But let's just say, for this example, that it's a rock mix. There'll be drums, bass, guitars, vocals, a few 'arrangement' layers, some rhythm tracks and guitar leads . . . that kind of a setup. Given this, I would start with the drums. They're always the loudest thing in the mix—aside from the vocal, of course—and, from personal experience, I just know where they should sit. In fact, drums are really important for me. I like big kick drums, and big snare sounds, and tom sounds, for the most part. Because of that, I always want to make those tracks sound really, really good. Once I do that, then I'll fit the bass in with the kick, so they go together. Then I'll work on the guitars, and work on fitting them around the drums and the bass.

In general, I start and solo the drums. Then I add the bass in, not listening to anything else for the most part. Then I'll add rhythm guitars. I'll work on those a little bit. Then I'll add the vocals in, and how they fit with everything else. It's worth noting, too, that each time you add another track, you'll have to go back and change your treatments on everything else. It's a balancing act: add a bit over here, you need to take away a bit over there.

So you start by soloing particular mix elements, then?

Generally, yes. If you try to mix with everything going at once, at least to start with, you'll find there's just too much to listen to. The process gets overwhelming really quickly then. I recommend starting with whatever is the focus of the track, and working outward from there. Get the important or core tracks out of the way first. Then you can fine-tune the details.

I'm privileged enough that I mostly work on an analog console. I know generally where on my meters my kick drums should sit, so I can fit everything around it without eating up all of my headroom. The key here is gain structuring, and making sure you're not going to blow up everything when you add the vocal on top. I worked hard to know where I want things to sit, in fact. It took a lot of practice. But now I can dial my kick drum into a particular level and tone, and I know everything will hang more or less fine around it. It's like a foundation stone, and I build around it.

Once I have a basic balance in place, and I feel like the track is really grooving, then I can get creative and worry about aesthetics and emotional content. This may mean I throw things slightly out balance here and there, for effect. But that's the emotional part I was talking about, where you're directing the listener's attention and feelings to help convey the underlying meaning of a song.

Let's say it's a jazz session, though. Are you still going to start the process the same way? Are you going to start with a kick and get your balance?

Not necessarily. Sometimes, if it's a jazz record or in a similar genre, I'll still work on the drums first and work outwards from there. However, in that case, I'm not worried so much about where the kick drum is sitting dynamically. In fact, on a jazz record, you often barely hear the kick drum; you hear the snare and the cymbals more than you hear the kick, because the overheads and room mics are doing most of the work. In that way, mixing is a bit 'looser'. With rock, you have to be so careful about levels and headroom because you know, at the end of the day, the mix is going to be squeezed for every last bit of dynamic space the mastering engineer can manage. So with jazz you know you have room to maybe sacrifice some level in favor of a deeper tone or feel. But you still need to balance things, and the kick is a solid place to start.

In fact, with genres like jazz, mixing entails a lot more trial and error for me. That's nothing inherent in the genre, though. I just don't mix a lot of jazz records, so I don't really have any common practice techniques that I rely on. I'm sure if I mixed jazz a lot, I would have a fairly rigid method worked out for it. Given how important I just said the ride cymbals are in jazz, I might start there.

In general, regardless of what genre I'm mixing for, I want to figure out what's important in the mix. If it's a solo instrument that is going to be really loud, then I hang all the mix elements around it, both in terms of their level and tone.

For a pop mix, on the other hand, there's a bit of a formula at work. Certain obvious things need to be where they need to be. In this case, I'll start with a big bass-heavy kick drum. The bass is really important for me, so I tend to start mixing there, whatever the music I'm working on. This is partially preference, but it's also a technical thing. Mixes get messy and overwhelming, really fast, in the bass region. Bass pushes a lot of energy, and often you don't even consciously hear it doing so. Bass frequencies are really big waveforms, and take a lot of energy to propagate, and take up a huge part of the speaker load. If you take a massive kick drum and then add a bass guitar on top of that, and then a bass synth, or any other bass element you might add, your bottom end will quickly blow up and get out of hand. A mix where the engineer hasn't filtered the bass can easily make your 2-Bus compression really go insane, and that's going to compromise some level and your mix will sound all the quieter and muffled because of it. So that's why I like to start with the bass, regardless of genre. Almost on reflex, I'll start by rolling some bottom end off a bass, because I want the kick drum to push the absolute bottom frequencies of a mix, and then I will fit the bass guitar around that. It depends on the mix of course, but that's where I like to start. It is all about organizing the frequency spectrum and giving each element a place.

Have you ever hung a mix around the vocal rather than the kick?

Not so much. I have done it in the past, though. And I have seen other mixers do it. And I've met and interacted with a lot of engineers who say that's their way of doing things.

It seems like that process would be similar, but also very different, from what you just described. Can you take us through your impression of a 'vocals first' mix process?

I don't think it's necessarily a conscious decision. It's just one of those things where that's what that particular mixer does, probably because that's just how they learned to do it. They may have come up being trained that way, being around people who did it that way. They consider the vocals the most important part of the mix right from the start, and they work in a way that, in their opinion, best serves that notion.

Also, it's worth noting that a lot of 'vocals first' mix methods aren't necessarily 'vocals first'. Often, they will treat the vocals relative to some other important mix element, so they can get levels that work. Especially with pop material, for instance, you'll see a lot of mix engineers start that process by focusing on the vocals in relation to the kick drum. Once they get those tracks balanced, they fit everything else around them. And that's because the vocals and the kick drum are both the most important elements of a pop mix. If you listen to the pop mixes, that's most likely what you're going to hear being emphasized.

I can see how that works, and I agree with it. It's just not how I do things personally. There's nothing wrong with that sort of diversity, in fact. There's no single correct way to mix a record. It's only the results that matter. Some people might say that my method is 'backwards', because once I drop in my vocals I have to go back and tweak all the tracks to accommodate the vocals. The problem is that if I were to start with a vocal track, I wouldn't personally know where it should sit. I need to hear the rest of the track to decide where the vocals should go. Is it going to be a really loud mix? Is it going to be really bright, as a whole? I need to answer these sorts of questions before I start on my vocals. But, again, that's just my personal way of doing things. And I do things that way not for any metaphysical reasons. I just 'came up' around engineers who mixed that way.

Part Three: Mixing for Markets

When you talk about how mixing reinforces the emotional content of a song, it reminds me a bit of the way that some cinematographers explain their craft. Would it be fair to say that what you do, when you're mixing, is similar to what a cinematographer does when they decide how to shoot a scene?

I do think mixing is similar to cinematography. Mixing is directing the audio image, making the track *flow*. The arrangement is largely determined during production, before a track ever gets to the mixing stage, but as a mixer you still have to work on how the track flows, so listeners feel different things in different parts of the song. Mixing is subtle in that regard. If you want a verse to feel very intimate, for instance, you will probably want the vocal to be very clean, then, and maybe add some subtle reverb and delay to enhance it a bit. When the chorus comes, and the song needs to feel very aggressive all of a sudden, you might distort the vocal slightly, and even double it. Maybe you'll add a lot of reverb, too, because you want it to feel less 'up close and intimate' than in the verse. Those are the kind of 'cinematography' things I was talking about, which can add support for the emotional content of a song, and which subconsciously guide the listener through the song in the process.

Another obvious example that I can think of, right off the top of my head, has to do with the way you treat screamed vocals in a mix. If a screaming vocal is mixed to sound really present and right in your face, that's going to produce a very different emotion than if those same vocals are mixed back in the mix. The mix with the present vocals is going to feel much more aggressive than the one that's mixed, so there's a bit of distance between the listener and the vocalist. You're going to have a different emotional connection with the song, given each mix.

Download and Listen
to track 9.1

At the same time that you're supporting the emotional content of a song through your mixing, you might also *generate* some emotional content of your own—use mixers to make listeners feel tension, relief, excitement or any other emotion that's available and you want listeners to feel. The guitar part in a particular section of a song might sound really interesting, for instance, so you try to make sure listeners hear it through your mixing. The guitar part could be going the whole time, in fact, and it's only in a particular place in the arrangement where you want to make it the main focus of the mix. So you bring it forward dynamically, maybe brighten it a bit, and then it sticks out of the mix all of a sudden at that point. Then the vocals are back in, so you bring the guitar part back down in the mix.

Do you do different things when you're preparing a mix for different markets? If so, can you give some specific examples? How might you treat the kick drum differently if it's metal vs. pop vs. rock? What different things might you do to the vocals given the same genres?

There are definitely things that you do differently when you're mixing records for different markets. I've already talked about how I would mix a rock versus a jazz record. The jazz market doesn't want to hear a 'thump, thump' underneath everything, like the rock market does. Jazz fans want to hear the intricacies of the recorded performances. Let's face it, jazz musicians are often amazing at what they do. So their fans don't want to hear a four-on-the-floor kick drum pattern the whole time. That would be totally distracting. So I don't emphasize the bottom of the kick drum the way I would in other genres. Sometimes, in fact, jazz drummers don't even play the kick.

Definitely, there are lots of things I do differently when I'm working on different genres. Let's start with rock. Rock bands and rock mixes have 'big' kick drums. The kick drum is usually prominent in a rock mix. The snare is big and 'snappy', the guitars are a wall of sound around the kick and snare, and the vocal is in the center, too, but it's not stupidly loud like it might be in pop. The vocal track is there, and you can understand it. But it's not EQ'd overly bright. In fact, it can be dull, or can have a megaphone effect on it, or it can even be distorted in rock. The market is fine with each of those different treatments, whereas in jazz those effects are often out of the question.

Download and Listen
to tracks 9.2 and 9.3

Of course, pop is often the 'flip side' of the rock coin. The kick drum in pop mixes doesn't usually have a 'tick' to it, like it does in rock, but it's still very prominent and what we engineers sometimes call 'woofy'. The kick drum kind of drives the pop arrangement, in fact; the whole track is often based on the sound of the kick drum. Sometimes there isn't even a bass guitar or bass synth in a pop mix, because the kick drum fills out all the low frequencies, and this is especially true when the kick is just an 808 with a very drawn-out decay.

So the kick is really prominent in pop. There's often not so many guitars these days, but there are a lot of synths that you need to fit around the mix. The vocal is stupidly loud and really bright, with lots of reverb and lush effects. Finally, the vocal is the center point of any pop mix, whereas in rock it's not.

Download and Listen
to track 9.4

With metal, on the other hand, you still want 'big' guitars. Mixing in that genre is even less about the vocal, often because the singers are screaming and you can't understand what they're saying half of the time anyway. The vocal track is not really loud in metal, though because of the screaming and shouting it can often seem like it's loud. It's present, of course, and you can hear it in the mix, but it's not the same level of loudness as in a pop mix, for instance. The kick drum in metal is not as big as a rock kick drum, but it's very 'ticky'. Half the time the metal performers are playing very fast and intricate kick drum parts, so the drum has to be 'ticky' or else you aren't going to hear those intricacies. You can't have a big kick drum going on when the arrangement calls for thirty-second notes—it just isn't going to translate.

Then you go to jazz or blues, and mixing decisions become more about balancing all the parts in the track together, as a whole. There's no big kick drum as a center point, and there's no big vocal as a center point, in those genres. Instead you have to feature everything, in a way, and bring individual tracks forward only during solo sections.

Are there any other things you might do differently when you're mixing tracks for different markets?

Actually, I'll probably choose different samples to swap out or pad the drums. I have 'rock' samples that are very 'big' and that sound like the quintessential rock kick drum and snare. For metal, I have very different samples, though. Those are very 'ticky' and 'attacky', and they have less 'body' than the rock samples, because the metal samples need to cut through the wall of guitar shredding that tends to characterize metal. And for pop, I have yet another set of samples. With pop, in fact, you can use electronic or totally sequenced kits. So the samples you use for a pop record often vary from track to track. But in general, my pop samples tend to be a bit more 'boomy' than the rock and metal samples, and less 'ticky' or 'attacky' overall.

What about vocal treatments? You mentioned vocals and level before, but what about how you process vocal tracks?

The way you approach mixing vocals, and the default vocal chains you use, will be different for different markets. The pop ideal is very bright, very compressed and very 'in your face' throughout the entire track. The rock ideal is slightly less compressed, less 'in your face', and isn't quite as bright as in pop and has more grit to it and can be dirtier. The modern country ideal is very bright, very 'in your face' like in pop, so it's kind of a mixture of the pop and rock ideals.

Part Four: The Future of Mixing

How long do you think the current model of mixing can survive in the present market?

I think mixing is going to suffer a very slow death, if at all. The big engineers that people actually hear about, I mean, for the most part, the most famous engineers are mix engineers. When people talk about record production, you only really hear about important producers and mix engineers these days. There just aren't that many tracking engineers making head-lines lately. They exist, don't get me wrong. But for the most part, mix engineers are more of a going concern in the industry.

Take a look at the back of a record in your collection. It will probably say, 'Produced by so-and-so' and 'Mixed by so-and-so'. But who knows who engineered it? A record's cover doesn't usually advertise who engineered it. The fact that it often advertises the person, or persons, who mixed it tells you all you need to know about the status of mix engineers in today's market.

So, yeah, I think mixing is going to survive for a long while still, even as money gets tighter and tighter. Sure, a lot of people record in their bed-rooms nowadays. But they still want what they record to sound good. And they'll pay to have their records mixed by someone to make sure that it does! That's where 'bedroom producers' seem to spend their money now. Then they try to get it mastered by whomever they can find, or they'll pay the mix engineer they hired a little extra money to have them slap a few things on the stereo bus and 'master' it. With mixing, almost everyone understands that it's something they have to get done, and that it can make a major difference in the way their record sounds. So 'bedroom producers' these days usually won't pay money to rent a studio—and the cost of rent-ing a studio for a day, even, can easily, and quickly, quadruple what it costs to hire a mix engineer to see a track through to completion!—but they will put their money into mixing. And that seems to be consistent.

Anybody can buy Logic or Pro Tools, and sequence and track some sounds. But they probably don't know how to make those sounds sit well together, sound optimal as a group. And they almost certainly won't have the experience to know how to make the mix decisions they make best serve the emotional content of a song. They're usually too fixated on just ensuring that their mix is commercially viable to worry about aesthetics, since so much expertise goes into producing even just a basic balance. They need a mix engineer, and they know it. And because of that, I think the current model of mixing, where you have a separate mix engineer and a separate phase of the production process dedicated to mixing, will con-tinue on strong for at least the next little while.

What, if any, general changes have happened in mixing since you started?

There's two different facets of this question that I could address. The first one has to do with sonics in general. Has anything changed about mixing sonically since I started? The answer is of course yes . . . the sonics are *always* changing and evolving in mixing!

You hardly notice it while you're doing it, to be honest. Sonics just slowly evolve, as everything does. For a while, you're doing things one way, but then you get sick of doing things that way, or you hear a better way when you're listening to the radio, coming home from a session at some ungodly early hour of the morning. So you slowly start to make your kick drum, say, 'tickier' when you're mixing. You get positive feedback on those mixes, and the next thing you know, that's how you treat your kicks.

The same thing will happen with how you treat your vocals, your guitars, your bass. It's really just a question of developing as an artist. You can't just deliver the same balances over and over again, or pretty soon your mixes sound like they were in a different era. But it's a slow process. And engineers tend not to be particularly conscious of the changes they're making to their craft as they're happening. They just know that they want to keep growing as artists, so they continually fine-tune their craft. And this happens at an industry level. So the sound of mixes is constantly evolving as individual mix engineers evolve the sound of the mixes they produce. It goes hand in hand with artists developing their sound as well.

Then there's the technical aspect of mixing. Things have changed technically, almost to a point where the way you mix now is entirely different from how you mixed even just five years ago. Or it can be, depending on which technology you decide to use. Plug-ins have gotten better at doing the things they're supposed to do, for example. That's a big one in the last ten years. In fact, some of the emulation plug-ins have gotten so good that even the most stubbornly out-of-the-box engineers have gone almost entirely in the box in the last few years. That's definitely a game changer.

And, then, those changes in tools and techniques change the way mixes sound in turn?

Of course. Engineers check out new plug-ins all the time. Sometimes they're impressed. It can be like a 'Eureka!' moment, in fact, stumbling on a plug-in that does something exactly the way you'd always hoped a plug-in would. You play around with this new thing for a few minutes, immediately realize its potential, and then they you start using it on everything for a while. And the process goes on and on and on. . . . The process of discovery never ends, really, so long as there are mix engineers who believe they can improve their craft.

That said, people do still like mixing in traditional ways. People still like mixing on consoles. Consoles aren't really going anywhere anytime soon, even if you need a bigger budget to work on one. But the older model isn't as popular as it used to be. People realize, now, that whether you work out of the box or in the box is mostly a question of personal preference, and even though each way of working has its pros and cons, they'll both get you equally professional results. There's no *need* to work on a console, is what I'm trying to say. But a lot of producers and engineers definitely still *want* to work on a console, and they see the benefit of doing so. In general, though, I think that a large-scale move to in-the-box workflows is a big part of what's changed, or is changing, in mixing these days, even if the out of the box remains strong. And you can hear this change in the way that mixes generally sound.

Note

1 In this chapter, the word 'market' could be interchanged with the word 'genre'.

Mixing In/and Modern Electronic Music Production

Andy Devine and Jay Hodgson

Historically, the production of electronic music has tended to be viewed through the lens of traditional pop/rock production techniques. While this may be fine in certain circumstances, upon closer inspection it becomes clear that many traditional practices employed by producers working in a pop/rock paradigm can take on very different definitions when used in the context of modern electronic music production. These potential differences in approach have, for many years, been underrepresented in academic research exploring concepts in audio engineering. In turn, this trend has resulted in a pedagogical gap in the design of teaching and learning programs where electronic music production is concerned.

The similarities and differences between the disciplines of traditional audio production and that of modern electronic music production are beginning to garner more attention academically. However, for the purposes of this book, we wished to focus on one area in particular, namely, mixing audio. Moreover, we felt strongly that it was of great importance to invite artists working in the area to join the academic discussion directly. As such, what follows comprises an ethnography of sorts, specifically focused on mixing approaches as discussed by electronic music producers. Our sample included numerous nationally and internationally recognized DJs and producers, whose work as musicians and composers we've long admired. We asked each participant in the study to supply biographies, which can be found at the beginning of the book under 'Chapter 10: Interviewee Biographies'. Our ultimate hope is that the following conversation will highlight which aspects of mixing electronic music are shared with traditional instrumental production and which aspects seem unique. We leave it to future researchers to follow up on these leads.

How would you define 'mixing'?

Rick Bull (Deepchild/Acharné/ Concubine):

Mixing involves finding the unifying elements (harmonic, melodic, dynamic) within a piece of sound/music, and ensuring an internal coherence and synergy between these elements, to produce a finished work ready for mastering. Mixing is an essentially intentional yet artificial process of engineering a physical 'space', power, dynamic, and ebb and flow, within any given work. It's a process of trimming some of the errant, or distracting, or undesirable elements in a recording, while referencing others, to reflect the intention of the author. Mixing is a great and powerful sonic fiction, placing a listener or dancer within an imagined sonic terrain, in such a way that the recording process might become more or less transparent, thereby allowing the listener to best inhabit the desired sonic imagination of the author. The goal of a successful mix is generally to transmute musical elements into a seemingly effortless musical narrative, and to use tools such as EQ, compression, panorama and relative volume to best support many multiple voices within an evolving audio narrative. Mixing seeks to create the illusion (generally) of transparency, when often it is far from transparent. Smoke, mirrors, voodoo, silence, tension and release . . .

Adam Marshall   I'd define it in two ways, depending on the context. First,
(Graze/New Kanada):   mixing in a production sense, I'd define it in the traditional way: of arranging and recording certain settings during production or final mix down. Second, I'd also see mixing in a DJ-specific sense, where it's a live performance of in-the-moment matching, cutting, fading and riding two (or more) records during a DJ performance. The main difference between the two would be that mixing in traditional production practice is usually a tightly controlled and planned operation, with a specific desired outcome, whereas mixing in a DJ sense is more of an in-the-moment experience where a lot of the magic (or chaos) happens (or occurs) when taking chances.

Pierre Belliveau   Mixing is finding the middle ground for all tracks to
(Gone Deville):   live together without stepping on each other too much. Mixing is about giving each layer its own space within the spectrum.

Noah Pred   Mixing is the process of balancing, blending and merging
(Noah Pred/False   multiple sonic elements into a coherent whole that
Image/Concubine):   fits the desired character of the producer—be it crystalline, murky or anything in between.

Phil France   Mixing is the stage of the process that you come to
(Phil France/   when you've got a track sounding as good as you can,
Cinematic Orchestra):   before mastering.

Ryan Chynces (Rion C):   Mixing is the art of combining distinct layers of sound, or sound recordings, to produce a unified whole.

TJ Train (Room 303   I'd say it's blending different instruments and sounds
/Night Visions):   together to allow the space for each instrument to be heard both independently and together.

Andy Cole (LuvJam):   Mixing is balancing audio elements to create the per-
                          fect output, where everything is clear—an audible
                          balance of sounds created purposely for your desired
                          audience or media output.

How did you learn to mix?

Rick Bull: In my case, I'm still learning to mix. At first, my tech-
                          nique was rather robust, apologetic and compensatory,
                          or additive. After years of trial and error, my approach
                          has become rather more robust, reductive (cutting the
                          fat, trimming the errant frequencies, keeping things
                          simple, direct and confident) and physical. Initially,
                          most of my mixes with were 'out of the box' (through
                          analog consoles and mixers), but over the years I've
                          learned a more varied series of approaches. Assets
                          I've found along the way include mixing on a vari-
                          ety of monitors—a combination of high-end monitors,
                          ear buds, computer speakers and in different rooms to
                          compensate for less-than-ideal studio situations. Ulti-
                          mately, a mix is about placing a work in a given con-
                          text (e.g., a nightclub or for home listening) and, with
                          this in mind, it's been really useful to 'test and trial'
                          my mixes in many different physical spaces. For exam-
                          ple, as I learned early on, a radio mix might sound
                          awful compared to a club mix in a certain genre. I've
                          learned a lot through mixing in compromised situa-
                          tions—e.g., mixing a club-track in a small flat at low
                          volume versus in a studio situation at higher levels.
                          I've mixed some of my best bottom-heavy records
                          using my fingers, physically resting on a monitor cone,
                          at super-low volume. Ultimately, I've been best served
                          realizing that sound is fundamentally a physical expe-
                          rience, and one best approached playfully, boldly and
                          intuitively, rather than academically.

Pierre Belliveau: I learned through personal research, investing time
                          and numerous discussions with other professionals.
                          Mostly, though, the most important thing you can do is
                          train your ears and learn to trust them.

Noah Pred: As with most of my technical expertise, I'm primarily
                          self-taught. Most of this learning occurred before the
                          advent of online tutorial videos, so I was truly on my
                          own. I learned through a great deal of trial and error,
                          assisted only by the occasional industry magazine or
                          instruction manual.

Adam Marshall: Everything I've learned is self-taught. When I was
                          growing up, there was not a lot of direct education to
                          be had in producing electronic music—but as a DJ,
                          I knew what I was aiming for and tried to learn the

skills to allow myself to create tracks that I, as a DJ, would want to play out in the clubs. I'd say in the early stages I learned the most from my peers.

Ryan Chynces: I came to music production from being a DJ, and therefore my mixing sensibility was informed by combining records in a live setting. I taught myself to mix in software such as Ableton and Adobe Audition, with the help of books on the topic, and lots of trial and error.

Phil France: I picked up as much as I could from studio time spent working on various albums and projects with the Cinematic Orchestra and working on various other albums and projects. I've still got a lot to learn about engineering, production and mixing. However, instead of trying to know it all before you start, a couple of people I respect a great deal mentioned to me that it is okay to figure things out as you're going along. So that's what I do.

TJ Train: I'm self-taught. When I was starting out, there was a ton of information out there on spectrum analysis, EQ'ing, compression, etc., including online books and videos. I used those and other resources.

Andy Cole: From an early age, I have always been aware of the balance of sound. Actually, my first experiences with mixing developed through learning to DJ mix—simple fades, panning and balancing bass/treble/high layers. I then further developed my mixing through my television work, where I would layer sound-FX onto my motion graphics. I learned not to 'peak' any sounds, thereby creating a good, clear balance. Then I have developed my skills further by using Pro Tools, Logic Pro and Adobe After FX.

Craig Bratley: I learned to mix through a lot of trial and error. I also read a lot of books on the subject, which led me to study for a BSc in Music Technology. I still read a lot on mixing today, as there is always something to learn. There are also some great tutorials online, such as Pensado's Place 'Into The Lair'.

How do you approach mixing in general? That is, what (if anything) are the things you do each time you mix a track? Or do you prefer to remain open about your production processes?

Rick Bull Generally speaking, my mix approach is very simple.
(Deepchild/Acharné): I try to minimize sonic variables—basic EQ'ing, compression and panning are my primary concerns when

mixing. I don't feel that I'm sophisticated enough as a mix engineer to create much of the nuance I'd like in a high-end mix down, so I tend to prefer creating mixes with a fingerprint character and with a sound I find personally pleasing. I don't think I have the sonic agility to be 'all things to all men', so I shoot from the gut—I shoot for 'simple', 'direct' and 'easily identifiable' (usually) as the sonic fingerprints of a Deepchild mix. In my case, this means a direct, warm and perhaps a little more bottom-heavy approach than is usual. My approach is hardly refined, but I do hope it is distinctive. I feel that many mix engineers end up producing under-par productions by overworking a mix, rather than allowing it to breathe. Ultimately, I tend to try to remind myself that less is more. It's been useful to ask, 'what can I edit out?' rather than, 'what can I add?' when mixing a track

Pierre Belliveau   Panning is pretty standard no matter what the track
(Gone Deville):   —it gives each element its own space in a mix. Also, finding which frequencies I want to prioritize for each layer is common for me. I also try not to compress anything unless I absolutely need to. Often, sounds can fit into a mix with nothing other than the right volume and panning applied. I'll still EQ most tracks, but not too destructively. To be honest, I'm looking forward to the day when I don't mix my own songs anymore, so that I can let someone else bring a new life and flavor to my tracks.

Phil France: When I'm happy that a track is finished, I've generally got a pretty good idea of how I want it to sound, and I will do all the stem bounces, and make sure all my levels are fine, and balanced, before I go into the studio for the mixing session. Music that an artist(s) is working on can really benefit from having a fresh pair of ears working on it for the mix, as the artist(s) is generally pretty close to it and may miss better ways of doing things, or ways of making it sound better. At the moment, largely due to the fact that my major financial investment is renting a studio space, but also because I'm a bit of a control freak, I do as much work as I can myself. I have a clear idea of how I want it to sound, usually, but also understand the importance of having another pair of professional ears listen, to potentially improve things and also to do a double-check. If there isn't the budget or time for working with another producer or engineer for the mix, I'll get it sounding as good as I can on my own, check the levels and then run it through some plug-ins I have. Basically, I want to achieve a full, warm, defined and fat sound across all the frequency ranges in the track. I'll run a track through a (stereo-bus) compressor, which helps to glue all the track together. I'll also try to remove any frequencies from the individual tracks that I don't like through equalization, before the mixing session. I'll usually remain open during mixing, and I'm happy to try new things. In fact, my approach to the mixing process is to try all the ideas that enter my head, to see if they work or not. It's trial and error, but trial and error within the parameters and guidelines of what technically works.

Noah Pred: My mixing technique has changed a lot over the years and will surely continue to evolve and adapt, along with technology and my knowledge of it. Currently, most individual tracks in my projects have a UAD Neve 88RS emulation on them for gentle compression, gating and EQ, along with optional dx-160 and saturation units, which I can activate as needed. Importantly, all of my track volume faders are attenuated to −7 dB by default, so there's always plenty of headroom, and I try to be vigilant about maintaining 'green levels' throughout every gain stage. Once an arrangement has taken shape, I'll group my tracks into sensible instrument or frequency-spectrum buses, at which point I'll apply further EQ, bus compression, and vintage excitation (often with a combination of UAD plug-ins and iZotope's Alloy 2). I've also got all of my effect returns collected onto their own group bus, with some basic EQ and excitation to keep my reverbs and delays from getting muddy.

Ryan Chynces   Each time I mix a track, I pick one of the multitracks
(Rion C):   to be dominant. I set it to zero, and then adjust the levels of all the other tracks relative to it. I keep the volume of that dominant track constant throughout the entire mixing process, but am constantly adjusting the volume (and track delay) of the other supporting multitracks in relation to it. I'm open about what multitrack will be the dominant (the standard of course is the primary 'four on the floor' kick beat), but if I have a great harmony or melody track, sometimes I'll build everything around that.

Andy Cole: I tend to listen over and over to each layer of sound in a mix, to be sure it's working individually over time, before I listen with all of my layers combined. I have learned by mistakes not to master my own files, not to use stereo-bus compression tools, and just to let my mastering engineers work their magic there! However, referring to your 'open' production process, I quite often throw in a selection of sounds, and just see if they sit well. Some do, some don't—I'm quite experimental in that sense. Over time, I have developed further programming skills that enable me to almost weave certain sounds in and out with fine attention to detail to the sound levels of the individual layers (i.e., when a sound should be added or removed, can be added or taken out, when a sound might sound better alone, to enhance the impact, and so on).

TJ Train:	I prefer to remain open. In general, I now like to focus 90% of my energies towards the creative process and 10% towards mixing. I feel that mixing is an entirely different art form from actually making music, and takes years of dedication and experience to do it right.
Craig Bratley:	I'm very open about the way I mix, as electronic music lends itself to trying different things. There are one or two things I always do, like having a low cut filter on every track except the bass and kick, but I never follow 100% the same approach.
Rick Bull:	I feel like I have so much to learn. My ears tend toward a specific sonic palette (and an interesting aversion to certain frequencies), but my grasp over some tools can also be a little heavy-handed, perhaps. I find it easy to get attached to a certain way of doing things, which may occasionally blind-side me to solutions to sonic problems which might be obvious to those less attached. For example, I remember (about ten years ago) working with a vocal in a track which was around a quarter-tone out of tune. After struggling to disguise this fact and making a mess of the vocal, I settled on a combination of sometimes apologetically auto-tuning the vocal and at other times letting it sound in its own de-tuned glory. I feel that within the mixing environment, there is no single 'best' way of approaching a sonic challenge, and that listeners will generally re-cast an errant element with repeated listens to become part of the desired fabric of a tune. Trying too hard to 'sterilize' or 'perfect' an imperfect mix can often lead to a very two-dimensional listening experience and a profound sense of distrust from the listener. Writers like David Toop and Brian Eno address this very astutely and point out how much mixing conventions also change over time. To use an extreme example, many of the mixes of Contemporary Christian music in the '80s sought to remove the sibilance, hum and physical 'detritus' in a recording to create one illusion of sonic 'transcendence' and 'perfection'. These mix conventions within that tradition are very different now, particularly as we have come to associate tropes like sonic 'authenticity' with certain heavily nostalgic processing techniques like tape saturation and compression. We hear through a haze of nostalgia, and we take comfort in the memory of certain technology. Creativity involves acknowledging these sonic assumptions and knowing when they might be worth subverting.
Pierre Belliveau:	I guess that, in some ways, I do prefer to remain open about the mixing process. In some ways, I guess. Boundaries and limitations don't always have to be a bad thing, especially when creating—it's good to not always draw lines around where you can and can't go.

Noah Pred: Some degree of openness in the mixing process is important creatively. There should always be room to try new things, to test new tools and to experiment with alternate techniques, particularly when your go-to approach isn't working. However, I think there's something to be said for having familiar guidelines, tools and signal flows configured for easy usage, if for no other reason than to streamline the process, and possibly help to give you your own 'sound'.

Adam Marshall: I don't tend to share my production processes because I am not convinced that they are 'correct'. I just know that they work for me, and have sort of become part of my 'style'. However, I think it's important for people to share their production processes if they want, as it's an invaluable source of learning for people that are just getting started.

Phil France: Being open means that you may find something that sounds better, but I've found that it helps to have a clear overall focus about what you want in mind as well. If something doesn't work quickly, I'll disregard it and then move on to the next idea. One of the things I've learned is that it takes me time to listen to something, so I'll review and double-check it on different speakers away from the studio during all stages of producing the work. I've learned that if I'm involved in a session with other professionals, and I'm in charge, it helps to have a clear idea about what I want and to be prepared and decisive. For me, in this context, the decision is always about you and your work, and how you feel and think about it. I'll always address the things that come to mind when I'm mixing, then articulate them and figure out how to try the change them, so I can 'A/B' the results. All my pieces have to 'settle', so there are no niggles or regrets about the mix in my mind.

Ryan Chynces: Changing up the mixing and production process opens up numerous possibilities. For example, mixing an entire track around a dominant melody opens up a different garden of forking paths, as compared to laying everything onto the kick drum.

TJ Train: Remaining open allows you to try new things and move in new and experimental directions. I believe that having a regimented production process restricts you as an artist and prevents growth.

Andy Cole: To be open is a good thing. I prefer trial and error— just throw it in the mix and see if it works. What's also helped me are my DJ skills, where over time (and through trial and error) you learn to mix/dissolve one sound into the next, which is also learned by trial and error. I treat my mixing as I would my sketch pad or notepad: try it, record it, test it. If it works, then great, and if it doesn't work, delete it or amend accordingly.

Craig Bratley: If you remain open, there are no parameters to stick to and you can have the odd, happy accident along the way.

Do you think mixing is done differently in what I'll call, for lack of a better term, 'electronic dance' genres than in more traditional markets (like rock and folk)?

Rick Bull: The joy and trappings of electronic dance mixing have traditionally been that it's far more immediate and physically functional than a lot of rock 'n' roll, by virtue of the medium in which it is experienced. There are certain conventions which are difficult to break within the form of a disco/house track, without compromising the physical power of a recording. Due to the fact that dance music has traditionally been listened to at high volume, in a club, the broad brushstrokes of mixing a kick, bass, snare and hi-hats together are relatively concrete, with mix creativity being explored through placement and dynamics of more subtle sub-mixed elements. I like the challenge of mixing dance music because it's generally so highly artificial to begin with; once the fundamentals are in place, it's exciting to work out how best to exploit what's left over. I've seen how some younger rock mixers can overcomplicate their process and end up with a mix down which sounds too bitty or overly complex. Dance music has a way of bringing things back to basic principles and grounding a listening experience in the body. I've found that these kind of lessons have helped me produce better mixes for non-electronic music. Dance music mixes have often felt like 'visual' experiences for me—broad sonic blocks to be placed in a clear space. Once the fundamentals (and in dance music, these fundamentals are fairly obvious) are in place, there's creative room to play with what remains.

Adam Marshall: Yes, I think so. With electronic dance music, it's assumed that the track will be played in a certain environment (a club system) as the pinnacle, so a lot of the mixing and production decisions are geared towards this certainty. For more 'listening' types of music, I think things need to sound good in a much wider variety of environments, and this must be taken into account when recording, mixing and mastering.

Pierre Belliveau:	I think mixing in electronic dance has the same concepts as mixing for other markets, but the priorities are not in the same areas. For example, in electronic dance music, a kick will take a much more central place in the mix than indie music, and indie may not be as comparatively 'hard' in the low end. At the end of the day, it's still the same science, no matter what. But genres also encompass different mixing and mastering techniques.
Ryan Chynces (Rion C):	It certainly is. For one thing, most of the multitracks that are used will be MIDI based, as opposed to audio based, and working with MIDI has its own unique challenges and opportunities. Another difference is that electronic producers aren't concerned with reproducing a live aesthetic. A rock band, for example, might want to mix their song so that it's more of a faithful reproduction of a live performance, whereas a dance music song is produced to *be* the live performance.
TJ Train:	Absolutely! In electronic music there are dominant elements such as the kick drum, bass line and perhaps a lead or two. In most other genres, it's more cohesive where all instruments play an equal role. Now, this perhaps isn't true for all electronic music but it certainly it is for a good majority.
Andy Cole:	Yes, possibly, but good sound is good sound! I have mixed folk elements with electronic styles, and the ethic is the same. I have not really mixed rock as such, but I can imagine it could possibly be more difficult due to its 'live' nature. That said, the same principles apply.

What, in your opinion, is the same about electronic and other mixing processes? What is different?

Phil France:	I would think that the basic principles are still the same—sound source, signal chain, levels, phase, equalization, compression, balance, etc. I think it boils down to whether you are recording a machine or a human being playing an instrument. I noticed from my time in the Cinematic Orchestra that electronic instruments, compared to traditional instruments, have a hell of a lot more level. I'd also suggest that certain bits of gear are more readily associated with particular instruments, genres and sounds.
Rick Bull:	Dance/electronic music tends to be defined by a certain physical functionality or literal sound system. Folk/rock

music can often get a little more hijacked by a perceived need for 'sound-authenticity', which I've seen throw some overly analytical engineers into a spin. I think that increasingly, as genres collide, this is less of an existential issue. However, I do feel that dance music has reminded me that it's all just a glorious fiction we are creating, and that independently of genre, we can have fun acknowledging this fiction without getting too hung up over it. Ultimately, a mix that 'gets things done' without regressing into contrivance is an enduring mix.

Noah Pred: While many fundamental techniques are shared across genres, mixing electronic music tends to have a unique set of priorities. There's generally less emphasis on the midrange, though 1k–5k should still be approached with care and precision. More attention needs to be paid to the low end, with good reason. A lack of low end can be a serious impediment for any club play, whereas too much low end can seriously muffle the rest of the track when compressed through a club system. As a result, extreme side-chain compression seems to be used considerably more often in electronic music to address these imbalances. Also, typical club systems tend not to translate extreme stereo activity very well, so electronic music has a strong focus on mono and stereo mid signals.

Ryan Chynces: A similarity would be trying to fill up the available sonic space, and aiming for a sense of proportionality between all the elements.

Is it common for artists working in 'electronic dance' genres to mix their own material, even while they sequence and compose it, etc.? That is, is it common for artists specializing in 'electronic dance' music to oversee every aspect of record production, save mastering? Why do you think this is, and how does it impact creativity in the genre?

Phil France: I think that most electronic producers are capable of getting a decent mix on their own. However, depending on what they think is right for the track or record, they or their record company may ask a specialist mix engineer to do the work. I'm thinking about Caribou and FKA Twigs (among many others) and mixing engineer/producer David Wrench, for example. I think dance music has its own DNA in how quick it was on the uptake to what music you could make with electronic instruments and computers. If you're putting things into a sequencer then you are learning about mixing and production by default, as opposed to, say, writing a song and singing it on your guitar. It has enabled the genre to blossom creatively because it was all being figured out as it went along. It was new and so there weren't any traditional generic templates— before it became commercially successful, that is. The means of production using computers can be relatively cheap compared to traditional recording studios, but the integrity and quality of the musical ideas is always the thing that will shine through.

TJ Train: I think that as they become busier with touring, etc., it becomes more common for artists to seek out a mixing and/or mastering engineer so they can focus on their creative process. When starting out, the learning curve usually involves trying to do their own mixing and mastering, because funds usually do not allow otherwise.

Rick Bull: As a basic function of economics, and admonishing the 'DIY' mythology of dance music history, I think that the great majority of electronic music artists pride themselves on being both composers *and* engineers. I've seen something of the opposite in a lot of rock/ folk artists, who tend to feel that their role is primarily as composers. It's an artificial divide, which I do think is closing. The folk and rock tradition was, until more recent decades, very much beholden to the power structures of major labels and studios, which have been invested in divesting many artists of their own sense of volition of power in producing finished records. The traditional industry model relies on the mythologies of stratified labor models in a production process (i.e., writers, mixers, engineers, pressing plants, distributors, etc.). As this traditional model collapses, and mix technology becomes democratized, I think that artists in general are realizing that they have the ability to grasp all of the aspects of music production from composition to finished product. There is no longer so much artistic stigma attached to 'doing it all', and extra assistance tends to be employed now more as a matter of utility than principally as a matter of pride.

Adam Marshall: Yes, this is common. This happens because, on the 'underground' levels that I usually work within, there is not the infrastructure to bring in outside experts for certain parts of the production process. This is due to the fact that ('underground') labels are usually self-funded, and therefore it's just not financially possible to bring in outside experts. It's DIY, but out of necessity.

Noah Pred: Empowered by technology, it seems the majority of electronic producers take pride in mixing their own material. Not only does it give them more control over the sonic character of their work (something that gets a lot of attention in their field), but it can also be an integral part of the creative process. Personally, I tend not to focus too much on mixing until the final stage of production, since mix balance can be significantly altered by changes to an arrangement. However, I know a lot of producers who front-load a good deal of mixing at the beginning of a project, ensuring every sound is honed into place before proceeding with further composing or arranging duties. For those who can work effectively in this way, it can lead to an impressive degree of sound design. On the other end of the spectrum, I've seen this approach prevent other producers from getting anything done at all.

Ryan Chynces: Since an electronic music artist generally hasn't invested the time into mastering an instrument and then performing it professionally on a track (like a guitarist would), that time and energy is instead invested into developing expertise in the entire production process.

Craig Bratley: It can vary. Mixing and writing are often part of the same process with regards to electronic music. It's quite common for an artist to mix their material as they are writing. Sometimes you need to have the right drum sound, or get the bottom end working together, before you can make a start. Even the sounds you choose can have an important impact on the final balance—selecting the right sounds in the writing process saves you work trying to find a space for everything when you're mixing. Other times, you might just write a track really quickly, and mix it later. And there are a lot of people who write tracks and get a mix engineer to polish them up later, as they might not have the production chops to get the track sounding exactly how they want it to.

Andy Cole: I can only speak for myself. I develop 'layers', then go back to add more 'layers'. Perhaps, over time, I may compose another layer, or add a vocal that comes up after listening to the piece. However, perhaps at the start of my production processes, I wouldn't 'compose' as such, but simply 'try', 'experiment' and 'see what happens', much like a visual artist would (i.e., put something onto paper, add to it, take things out and so on). I guess you may start with an initial idea in your head, but the outcome may be completely different from where you started. I guess that is the beauty of the electronic/digital mixing processes—you can cut, copy, paste, save, delete and so on. In terms of mastering, though, I prefer for a mastering engineer to create final masters. If my mix elements are clear and balanced, then it lets a more experienced person complete the project. That's not to say that I always work with a mastering engineer. I have created my own internal masters, for personal use, and they do sound good, but it's also always good to have another set of ears hear something before you finish.

Noah Pred: Particularly with electronic producers' near-universal embrace of creative platforms such as Ableton Live, Reason, and Bitwig (as opposed to more traditional tracking-oriented platforms such as Pro Tools, Logic or Digital Performer), composition and production are so inherently intertwined as to be in most cases indistinguishable.

Phil France: Yes, I work on a piece right until the end of the mix session. You might notice something during the mix that shouldn't be there (for example, an audible click on an audio edit), for example, which needs fixing. If you're happy with the mix, you also need to take it away from the coalface, so to speak, and listen on different speakers until you are entirely happy that it's good enough.

Ryan Chynces: Absolutely—100%! I'd also say that 'producing' is part of the composition process in electronic dance music, since new sounds, combinations, etc., emerge during the production process depending on how heavily (and creatively) various effects are applied. While rock artists may 'go lighter' on processing effects during the production process (to maintain that 'live performance believability'), electronic music artists are under no such constraints.

What tools do you use to mix your music?

Rick Bull: I've moved from various outboard solutions, to mixing entirely 'in the box' these days. This has primarily been born of a desire to work simply, quickly and in a variety of environments—from cafes to airplanes, studios to in bed, in a way that a simple domestic laptop can support. I work with minimal third-party plug-ins, entirely from within the Ableton Live environment. I travel so much and frequently that it's been in my interest to tailor my process accordingly, and so I try to work with and master whatever tools are 'at hand'. Where necessary, I'll solicit third-party help and studios. I use minimal control surfaces (even when they are 'on hand'), as I prefer to really try to 'learn' the most basic of tools efficiently.

Pierre Belliveau: Headphones, monitors, spectrum analyzer . . . that's pretty much it, I guess. And the good old car test. That's a must—no joke!

Noah Pred: I do all my mixing in Ableton Live 9.5, with a combination of Ableton's audio devices, UAD-powered plug-ins, and plug-ins from Eventide, iZotope, Wave

Arts and Waves. At this point, the UAD emulations probably feel the most indispensable, due to their faithful harmonic saturation characteristics and signal path modelling. That said, I think an important part of mixing is knowing the principles to make do with whatever tools you have available.

Adam Marshall: I alternate, depending on what equipment I have access to. For initial mix downs, I usually use Ableton. But for final mix downs, I enjoy summing my mixes through outboard hardware mixers (when I can).

Phil France: Logic X, Soundtoys, Lexicon and UAD plug-ins.

Ryan Chynces: Ableton Live, headphones and studio monitors.

TJ Train: When I do any sort of mixing on my own, I use the Fabfilter plug-in suite, and UA plug-ins for stuff like compression and EQ.

Andy Cole: Logic Pro for production. Allen & Heath Xone 62/92 mixers for DJing.

Craig Bratley: I use the same tools as everyone else. I have a hybrid studio so I use the best of both worlds—a bit of EQ, compression, reverbs, delays and so on. Nothing esoteric, really. I also use a summing mixer.

Which (if any) of the tools you use for mixing would you say are indispensable to your process? Why?

Craig Bratley: EQ, as it allows me to make space for the individual elements.

Pierre Belliveau: Headphones. They provide a direct reference, irrespective of whether you have good room acoustics. The acoustics in my room are not good, so headphones are a must.

Rick Bull: A solid set of monitors are vital. I use a combination of Genelec 2020As, Yamaha HS80Ms, cheap Sennheiser earbuds and Sennheiser HD25 headphones. I've learned to hear through these fairly readily available tools, to produce a passable mix. Ultimately, I feel that if you can learn to mix on really simple tools, then having access to more 'high end' tools feels more effective. I'm personally really a 'nuts and bolts' guy. I have a small handful of third-party plug-ins which I find indispensable, primarily the PSP Vintage Warmer compressor and Xenon limiter. These are tools I've just stuck with over the years, and ones that I understand the sound and functionality of. They are robust, not particularly subtle and intuitive tools. There are some gorgeous multiband compressor and EQ suites I'd love to get my hands on, when finances avail themselves. But for now, the bundled tools within Ableton Live, and a few third-party utilities, have helped me to forge my signature sound—and it's one, in general, that I'm happy with.

Adam Marshall: When I perform live, my analog drum machine is invaluable because it produces bass tones that could not be (easily) replicated from my laptop. In fact, in most productions I do, I find adding an analog element to the mix really fills things out sonically. I am not married to any specific tools, but I do find it works well for me to set limitations and just get to work with what I have, instead of always waiting until I get the ultimate setup. Wait for that, and you'll never do anything!

Ryan Chynces: Headphones and studio monitors working in tandem are important for achieving optimal sound on a maximum number of playback devices. A program like Ableton Live is, of course, essential, since that's how the mixing is done. However, within Ableton, I'd say that the essential effects are the limiter and track delay. Track delay is very important when you're pushing effects like the limiter hard, since a heavily applied limiter on a given multitrack can make it seem to lag behind the other multitracks. Therefore, being able to nudge it forward a few milliseconds to compensate is vital.

TJ Train: Fabfilter. I use the filter/EQ on just about all of my music because of the smooth nature of their plugins. They have a very transparent and 'non-computer' sound to them, which I love.

Andy Cole: Logic. It allows me to record elements, live instruments, live mixes and samples, and then to work on each layer directly to create the best balance of sounds.

Summary and Conclusion

In the conversation transcribed and arranged above, a number of themes regularly surface, and a general consensus on approaches to mixing audio in electronic dance music contexts becomes readily apparent. Indeed, it seems clear that the artists above approach mixing in a manner largely peculiar to electronic music production. Mixing seems interwoven with the production process at large in our sample's workflow, while in other genres it remains largely a discrete activity carried out at a time and place removed from the compositional, recording and editing activities involved in a record's production. This seems to have led, in turn, to an equally unique conception of mixing's general function in record production.

According to the artists above, mixing emerges in electronic dance music genres as a 'creative' more than 'corrective' process. In some cases, in fact, artists seem to treat mixing as a creative process wholly indistinguishable from—that is, holistically ingrained within—all of the broader composition and production processes that go into making a record.

'Liveness', albeit a 'liveness' peculiar to electronic dance music genres, also seems to have played a crucial role in the way our sample developed their mixing practices. Initial experiences of mixing audio in a live DJ/performance context seems to be an almost universal early experience here. This 'live', 'on the fly performance' approach to mixing audio lends itself to an open approach to the process, leading to an openness not just with the products of mixing but with regards to the process itself. The artists we spoke to seemed willing to experiment not just with how a mix should sound, in other words, but also with what techniques and tools belong to the 'mixing stage' proper. Moreover, the fact that electronic music productions are created with the dance floor and DJ performance in mind also seems to play a guiding role in mix practices. Traditional rock/pop productions often aim to convey the spirit of performance, while the artists we spoke to seemed to suggest that electronic music productions *are* performances in and of themselves.

Another point to consider is that the majority of electronic music productions discussed above took place exclusively 'in the box', that is, within a DAW environment. Even when artists 'came up' in analog environments, they claim to now almost always work in primarily digital contexts. This has a number of crucial implications for mixing. As Rick Bull noted, the technical democratization of production tools has promoted a similar democratization of production roles, as artists previously cut out of production and engineering processes now find themselves equipped to participate and oversee those processes completely (if they so desire). At one time, songwriting was the preserve of the artist, after which the production process would then be relinquished to professional recording engineers, producers and mix and mastering engineers, all playing discreet roles in often-discreet specialist facilities. With advancements in computing power over the last two decades, coupled with the relatively affordable prices of most computer recording software, the means of production is now freely available to artists willing to make a moderate initial investment. In turn, this has led to a situation where artists also have the tools to 'teach themselves' previously esoteric aspects of the production process, leading to a further willingness to experiment with aspects of production once considered the exclusive purview of specialists.

Obviously, much more could be made of the conversation above than we have room to do here. Our goal in compiling this conversation was simply to begin the process of examining the unique position of mixing in electronic dance music genres and to allow the artists doing that mixing a voice in broader academic considerations of their crafts. We would simply conclude by expressing our sincere gratitude to the artists who participated for allowing us this glimpse into their artistic practices and considerations. Clearly much more scholarly work needs to be done to fully grasp the various nuances of electronic dance music production practices, and it is our sincere hope that this chapter provides readers with a useful entry into that process.

11

Groove and the Grid

Mixing Contemporary Hip Hop

Matt Shelvock

In hip hop, a complicated relationship exists between production and mixing practices. Celebrated producer/mix engineers including MixedByAli, Noah '40', JustBlaze and Dr. Dre, among others, for example, demonstrate that the lines between mixing and production are often blurred within this genre. According to Derek Ali, aka MixedByAli:

> The guys at TDE [Kendrick Lamar, Ab Soul and Jay Rock] do things on records sometimes because they know I'm going to come behind them and do something that's gonna make it better. Over the years I've been so experimental with my mixes that they'd come in and try new things with their voice just to see what I could do with it. They use their voice as an instrument that can be added to.
>
> (Ahmed, 2014)

Here Ali demonstrates that his mixing style is so thoroughly experimental that it may resemble the production process at times. Indeed, his sonic thumbprint is particularly audible on the vocals of songs such as 'Cartoons and Cereal' (2012) and throughout *Good Kid, M.A.A.D City* (2012). Additionally, Jay Z's in-house producer JustBlaze demonstrates his own complicated relationship with production and engineering practices by describing his creative process:

> As far as how my process works, sometimes what I'll do, for example when I'm working with Saigon, is we'll start in the B room then we'll start to mix completely in the box. Once we get it to a point where it sounds good, or it sounds good to a certain point but we want to run it through the analog gear, we'll make stems of everything. Then we can bring those stems to the A room and finish the mix from there. That's pretty much what we did with the Jay-Z record. We started in the B room, all in Pro Tools, and did as much of the mix as we could in there because we weren't sure which record was going to be the single.
>
> (quoted in iZotope, 2016: https://www.iZotope.com/)

JustBlaze uses a multistage approach for crafting records by separating pre-production, production and mixing processes. He continues to say,

"I think hip hop for so long has been producer-driven—not that other music isn't—but in hip hop the producers are stars" (quoted in iZotope, 2016: https://www.iZotope.com/). Indeed, the producer is a central figure within hip hop music, but so much of what constitutes production and pre-production within hip hop requires fluency in audio mixing and its constituent techniques. As such, this article will consider several techniques occasionally associated with the arrangement and production of recorded music within the purview of mix practice. This attitude—as demonstrated through the above anecdotes and examples throughout this chapter—reflects the current culture surrounding hip hop production.

Generic Overview

The sonic characteristics of hip hop are distinct in comparison to other genres. Of these characteristics, perhaps the most notable is the presence of a rapper or emcee (or MC). Rapping is a type of spoken-word vocalism that is ubiquitous within the genre. Phrases within a rap verse adhere to a recurring rhythmic structure, or *flow*, from which the emcee will vary for the sake of maintaining compositional interest. Phrases within a verse also tend to end with a terminal rhyme—although internal rhyme is also possible. The general originality and cleverness with which emcees deliver these rhymes defines a large part of a performer's identity.

Emcees A$AP Rocky and Eminem demonstrate that there exists a tradition of appreciation among practitioners for the sonic history of hip hop. On artists such as himself, Kendrick Lamar and J Cole, A$AP Rocky explains, "We're hybrid children, we grew up students of hip hop. I really enjoy every element of hip hop. Everyday I'm learning about old school cats that have the same kind of flavor that we do . . . I'm just a student. I grew up on all that stuff" (Sencio, 2013: http://hot937.cbslocal.com). Eminem echoes this sentiment by stating,

> being a student of hip-hop in general, you take technical aspects from places. You may take a rhyme pattern or flow from Big Daddy Kane or Kool G Rap. But then you go to Tupac, and he made songs . . . Biggie told stories. I wanted to do all that shit.
>
> (*Rolling Stone*, 2013)

Appreciation for the sonic history of hip hop extends beyond lyrical homage, however, as practitioners often pay tribute to both past hip hop instrumentals and other foundational elements of the genre. The hip hop producer's primary means of engaging with previously recorded material this way is through the act of *sampling*. Sampling dates back to the formation of dub music in Jamaica in the 1960s. The music of Osbourne 'King Tubby' Ruddock, Lee 'Scratch' Perry and Errol Thompson, for example, would sample so-called instrumental *breaks* from B-sides of 45 RPM records and compose around them by further emphasizing drums, adding signal processing (especially reverb) and sound effects, and other instruments. By the 1970s in America, DJs such as Kool Herc, Grand Wizard Theodore, Grandmaster Flash and Jazzy Jay had adopted the practice of sampling breaks. These DJs formed the backbone of hip hop music through this quasi-compositional method, and expanded upon the technique of sampling breaks by adding vinyl *scratching* to the available sonic repertory (Hager, 1984; Fricke and Ahearn, 2002; Chang, 2005).

Another distinctive characteristic of hip hop, in addition to the presence of a rapper and sampled audio, is the general emphasis placed on drums and low-end frequencies within the culture of listening. The kick drum and bass tend to sound extremely prominent in hip hop mixes. Clarity of low-end frequency content is of the utmost importance to hip hop fans, who famously, for example, popularized the use of subwoofers within car speaker systems. Additionally, in 2012, Dre Beats held a market share of 64% for headphones valued over $100, and was valued at $1 billion as a company (Dorris, 2013; Neate, 2013). I clarify this fact because mix engineers routinely consider the best possible sonic representation for the audio they are working on. In order for this *best fit* approach to work, the playback systems of music fans must be considered.

Crafting the Instrumental

As stated, creative workflow within hip hop often blurs lines between production and mixing activities. This primarily owes to the fact that contemporary hip hop is built through the digital audio workstation (DAW). Even producers who originally worked primarily within the analog domain or on hardware, such as DJ Premier, currently rely on a DAW from pre-production through post-production. Premier explains his admiration for the DAW by stating:

> I think Pro Tools is a great gift for all of us who have dealt with tape. It can do a lot of things that I couldn't do in analogue; it allows you to mess up, and re-mess up and redo and undo, and so on. I am just in a whole different world and frame of mind these days and Pro Tools just enhances me as a producer and a person.
>
> (Tingen, 2007)

Here, the ability of the DAW to quickly perform edits, retake passages and mix on the fly clearly aids Premier's creative process. DAWs, when used in this way, can act both as a sketch pad for rough ideas and as a program for finalizing production, mix and mastering decisions.

This section discusses mixing hip hop with the assumption that DAWs remain an integral part of the beatmaking, production and mixing processes for contemporary hip hop music. 'Crafting the Instrumental' describes methods for mixing hip hop instrumentals by first discussing a number of standard generic components including samples, percussion and bass. This information is accompanied by suggestions for implementation within a mix composite. After discussing a process for mixing an instrumental track, the proceeding section ('Mixing Vocals') will cover strategies for mixing vocals. In hip hop, often vocalists work from a pre-composed instrumental while recording. As such, this article will first describe a generic workflow for finalizing an instrumental before adding in vocals for a final mixdown.

Sampling

To *sample* refers to the act of incorporating pre-recorded material into a new musical instantiation of some type. Producers can sample from any number of musical elements including (but not limited) to basslines, drum breaks, fills, background vocals, strings, gospel choirs, entire songs and any other conceivable audio source. Production duo Christian Rich, for example, sampled a yelling sound on Earl Sweatshirt's 'Chum' at 00:16 (2013). The pair recount the following story regarding their search for this sample in an interview with LRG clothing company:

> We were fascinated with this song from the 90s—this Wutang hip hop song. It took us 10 years to find the song [they sampled originally], and when we finally found the song we liked this scream in there. We were like, "Yo, we gotta find that album, or that song, and get that scream." We found the scream, and that's the scream in the beginning.
>
> (quoted in LRG Clothing, 2013)

This anecdote demonstrates the importance of sampling to past and present hip hop, and also makes evident the aesthetic significance assigned to short peripheral sounds within the genre's sonic landscape. Additionally, here Christian Rich also exemplifies the fact that sampling is so entrenched within the sound of hip hop that producers intentionally incorporate samples used by celebrated producers of past eras.

There are several approaches taken by mixers, beatmakers and producers who sample previously recorded music, and these approaches can be classified according to the compositional purpose of a given sample. The sample discussed above from 'Chum' (2013), for instance, occurs only in the introduction. This type of approach is common in hip hop, where a sample may be used intermittently as a sound effect or percussive element. Samples can also, however, provide a track with recurring harmonic and melodic content. This is exemplified by the piano sample that plays throughout 'Chum', where this repeated segment forms the song's harmonic structure and content.

It is perhaps this tendency for sampled material to account for the rhythmic, harmonic or melodic content of a song that is most controversial for some people. For copyright owners, the issue is clear: they desire monetary compensation for their intellectual property. For musicians and critics, on the other hand, there seems to be a complicated quasi-moral issue surrounding the authenticity of incorporated sampled audio within a piece. For some music fans and critics, sampling is castigated as a non-legitimate form of creativity. Noah '40' Shebib (OVO Sound, Drake, Lil' Wayne, Action Bronson, Beyoncé) comments on his own use of samples by stating:

I use em sort of as a tool I guess, you know? I'm a fan of publishing, so I try not to sample too much, but it's a tool in your arsenal. There is something to be said about not sampling, and if you say "I'm not going to take that from someone else," I say "great, congratulations—I'm happy for you." But, you can sample, and what if a sample makes the song better? What I try to do is grab something, take a loop out of it, then flip it, reverse it, distort it, or chop it up. Then I ask myself "Did I create something?"

(quoted in Pensado's Place, 2014)

Here Noah demonstrates that he views sampling as one potential vehicle for achieving a desired aesthetic goal. Additionally, he explains some of the techniques producers use to craft and manipulate sampled music: cutting, reversing, processing and re-arranging. These tasks are accomplished through either loading vinyl, tape, CD or mp3 audio onto an MPC-style hardware sampler, or the DAW. With a hardware sampler, *slices*, or small sections of audio data, can be assigned to discrete channels and triggers for automatic playback and processing. Through utilizing this machinery, producers can easily mix and match slices of a pre-recorded piece.[1] Within a DAW, producers can utilize audio editing tools in order to accomplish the same tasks. For instance, most DAWs feature editing tools that resemble the control panel for Microsoft Paint or Adobe Photoshop. With these tools, audio can be selected, deleted, moved, time-stretched or pitch-shifted, among other similar operations.

Time Stretching and Pitch

There are, however, drawbacks to standard sampling tools. Time stretching can be particularly problematic when stretching audio beyond approximately 5% (faster or slower), for instance. Beyond this amount, noisy transients and harmonically rich segments from the source material produce *wavery* artifacts (Senior, 2011). *Beat slicing* is a potential solution for instances where a given time-stretch algorithm causes too many audible artifacts to occur. The technique of beat slicing refers to the act of cutting pre-recorded audio at rhythmically significant moments and repositioning these slices within a DAW time grid.[2] If moving to a faster tempo, individual slices must be shortened in order to provide ample time for the next slice to occur. When moving to a slower tempo, slices must be stretched in order to accommodate the extra silence that occurs as a result of the expanded time grid. Automatic beat-slicing tools also exist. iZotope's Phatmatik Pro and Propellerhead's Recycle constitute two examples of such software.

Once a sample has been time-stretched or beat-sliced to fit a track's tempo, recordists must consider the pitch of the sample. Pitch tends to be more easily altered than rhythm, but remains an important consideration nonetheless. Many DAWs have built-in pitch alteration capabilities, wherein users can alter the overall pitch of a sample in cents. Other tools such as Celemony's Melodyne are quite commonly used, for example, which allows both manual and automatic pitch correction.

Noise Removal

Sampling originally employed the use of vinyl discs, and although the practice of so-called *crate digging* for inspiring recordings is alive and well today, producers and artists may also choose to sample from tape, CD or mp3. While digital media tends to sound full and clean, other types of media may be bandwidth-limited or contain audible hiss and crackle. Hip hop tracks may sample from either of these sources and can thus suffer from a multitude of noise issues. Some hiss and crackle can be considered acceptable, and may even be masked by the other instrumentation in a full mix. Kendrick Lamar's 'Rigamortis' (2011), for example, features this type of vinyl crackle throughout the song. The presence of this type of noise is not appropriate, however, for every situation, and occasionally noise-removal operations will have to be performed on the sample in question.

For hiss and crackle, simply filtering out high-end frequency content can fix issues at times. This solution, however, will also cause the sample to sound *dull* in comparison with the original source. As such, simple filtering may also be a poor choice for removing undesirable high-frequency noise. In cases where filtering causes the sample to sound overly dull, instead opt for multiband dynamics processing. Users should target the offending frequency region only (e.g., above 3 kHz–5 kHz) with a fast-acting expansion tool. Ratio and threshold can be tweaked to reduce any offending noise contained within the region, but should only be increased insofar as the source material requires. The advantage to using multiband expansion in this way is that the end result should be much more *transparent* sounding than simple filtering.

Another type of noise that occurs during the sampling process is the accumulation of ground hum at 50/60 Hz. Noise removal via comb filtering works well here, and can be performed through Tone Boosters' TB_ HumRemover, for instance. This approach necessitates caution, however, if musical pitches match the hum frequency (50/60 Hz). In this case, musical pitches will be attenuated along with the undesirable noise.

Groove and the Grid: Blending Samples with Percussion

Percussion instruments are integral to the hip hop aesthetic. Both the timbre and the rhythmic composition of these instruments must be considered by those who mix hip hop beats. Additionally, temporal alignment between key transients in a track's main sample and any added percussion must be established. Editing and sample delays can be used to align the rhythmic events of sampled material with additional percussion this way. These tools, however, can also be used to *humanize* grooves by creating small rhythmic irregularities.[3]

When rhythmic conflicts occur between different types of samples, or samples and MIDI instruments, editing is likely the best solution. For easy comparison, users should place conflicting tracks side by side within a sequencer. Zoom in on the waveform visual so that transient information is visible on both tracks, and proceed to nudge the tracks forward or backward on the time grid until the transients are aligned. Once alignment is achieved, it may be desirable to emphasize the transient of either instrument. To do this, simply use a sample delay to cause the desired transient to occur first. If one wants to mask the snare drum present in a given sample, then an additional snare instrument could be added and given a negative delay setting (i.e., −3 m/s). This would cause the additional snare to sound first, thus masking the transient of the snare within the sampled source material. While spectral treatment may still be necessary, this type of masking presents one starting point.

Hip hop features a strong focus on rhythmic events and *groove*. Groove can be altered—as in real-world drumming—by causing instruments to occur behind the beat by small (but varying) time intervals. This routinely happens in hip hop via manipulating *swing* on Akai MPC style samplers. The swing parameter determines how far behind a given sonic even occurs from its original rhythmic value. MPC designer Roger Linn describes his approach to the swing algorithm by stating:

> My implementation of swing has always been very simple: I merely delay the second 16th note within each 8th note. In other words, I delay all the even-numbered 16th notes within the beat (2, 4, 6, 8, etc.) In my products I describe the swing amount in terms of the ratio of time duration between the first and second 16th notes within each 8th note. For example, 50% is no swing, meaning that both 16th notes within each 8th note are given equal timing. And 66% means perfect triplet swing, meaning that the first 16th note of each pair gets 2/3 of the time, and the second 16th note gets 1/3, so the second 16th note falls on a perfect 8th note triplet. The fun comes in the in-between settings. For example, a 90 BPM swing groove will feel looser at 62% than at a perfect swing setting of 66%. And for straight 16th-note beats (no swing), a swing setting of 54% will loosen up the feel without it sounding like swing. Between 50% and around 70% are lots of wonderful little settings that, for a particular beat and tempo, can change a rigid beat into something that makes people move.
>
> (quoted in Scarth, 2013: 1)

The swing algorithm described by Linn has been employed within hip hop beatmaking since the 1980s, and continues to be relevant within the DAW. Ableton, for example, allows users to apply humanization from the so-called *groove pool* to both audio and MIDI tracks. Additionally, the software offers a large collection of MPC style grooves, which are intended to emulate the MPC *swing* algorithm. If groove quantization does not provide enough humanization, or perhaps where a different type of rhythmic approach is desired, beatmakers can simply key in percussive events manually.

Low-End Theory: Drums and Bass

Low frequencies are extremely prominent in hip hop mixes. In fact, the listening culture of hip hop demonstrates its fascination with bass through the adoption of its preferred playback technology: subwoofers, boomboxes and large headphones. Drum machines—particularly the Roland TR-808, but also samplers by Akai, E-mu and Alesis—comprise a large portion of hip hop's low frequency content via the kick drum. In contemporary hip hop production, samples of drums may be loaded into a DAW rather than a drum machine, but both approaches are common. Producers, beatmakers and mixers often sample classic drum machine sounds through DAW tools, but also sample live drums. In addition to methods for creating and mixing kick instruments, the following section will also discuss both methods for generating bass and mixing bass lines.

808s and Heartbreak: Mixing Sine Wave–Based Kicks

The sound of the 808 is ubiquitous within hip hop. The classic drum machine can be heard on Afrika Bambaata and the Soulsonic Force's 'Planet Rock' (1982), for an early example. More recently, Kanye West's affinity for the 808 is made clear by both the sonic characteristics and the name of his 2008 album *808s and Heartbreak*. West's production style also relies on the classic machine: the percussion in Lil' Wayne's 'Let the Beat Build' (2008), for instance, was built using only 808 percussion and a single sample in addition to the vocals.

Samples of 808 drums are ubiquitous within the production world. MIDI synths, as well as hardware and software samplers, often come pre-packaged with a set of 808-inspired sounds. Additionally, producers may opt to design their own sounds that resemble the 808. This can be achieved with any synthesizer that contains an isolated sine wave oscillator. In Native Instrument's Massive, for example, users can create a simple 808 kick using the following settings:

> Oscillator 1: 'Sin- Square'; 'Spectrum'; Pitch: −24; WT Position: 0%; Intensity: 100%; Amplitude: 100%
> Oscillator 2: Off
> Oscillator 3: Off
> Filter 1: Filter type: Lowpass 4; cut-off: 100%
> Envelope 1: modulates 'pitch' parameter of oscillator 1 +24 (2 octaves); attack: 0%; decay: 0%
> Envelope 4 (universal envelope): attack: 0%; decay: 35%

The above patch will produce a classic 808-style kick from a simple sine wave oscillation.[4] The resultant sound, however, despite sounding *clean,* lacks harmonics and may be difficult to hear through a dense mix.

Sine-based kick drums, like those produced in the above patch, can be treated to sound more prominent through boosting or adding additional overtone content. A number of strategies may be used to enhance the overtone content of the 808. Saturation, bit-crushing and harmonic enhancers constitute three methods for achieving this goal. The most common of these, perhaps, is saturation. Subtle saturation can provide slight compression and enhance the harmonic content of the 808 kick in order to cause it to appear perceptually *forward* within the mix. In order to accomplish this task, a distortion plugin such as Sound Toys' Decapitator may be used. Decapitator features a *thump* setting that works in conjunction with the hi-pass filter to add an amplitude boost at the desired setting. For example, if the hi-pass filter is set to 60 Hz, Decapitator will boost 60 Hz in addition to providing a filter here. Users can implement this technique in conjunction with multiband compression in order to extend the decay of the harmonics produced. A sonically related technique to saturating and compressing a signal is to generate distortion via bit-crushing. Bit-crusher plugins emulate the distortion produced when reducing the word length of a given audio file, and can be heard within the production style of !llmind on *Human* (2015), for example. Another method for increasing harmonic content within an 808 kick is to use methods for *enhancement*. For example, Waves' MaxxBass works by analyzing the low-end fundamental content of an audio source, and then creating new harmonics above that fundamental which can be mixed back into the original signal. One might consider running two instances of Waves' MaxxBass on a sine kick. The first instance will generate harmonics related to the fundamental of the kick, and the second instance will generate additional overtones from the harmonics generated in the first instance.

Other Kicks

808-style kicks are often blended with other synthesized or sampled drums. 909 and 808 sounds are often layered, for example, and can be seen within the free sample library released by hip hop producer Just Blaze in 2015. His sample library contains, for instance, two kick drums used on Jay Z's *The Blueprint* (2001), each consisting of a blend of synthesized and sampled layers. In addition to synthesized 808 and 909 drum sounds, kick drums can be sampled from other recordings or taken from sample libraries. Just Blaze's sample library, for example, contains such sampled material. The 'KICK_EASY.wav' sample contains an audible vocal sound left over from the source material (Camp, 2015).

Bass

Bass, while a prominent feature of both classic and contemporary hip hop, most typically remains secondary to the kick drum. A number of bass-producing methods may be used in hip hop production such as sampling, live tracking and synthesis. In cases where samples feature prominent bass lines, producers and beatmakers occasionally opt to use the recycled bass line in the new arrangement. Often, as a consequence of splicing and re-ordering samples, a new bass line forms from the edits made to the original source material.

For songs where the sample does not already contain a bass line, or where a song does not already contain a foundational sample as discussed in section 2.1, a sampled bass line may be used. In this case, a bass line from a sample library or previously recorded song is implemented within a given session and often receives treatment in the form of editing and repositioning in order to establish a more original feel.

Perhaps the most common method for implementing bass within hip hop, however, is synthesis. Sine-based bass instruments are quite common and—as with the kick drum—evolved from widespread TR 808 usage within the genre's earlier stages. A simple sine bass can be created in Native Instruments' Massive, for example, by inputting the following settings:

> Oscillator 1: 'Sin- Square'; 'Spectrum'; Pitch: −24; WT Position: 0%;
>   Intensity: 100%; Amplitude: 100%
> Oscillator 2: Off
> Oscillator 3: Off
> Envelope 4 (universal envelope): decay: 65%; level: 0%
> Voicing menu: Max: 1
> Insert 1: S shaper; Dry/Wet: 20%; Drive 50%
> FX 1: C-tube; Dry/Wet: 20%; Drive: 50%

The above settings will produce a deep sine wave bass texture that is suitable for sub bass, with added distortion to generate harmonics.[5] Additionally, any of the mixing strategies discussed in section '808s and Heartbreak' for enhancing the overtone content of the 808 kick drum may be replicated on 808 bass tracks. Mixers and producers should be careful, however, that the two powerful sounds do not compete for sonic territory.

Finalizing Kick and Bass

As mentioned, kick and bass drums often compete with one another for sonic territory. Within hip hop and related genres, the generalization can be made that the kick drum often occupies a higher frequency territory than the bass. Depending on subgenre, different approaches for blending the two instruments may be used. Trap, for example, features exaggerated sub bass frequencies as typified within Travis Scott's *Rodeo* (2015). A more tame approach to sub bass is heard on Dr. Dre's *2001* (1999), on the other hand. As sub bass (20 Hz–60 Hz) tends to occupy much of the available energy within a mix composite when exaggerated, kick drums can have a hard time competing for audibility in trap music such as Scott's. In order to aid the kick drum in cutting through in this instance, typically high-pass filtering will be applied to attenuate sub bass frequencies for this instrument, and in so doing removes frequency information that may compete with the bass.

In addition to filtering, the bass can be treated through lateral dynamic reduction (Hodgson, 2011). This method attenuates overall volume of the bass by reducing its dynamic range through compressing via an external input. This method makes use of the *side-chain* feature available on many hardware and software compressors. Here the kick drum is used to provide such an input, and a compressor on the bass channel will react to the incoming kick signal.

Export/Bounce

The previous sections have focused on methods for mixing and selecting samples, drums and bass. Other peripheral instrumentation of any type can be added as well, but these sources are diverse and often play a more trivial part in hip hop mixes than the sonic aspects discussed above. Once an instrumental is completed, it is sent to an emcee for the addition of vocal tracks. Before the file is exported, engineers may consider applying light compression or limiting and EQ to the stereo bus to aid in delivering an *energetic* sounding background track to the artist. This step is not necessary, but may be appreciated by the vocalist.

Mixing Vocals

As mentioned earlier, a sonically distinct feature of hip hop music is the presence of rap vocals. It is crucial that engineers craft rap vocals to sound intelligible within the mix—often a difficult task—as the vocals may clash with the rich midrange and upper midrange exhibited within funk and soul vinyl samples, snare drums, gospel organs, pianos, strings and other idiomatic textures. On achieving an acceptable balance between these sounds, Grammy-nominated mix engineer Matthew Weiss (9th Wonder, !llmind, Snoop, Sonny Digital) explains:

> Quintessentially, hip hop is all about the relationship between the vocals and the drums. The number one contestant with the voice is the snare. Finding a way to make both the vocals and the snare prominent without stepping on each other will make the rest of the mix fall nicely into place.
>
> (Weiss, 2011)

Here, Weiss provides a concise characterization of hip hop as a genre that favors drums and vocals. As such, mix engineers direct attention towards the sonic impact of these two instruments during playback. The importance of an artist's vocal *sound* is demonstrated through artist–engineer pairs such as Kendrick Lamar and MixedbyAli, as well as Jay Z and Young Guru, among others. Young Guru remarks on his relationship with rapper Jay Z by stating:

> Even when Jay's working with another producer who has his own go-to engineer, Jay takes me along to engineer his vocals. This is to do with my knowledge of his way of working and the comfort and trust factor between us. We do more than just recording: he also bounces off ideas with me, so there's a kind of synergy about the records we create together.
>
> (quoted in Tingen, 2009)

This anecdote demonstrates that emcees may even employ the same engineer repeatedly once an artist develops a level of trust in such an individual. Given the importance of vocal sound to both mix engineers and the overall aesthetic of hip hop music, the following section will discuss strategies for maintaining vocal intelligibility within hip hop mixes.

Types of Vocal Tracks

There are five classes of hip hop vocal tracks, each defined by its aesthetic function. These include (i) the lead vocal, (ii) answer tracks, (iii) ad lib tracks, (iv) emphasis dubs and sometimes (v) a supporting vocal in the form of a hook or background singing. As in other genres, a lead vocal acts as the primary compositional vehicle for delivering lyrics and textual meaning. In hip hop, this lead vocal is almost always delivered within the paradigm of so-called *rapping*. The lead vocal (i) may rap continuous *verse* sections, or may separate sequential verses with a *hook*. Hooks (v) may be rapped or sung by the emcee or another individual in a more traditional pop or R&B fashion. Emcee Action Bronson's 'Baby Blue' (2015), for instance, features several verses separated by a *sung* vocal hook. In this case, the rapper opted to sing the hook himself, but it is equally common for another singer or sample to fulfill this role. Hip hop also commonly features a variety of background vocal tracks, such as an *answer track* (ii). These tracks can support the lead vocal in any number of ways. Most often, these answer tracks will offer some type of agreeing sentiment with the lead vocal. For example, in Kendrick Lamar's 'Swimming Pools' (2012), the repeated phrase 'drank' supports the lead vocal as follows (00:13–00:26):

> Pour up (Drank), head shot (Drank);
> Sit down (Drank), stand up (Drank);
> Pass out (Drank), wake up (Drank);
> Faded (Drank), faded (Drank).

In addition to answer tracks, another type of common supporting vocal in hip hop are ad lib tracks (iii). These tracks establish a quasi-improvisatory feeling within the music through the provision of seemingly *off-the-cuff* lyrics and vocables. In both classic and contemporary hip hop, often the intros and outros of songs contain such ad libbed material. An example of this type of vocal can be heard in Statik Selektah's 'Carry On' (2014) (00:13–00:24). Here emcee Joey Bada$$ delivers introductory ad lib material while Statik Selektah provides instrumental support and scratching. The least audible, but certainly not the least significant, type of vocal track featured in hip hop recordings are *emphasis dubs* (iv). These tracks simply reinforce rhymes contained within the lead vocal.

Lead Vocal

When mixing lead vocals, there are few (if any) universally constant methods used by engineers. A variety of approaches for mixing vocals exist for reasons such as physiological differences in human voices, differences in tracking methods and differences in aesthetic goals from project to project. Difficulty arises when one attempts to prescribe a *best fit* approach for mixing vocals, particularly where specific numeric descriptions are involved. There are, however, a number of general aesthetic tendencies of hip hop vocals that can be described. The following sections will proceed this way by discussing the characteristics of exemplary hip hop vocals within the scope of a mix composite. A number of general provisional signal processing maneuvers will be described in the following section, but these examples merely provide starting points for readers.

Consistency

A key factor in maintaining vocal intelligibility throughout a given vocal track is the overall consistency of amplitude. While skilled vocalists may be conscientious of dynamic range throughout the tracking process, additional processing is often required to achieve a more consistent amplitude level for the duration of a vocal track. Engineers can automate volume manually or use a plugin such as Waves' Vocal Rider in order to tame overall amplitude, or *macrodynamic*, issues within the vocal. Where slightly more invasive peak taming is required, compression may also present a good choice. A compressor with medium attack and release times, and a light ratio and threshold, should address such peaks without providing an abundance of timbral coloration.

In addition to macrodynamic shaping, a compressor can be used with a *microdynamic* strategy in mind. Where the macrodynamics of a track refers to its overall amplitude scheme throughout its duration, microdynamics refers to the dynamic characteristics of individual events within the track. For instance, hard consonants at the beginning of words can establish small amplitude spikes within individual words or phrases, and in so doing create a microdynamic imbalance between the initial spike and any proceeding lyrics. If problems such as this arise, a potential solution is to apply compression with a light ratio and fast attack-and-release settings.

Balance with Instrumental

Once dynamic consistency has been established within a vocal, mix engineers should begin to consider the dynamic and spectral balance between the vocal and the instrumental tracks. As the human voice is a diverse instrument, there is no *one-size-fits-all* description of how to achieve this balance; however, a number of sonic considerations will be suggested in this section.

Lead Vocal and Instrumental: Low End

A common suggestion made to beginner mix engineers is to high pass nearly every non-bass instrument around 100 Hz. Indeed, this type of strategy aids in clearing up *muddiness* in a mix by removing extraneous low end from non-bass instruments. Spoken male voices, however, range in fundamental frequency from 85 Hz to 155 Hz (Titze, 1994); ergo, a high-pass filter set to 100 Hz may attenuate some important spectral information. Additionally, in rap music there is a tendency to provide down-tuned vocals at times. This trend was heavily popularized by Houston-based chop and screw artists and can be heard on Z-RO's '25 Lighters', for example. Brooklyn-based A$AP Rocky has re-popularized this trend in a more mainstream way as heard on 'Purple Swag'(2011), 'Bass' (2013) and 'Goldie' (2012).

Rather than using a predetermined high-pass filter setting, mixers should instead rely on their ears by sweeping through the low range from 50 Hz–150 Hz. The vocals should remain full sounding in order to fulfill the aesthetic goals of hip hop, however. If both a full-sounding vocal and a clean-sounding mix cannot be achieved, recordists must revisit instrumental tracks in order to remove competing low end.

Lead Vocal and Instrumental: Mids

Human hearing is attuned quite well to midrange frequencies, and as a result mixes often contain a number of voices within this region. Common examples of these sounds in hip hop include snare drums, guitars, organs, pianos, strings and choir vocals. Given both our acuity in hearing frequencies in this range and the availability of instruments with fundamental frequencies contained within the midrange, there is a high likelihood for instruments to mask one another within this range. For example, snare drums often must sacrifice some midrange content or overall level in order to make room for the vocal. In hip hop, this is perhaps the most common mix issue within the midrange that must be addressed.

Additionally, different strategies for mixing midrange are necessitated by different approaches to instrumentation. A Tribe Called Quest's '1nce Again' (1996), for example, features a snare that is slightly louder than the vocal. In order for the vocal to remain intelligible within the mix composite, some low frequencies are removed and the higher midrange frequencies are emphasized and perhaps lightly saturated. This allows for a higher frequency territory to be emphasized in the vocal than the snare, and aids in maintaining listener separation between the two tracks.

Lead Vocal and Instrumental: High End

Vinyl samples often sound fairly saturated, and may contain lush strings and loud percussion. Ab Soul's 'Mixed Emotions', for example, samples 'Merv's Theme (Theme from 'The Merv Griffin Television Show')' (1965).[6] Upon inspecting the original sampled material, one hears loud percussion and strings. Both sounds elicit a high-end *shimmer*, and sound quite saturated—perhaps owing to the analog recording processes used in 1965. In order to fit within Ab Soul's track, the original sample has been processed to remove some of the high-end frequency information, and in so doing causes the instrumental sample to sound noticeably less *exciting*. The end result is that Ab Soul's voice can cut through the mix and remain both intelligible and more exciting for the listener, because the corresponding frequency information associated with these characteristics is no longer masked by the sample. If vinyl samples, or other instrumentation, maintain too much high-end *harshness* or *excitement*, listeners may ultimately have difficulty focusing on the lead vocal and instead may shift listener attention to the harsher or more exciting sound.

Lead Vocals: Notes on Ambience and *Air*

Air often refers to frequency information above 10 kHz. This region can be boosted quite simply via EQ, but this may yield artificial results depending on the amount of additional gain required and the amount of processing applied to the vocal already. Another, less invasive strategy in many genres might be to add reverb, but hip hop features a conspicuous absence of reverb on vocal tracks. As mix engineer Jaycen Joshua explains, "everyone knows that reverb is the kiss of death on rap vocals. Reverb and rap don't mix" (Tingen, 2010). This primarily owes to the vocal delivery of rap, which is often faster than other styles and more rhythmically intricate than pop vocals. Additionally, reverb causes vocals to sound as though they are occurring from a distance. Rap, however, often places an emphasis on aggressive delivery and *in-your-face* attitudes. As such, standard reverb settings should be avoided when trying to enhance *air* within a track. Where reverb is used, however, a short, quiet reverb with a wide stereo image is often best. Another viable technique may be to apply a small amount of multiband compression to frequencies above 10 kHz, or perhaps a combination of a slight treble boost and compression with a slow attack and fast release.

Background Vocals: Answers and Emphasis Dubs

In rap music, background vocals either answer the lead the vocal or provide emphasis for rhyming, or quasi-rhyming, words. Busta Rhymes demonstrates the vocal answer technique, for example, in 'Woo hah!! Got You All in Check' (1996 [00:20–00:26]). The hook of this track features Busta Rhymes stating, "when I step up into the place aye oh step correct", which is answered by a separate vocal that exclaims, "woo hah (woo hah!)!" This additional vocal is also heavily layered for emphasis. Vocal answers are not always so prominent, however. For instance, in the Beatnuts' 'Watch Out Now' (1999 [01:18–01:20]), JuJu and Psycho Les state, "It doesn't take much for us to let the metal holler", for which an overdubbed vocal repeats the word "holler". The overdubbed repetition of "holler" is quieter than the first, and has been treated to exhibit less presence overall. Depending on the lyrical and compositional function of an answer track, it may be more subdued, as with the Beatnuts' example, or more of a focal point, as with Busta Rhymes.

Emphasis dubs provide an additional type of background vocal in hip hop. These overdubs reinforce important rhymes and phrases rapped by the main vocal. An example of the technique of emphasis dubbing may be found in Dr. Dre's 'Still Dre' (2001 [01:30–01:38]), where he states "And even when I was *close to defeat*, I rose to *my feet*. My life's like a soundtrack I *wrote to the beat*". The italicized sections are doubled with an additional voice. This voice is quieter than the lead, with less harmonic complexity and dynamic range.

Background vocals such as answers and emphasis dubs demonstrate a few sonically consistent tendencies within hip hop. With the exception of songs where an answer lyric may be integral to the hook, background vocals are processed to sound both smaller and less exciting than the lead hook. In order to obfuscate these vocals slightly, a combination of filtering and reverb may be used. While reverb is the so-called 'kiss of death' for rap lead vocals according to Jaycen Joshua, subtle amounts of reverb on background vocals can be acceptable and will cause them to sound more distant (Tingen, 2010). Both low- and high-pass filtering may also be used: high-pass filtering will excise redundant low-end frequencies that already exist in the lead vocal, and low-pass filtering will remove some *excitement* and *air* from the background vocals. By removing high-end frequency content and adding gentle reverb, engineers can effectively *dull* the background vocals, and will thus allow the lead vocal to maintain prominence. Additionally, panning is used to remove these peripheral vocals away from the center of the stereo image. Upon listening to 'Still Dre', for instance, listeners will hear that emphasis dubs tend to be panned towards the left perhaps 25% to 40%.

Conclusion

Mixing, production and beat construction may officially occur at different stages within hip hop music making, but each stage tends to employ several methodologies typically associated with traditional *mixing*. In fact, as a music genre, hip hop is inseparable from the recording studio as a by-product of the genre's evolution from an underground phenomenon based on sampling breaks. The DIY attitude present during the genesis of the genre persists to this day, as evidenced by the popularity of samplers, triggers and non-professional studios on shows such as MassAppeal's *Rhythm Roulette*, where celebrated hip hop producers construct beats from music discovered while *crate digging*.

Core techniques associated with hip hop beatmaking, pre-production and production are discussed throughout this chapter insofar as each engages with the practicing of mixing recordings. The practice of mixing hip hop, of course, is complicated by the fact that the exemplary mix engineers discussed throughout this article, such as MixedByAli, may add samples, synthesis or other compositionally significant material to a project as a matter of course. While a number of crucial techniques and processes are covered throughout this article, hip hop strongly favors individualism and creativity as core aesthetic tenets. As such, those involved in making hip hop music should feel free to experiment with, and expand upon, the techniques provided throughout this chapter.

Notes

1 Demonstrations of this process can be seen on many episodes of Mass Appeal's *Rhythm Roulette*—a YouTube series that features celebrated producers and beatmakers such as 9th Wonder, El-P and Kirk Knight. The show can be found at https://www.youtube.com/user/massappeal (Accessed January 2016).

2 Different strategies may be used depending on both personal taste and the envelope of a given sound. Beatmakers may deem it more appropriate to edit samples at the zero crossing, or closer to the peak of the transient.

3 Such irregularities are too precise to represent with standard notation. For example, occasionally drummers may offset a snare drum intended to occur on beats *two* and *four* by a few milliseconds. The drum in this case is still notated to occur on beats *two* and *four*, even though the resultant groove has been effectively altered. This type of drumming can be heard on Stax records and is a key feature of music played by Booker T and the M.G.'s, for instance.

4 This patch can be replicated on other synthesizers. To aid with this process, an explanation of a few Massive-specific options may be helpful. The "sin-square" wavetable option provides both sine and square wave input to oscillator one, and the "WT Position" set to 0% ensures that only the sine wave can be heard. Additionally, Massive features four envelope filters. Envelope filters 1–3 can be assigned to modulate other parameters within Massive (but remain otherwise inactive), and Envelope 4 is a universal filter that all oscillators are connected to by default. The 808 patch provided contains a discrete fundamental pitch that can be altered by keying in different notes. If users want to hear more of the fundamental, simply turn up the decay on envelope filter 4.

5 Where more punch is required, a duplicate version of this track can be added. This duplicate should alter incoming MIDI data to sound an octave higher than the original, and should also be filtered to eliminate spectral conflicts or masking with the initial bass track.

6 From *A Tinkling Piano in the Next Apartment* (1965) by Merv Griffin.

Bibliography

Ahmed, I. (2014). How to Make It in America: TDE's MixedByAli on Becoming an Audio Engineer. *Complex CA*. [online]. Available at: http://ca.complex.com/music/2014/05/mixedbyali-tde-how-to-be-studio-engineer/ [Accessed: 18 January 2016].

Camp, Z. (2015). Just Blaze Releases Early '00s Drum Samples, Making Good on Meow the Jewels Promise. *Pitchfork*. [online]. Available at: http://pitchfork.com/news/62149-just-blaze-releases-early-00s-drum-samples-making-good-on-meow-the-jewels-promise/ [Accessed: 18 January 2016].

Chang, J. (2005). *Can't Stop, Won't Stop*. New York: St. Martin's Press.

Dorris, J. (2013). How Beats by Dre knocked Out Better Headphones. *The Age*. [online]. Available at: http://www.theage.com.au/digital-life/digital-life-news/how-beats-by-dre-knocked-out-better-headphones-20130913–2tola.html [Accessed: 18 January 2016].

Fricke, J. and Ahearn, C. (2002). *Yes Y'all*. Cambridge, MA: Da Capo Press.

Hager, S. (1984). *Hip Hop*. New York: St. Martin's Press.

Hodgson, J. (2011). Lateral Dynamics Processing in Experimental Hip Hop: Flying Lotus, Madlib, Oh No, J-Dilla and Prefuse 73. *Journal on the Art of Record Production 5*. [online]. Available at: http://arpjournal.com/lateral-dynamics-processing-in-experimental-hip-hop-flying-lotus-madlib-oh-no-j-dilla-and-prefuse-73/ [Accessed: 18 January 2016].

iZotope.com. (2016). *iZotope Interview: Just Blaze*. [online]. Available at: https://www.iZotope.com/en/community/artists/engineers/just-blaze [Accessed: 18 January 2016].

LRG Clothing. (2013). *Off the Record: Christian Rich*. [online]. Available at: https://www.youtube.com/watch?v=c4nyHF23gnI [Accessed: 18 January 2016].

Neate, R. (2013). Dr Dre Beats Valued at More Than $1bn Following Carlyle Deal. *The Guardian*. [online]. Available at: http://www.theguardian.com/music/2013/sep/27/dr-dre-beats-1bn-carlyle-sale [Accessed: 18 January 2016].

Pensado, D. (2014). Noah '40' Shebib—Pensado's Place #151. *YouTube*. [online]. Available at: https://www.youtube.com/watch?v=ESUHhXgIaos&index=19&list=PLh0d_FQ4KrAVF5LM2sCC_nncdjk7jCK7G [Accessed: 18 January 2016].

Rolling Stone. (2013). *Eminem Reborn: Inside the New Issue of Rolling Stone*. [online]. Available at: http://www.rollingstone.com/music/news/eminem-reborn-inside-the-new-issue-of-rolling-stone-20131120 [Accessed: 18 January 2016].

Scarth, G. (2013). Roger Linn on Swing, Groove & the Magic of the MPC's Timing—Attack Magazine. *Attack Magazine*. [online]. Available at: https://www.attackmagazine.com/features/interview/roger-linn-swing-groove-magic-mpc-timing/ [Accessed: 18 January 2016].

Sencio, B. (2013). "Student of Hip-Hop" A$AP Rocky Talks His Unique Musical Style. *Hot937.cbslocal.com*. [online] Available at: http://hot937.cbslocal.com/2013/03/18/student-of-hip-hop-asap-rocky-talks-his-unique-musical-style/ [Accessed: 18 January 2016].

Senior, M. (2010). Mix Rescue: Preslin Davis. *Soundonsound.com*. [online]. Available at: http://www.soundonsound.com/sos/feb10/articles/mixrescue_0210.htm [Accessed: 18 January 2016].

Senior, M. (2011). How to Build Tracks around Sampled Tunes. *Soundonsound.com*. [online]. Available at: https://www.soundonsound.com/sos/dec11/articles/steal-the-feel.htm [Accessed: 18 January 2016].

Tingen, P. (2007). DJ Premier. *Soundonsound.com*. [online]. Available at: http://www.soundonsound.com/sos/jul07/articles/djpremier.htm [Accessed: 18 January 2016].

Tingen, P. (2009). Secrets of the Mix Engineers: Young Guru. *Soundonsound.com*. [online]. Available at: http://www.soundonsound.com/people/secrets-mix-engineers-young-guru [Accessed: 18 January 2016].

Tingen, P. (2010). Secrets of the Mix Engineers: Jaycen Joshua. *Soundonsound.com*. [online]. Available at: http://www.soundonsound.com/sos/aug10/articles/it-0810.htm [Accessed: 18 January 2016].

Titze, I. (1994). *Principles of Voice Production*. Englewood Cliffs, NJ: Prentice Hall.

Weiss, M. (2011). Tips for Mixing Rap Vocals. *The Pro Audio Files*. [online]. Available at: http://theproaudiofiles.com/mixing-rap-vocals/ [Accessed 18 January 2016].

12

The Mix Is. The Mix Is Not.

Robert Wilsmore and Christopher Johnson

> [E]stablish a logic of the AND, overthrow ontology, do away with foundations, nullify endings and beginnings.
>
> —Deleuze and Guattari (1987: 25)

Upon presentation of the mix, the one-is-not is made apparent. In our continuing exploration, as producers and academics, we will expand our discourse of the conceptual group The And and their singular composition 'The Song of a Thousand Songs'. We will take a new line of flight in discussion of the mix, which suffers similar ontological and existential concerns, but we will also keep our feet on the ground and explore a song that we have been involved in at a gritty level. Our aim, if we have one, is to show how different perspectives on the mix as multiple reveal different authenticities, different values and how these are constructed. Most of all, we just want to play between one and many, between idea and audio, between heaven and earth.

As our starting points we use Alain Badiou's exploration of 'the one-is-not' as it is discussed in the opening meditations of *Being and Event* ([1988] 2007, hereafter abbreviated BE), although we will leave aside the mathematical set theory used to establish his arguments and borrow instead some approaches to considering the multiple that will be of use to us. Second, we make reference to the multiplicity, the one and the many, as it is thought conceptually in Gilles Deleuze and Felix Guattari's *A Thousand Plateaus: Capitalism and Schizophrenia* ([1980] 1987, hereafter abbreviated ATP). We can, if it helps, take a step back and simplify our concern for a moment in order to make our starting point clear, while asking that the simplification not be taken as a fitting reduction that expresses all our discussion.

The simplification is this: *the* is singular, *mix* is multiple, *the mix* then is singular multiple. The definite and totalizing 'the' performs the act of singularizing (the making of the singularity) upon a concept that is multiple (mix), for mix designates the former presence, the constitution, of more than one. The mix refers to a past, a history, a former state (its multiplicity, its state of being multiple) when the mixing was being done. At some point the verb thickened, set and became the noun (did we allow the verb to set or did we just run out of time and let it congeal?).

Mix differs to one. Mix already concedes to a whole, a sum of its parts, which is not the same as one. Its not-oneness is a predicate as how can it be a mix of just one thing? To be pedantic, we would simply refer to that operation as being 'more of one thing' rather than a mix. We might stop there and say that our investigation is over; the one might be fought over with regards to the existence of a constitution, but the mix never claimed to be anything other than a result of plurality of some type, but we wish to consider the mix through its coexistent one-and-many multiplicity and its apparently paradoxical singular-multiple. The Deleuzian multiplicity, the rhizome, is one and many, the binary one or many is not in operation here and we have plumped (for function and as well out of pretentiousness) for the word 'themix' to capture this. For Badiou, that one is multiple is a starting point taken from Plato's *Parmenides* (1931, hereafter abbreviated P) that includes a dialogue on 'the one is not' but then this one-is-not is explored as different types of multiple, the *inconsistent* multiple (pure multiple) and the *consistent* multiple (the composition of ones). On discussion of Plato's *Parmenides*, Badiou writes in the second meditation "that in the absence of any being of the one, the multiple in-consists in the presentation of a multiple of multiples without any foundational stopping point" (BE 33). While the consistent multiple is the "composition of ones" (BE 35), "the multiplicity of composition which is that of number and the effect of structure" (BE 25). With these concepts in mind, we will speak of these modes:

1. *themix* as the-one-and-the-many,
2. *inconsistent mix* as pure mix (the multiple without ones, pure multiple),
3. *consistent mix* as mix that is composed of ones.

This gives us three ways to view the mix: as one-and-many, as multiple only, and as many ones. To keep some affinity to Badiou, we will often tie that last two together in sequence, as if they formed through this operation a single multiple, a "situation" in Badiou's terms. From these we can identify the conflicting authenticities that concern us with regards to the mix, particularly as our approach chooses to recognize that all modes can co-exist, although at times in our thoughts (and the listener's experience) one mode will dominate over the others, and often observers will place their truth according to what mode they favor.

We are not in a battle for the metaphysical high ground here (we'd lose anyway), and we can but say sorry to the philosophers whose terms we have remixed for our own purposes. We think we are probably closer in our use of rhizomatics to that set out by Deleuze and Guattari than we are to how the multiple "splits apart" (BE 25) in Badiou, but it is the splitting apart of the multiple that interests us, in particular the differing results that are seen in the split. We are aware that Deleuze and Guattari would probably not be that keen on how we apply their concepts to pop music (too striated and not smooth enough), even though they did shout at us that 'RHIZOMATICS = POP ANALYSIS' (ATP 24), but we need ways in which we can view the mix that help us to see what operations occur that produce the event, and these various points of recognition and observation allow that analysis to take place to reveal the authenticities and inauthenticities that dwell within mixes. No further apologies then; we have adopted and adapted these ideas for functional reasons regardless of how far they may have been shifted from their philosophical homes. Besides, we're not claiming to be in total control of what we are trying out here, but try we will. We are about to dive from the stage unsure of whether we'll be heroes carried aloft by the crowd or zeros with broken legs as we hit the ground.

So far we have spoken only in general terms with regards to the mix and to singular-multiple, and much of this might also hold true for the crowd, the grass, the band (The And), the many, the "pigs, stars, gods . . ." (BE 30), etc. In order to offer something more than a simple hypothetical substitution of one with mix, we will engage at ground level among the "mud, hair, dirt" (P 49) with the musician in the studio and their recorded product and the relinquishing of their produced thing to the control of the musician who will mix it (with other things) so that we keep on connecting "and . . . and . . . and" (ATP 25). Producer number 1 (she who produces the sound—the musician in the studio) to producer number 2 (she that will mix) to producer number 3 (the listener that turns the treble up on the radio). These entities are as real and as tangible as we may be able to draw upon in order to tie the abstraction to the world of the music producer, and we have to escape the philosophical abstractions at times and come down to earth where abstracts can sometimes look rather awkward, like a fish out of water. We cannot take the unveiling of the onion skin layers of the one (that is not) down to an un-further-able Planck length, we cannot zoom in all the way to a quark that is not a mix of anything but a true one (if indeed it is, we don't know, we're not quantum physicists). But we can speak of connecting the velocity of striking an instrument to the distance to the microphone to the line between breakpoints on the volume automation on the computer screen as contenders for ones or multiples that may constitute the count-as-ones of the mix. We can speak of being lost in the music that gives inertia to presentation which delays the count-as-ones. We can speak of the collaboration of the player, the mix engineer, the listener. We can speak of mixing that thickens, sets and becomes the mix (noun = verb stasis).

Besides, the Socrates of Plato's *Parmenides* is concerned to ask if the *Idea* is one or many, the discussion of the one and the many of real things is tossed aside as being so obvious as to be hardly worth considering:

> If a person wanted to prove of me that I was one and many . . . he would say that I have a right and a left side, and a front and a back, and an upper and a lower half, for I cannot deny that I partake of the multitude. When, on the other hand, he wants to prove that I am one, he will say, that we who are assembled here are seven, and that I am one, and partake of the one. In both instances he proves his case.
>
> (P 48)

But we will indulge ourselves at ground level and talk about real things as well as playing around with ideas up in metaphysical hyperspace, and we will consider a mix we have been involved in, and if we "cannot go beyond the trivial statement that differences exist" (BE xii) then so be it, we are content to dwell in the differences and see what they may reveal.

A Conceit

It is perhaps the right thing to do at this point to put forward our agenda, our clash of interests, our conceit, in this discussion of the mix. We propose our version of the discourse of the one and the many in the form of the conceptual song 'The Song of a Thousand Songs'. We see *all* of popular music, all songs, as being *one* song, we have removed all the (non-musical) borders (composers, song titles, etc.) that force songs to separate so that we can enjoy the ongoing symphony of the thematic development that is popular music. All contributors to this song are part of one band, The And, any internal differentiations such as The Rolling Stones or The The are merely localized calibrations within the multiplicity. We are only retelling an old story, one that simply agrees that rock n roll will never die. But it had not occurred to us to consider the mix of 'The Song of a Thousand Songs' or indeed the mixing engineers or the mixing of it as it is not yet, and may never, set, although it has congealed in some places and has rather overhardened in others. It might best be described as 'lumpy'. It is always becoming, it never becomes. It is always mixing; it is never the mix. The mix is not.

We will consider our conceit later but for now we will explore our approaches to the mix and offer observations of studio practice though a song that the world might recognize as a song rather than the disputable singularity of 'The Song of a Thousand Songs'.

The Ukulele at the Start of 'The Bins'

Just as Plato makes plain the obviousness of a person partaking in the one and the many and quickly discards it, so we will throw away some of the multiples that could be considered here and focus on particular arguments to the exclusion of others. 'The Bins'[1] (Johnson, 2016) starts with a single ukulele playing chords, a repeating pattern that makes up the phrase. We can break this down to the chords, then each chord to the notes of the chord, then each note to its fundamental and harmonics and its timbres, durations, etc., that are parts of its count-as-one, but we will focus on particular events as we see fit and discard others so that we can at least shed light on something rather than keep digging and become lost in the hole.

The ukulele in 'The Bins' was indeed one ukulele, and its sound to us as listeners is of one instrument being strummed to produce the chord. The means of producing this oneness, though, is different. When strumming and recording the chords in the studio, it was evident to us that the tuning was outside what was acceptable to us with regards to signifying a solitary ukulele being played somewhere on a beach ('The Bins' is a song about the summer). In this case, a slight out-of-tune-ness was acceptable, even desirable, in signifying some authenticity around the boy-on-the-beach normality of the event, but this was not captured in our first recordings (one ukulele strumming the chords); it was too out of tune for our liking. In the end, we recorded each individual line on each string on separate channels. Each channel had a pitch correction auxiliary attached with a relatively slow response setting to allow the individual quality of the ukulele's character to come out before being moved to a standard pitch set. That is, the notes have a charming out-of-tune character on the attack but then soon become 'in tune'. The four channels (one string recorded on each) were then mixed together to produce the one strummed ukulele. A number of these mixes were done with the intention of sounding as two ukuleles being played and presenting one in the left and one in the right of the audio spectrum. In the studio, we decided against having two ukuleles and plumped for one in the final version.

So what of this entity, of its ontology (if it has being), of its multiplicity? We are reduced to earthly perception at this point, to the phenomenology of the event. Upon perceiving the event, the chord, we can say 'there is one' (we can hear one chord, one ukulele), we perceive (or simultaneously perceive) that this is multiple 'there are pitches' (plural). It carries the one and the many of themix. Upon presentation, the chord is pure multiple, nothing but mix to infinity, inconsistent mix. Upon presentation within the count-as-one of the chord, the chord presents the notes; the count-as-one has ones (consistent mix). What perhaps makes the mix of interest is that it presents both its history and its present (at least with regard to its tuning, if not to the performance of the chord which was played one note at a time).

We have written of the construction of the chord and its tuning and have implicated ourselves in the acts of authenticity and inauthenticity in the doing of it, so let us expose the semiotic of this further. Within the consideration of the consistent mix of a chord, the chord is made of individual notes (composed of ones). In this mode, we reveal the tempered authenticity of the mix. Its authentic attack has the notes at the pitches as they are played at the moment of striking the strings, but we are not keen to maintain those pitches, so our pitch correction auxiliary quickly moves them to a familiar tempered scale, which is clearly inauthentic with regards to it no longer being true to the pitch of the instrument. Even then, the notes are short lived; ukulele notes do not sustain for long anyway, so the now-tempered remainder of the note has a very short existence, but it is long enough to make a difference that can be registered by the listener. This is the very thing of Western cultural appropriation—we have taken a 'native' instrument (as we might describe it in colonialist terms), the ukulele, full of quirk and charm, and allowed it at its surface, its instant presentation, to exhibit this beauty. But of course it is but a split second before we drag each pitch into the straightjacket of Western pitch constraints (just as supermarkets quickly dispose of any vegetables that are too crooked and simply not straight enough to be sold to a public that apparently can't cope with wonky veg). This is, however, 'The Bins', a summer characterized by melted Tarmac, fly-ridden bins and de-icer drunk drunks, its social realism charmingly nostalgic in its depiction of a summer in an England several thousand miles from the tropical beaches that might be inhabited by flower-clad ukulele players. Subtle as it may be, the authentic–inauthentic co-existence within a single chord is also that of the representation of summer itself within the song. Our British trappings of summer might normally include Hawaiian shirts and Bermuda shorts, just as our national dish seems to be the chicken korma (and we are not sure if we have shed our colonialism and allowed our cultural multiplicity to exhibit the equality it deserves or whether this is a way of trying to maintain some kind of ownership).

We have now hit the ethics of the mix. To have the notes bang on their tempered scale from the attack would rid it of charm, dehumanize it, but to keep it on its actual pitch from start to end would make us cringe (rightly or wrongly). Neither singular approach would produce a result that we would consider acceptable, so does this make the producer inauthentic or even unethical? Should we accept the pitch as it is and proclaim ourselves authentic? Let us go for another option that says that the producer's tools are as much the instruments of sound production as the musical instruments are and not prioritize one over the other. Hence in this case we have captured a bit of each, a new hybrid, a mix that is a one (or a count-as-one), not a ukulele-and-Logic-software instrument but rather a ukeLogic, an instrument that plays charming notes on the attack that then smooths out the pitches into mathematically exquisite proportions that resonate with the physics of our version of the universe. We are happily one and many, confidently rhizomatic in removing the imposition of the authenticity barriers that so concerned us ethically only a paragraph ago.

So what do our modes show us? They result in nothing more than what we already knew, that we are already many, that we are authentic and inauthentic, but what it does show us is *how* this is achieved, and how different values may be drawn from the same mix. From the point of view of the-mix, we are comfortable in acknowledging the many influences that make the whole. The pure, inconsistent mix is happy with its presentation for it seeks no further analysis. The consistent mix draws our attention to the chords, the ones of the count-as-one, only at this point when we consider the consistent mix do we look into the ethics of divided notes within the chords. Not surprisingly, there exists conflicting states in the presentation of the mix. The inconsistent mix is happy, the mix is knowingly content, but the consistent mix has issues. With this exposing of the complexities of the multiplicity, it is no wonder that different opinions of the mix exist; we merely choose, knowingly or otherwise, which mode or combination of modes we favor when we put forward our judgment.

Comp and Circumstance

Even within our times as producers, we have noticed a significant shift in the point at which contingency is at its greatest in the production process. In the times of tape, we worked with the performers to get one good take, or maybe two or three that we knew would be what we would be working with at the mix stage. This has changed considerably; we might now have many *many* takes, multiple takes of fragments, and keep going until we feel we have enough to the point where we have it covered more than plenty so that we (or whoever is comping and mixing) will be able to select from the great riches of the takes. What a pain this has turned out to be—we have stored up for the producer the unenviable task of sifting through the many ones (takes) to find the best pieces in order to construct the best count-as-one. The job has become that of the forensic detective carefully sorting through all the material evidence that has been collected until they find a truth (or at least something that will hold up in court). The truths we orbit are those of the three modes described and our intuitions based on the varying weightings of these. At this point, we can observe this moment of contingency through these modes, bearing in mind that the presented multiple at the end of this process is designed to erase all the evidence of the existence of the multiple takes. The very reason for the many takes is to produce the one perfect performance; without the multiples, the perfect multiple (an ideal pure mix) would not be possible. The final count-as-one, a track in this case, be it vocal line, guitar solo or (Pythagoric) snare, should leave no sign of its former multiplicity (if the producer is slick enough). This has a different analysis of the phenomenology of the mix when compared to the multiplicity of the ukulele case study. In the ukulele example, the decisions lead to an audible compound of authentic and inauthentic tunings, but here nearly all of the takes have been rendered inaudible, having been erased from the final version. In terms of perception, the many takes do not form part of the multiplicity because they are not presented, having been erased before the event of presentation. If we are not able upon hearing the line (the vocal, the guitar, the snare) to realize its Frankenstein-like construction, then the takes are absent from both inconsistent and consistent mixes. We will be aware of the makeup of the line as notes (ones joined together), etc., but not the multiple takes that have vanished in the comping process. Rhizomatically, this leaves an abhorrent trace. This comped line has as its history multiple lines of flight that have now been reduced and fixed by a trace. If we had kept all the takes, we could show on the screen visually how the trace, the audible line, jumps from one take to the next, muting out the track it has just left so that the operation of the count-as-one is exposed as the muting and unmuting journey across the tracks to create the final trace. The map from which the trace came has been destroyed.

But that is the producer's job, isn't it? We have just made the art of mixing the comp look like a criminal act, a fascist act, a Frankenstein's monster that erases those parts that we do not wish to participate in the event, only to leave the one (that is not one) for presentation. And isn't that one of the well-established criticisms of our art anyway? It suggests that there is a true one, however flawed it might be, in the singular performance of a line (one of the takes) that we have now technologized out of existence, that we have obliterated the authenticity of the singular line of flight. This argument prioritizes the performer above the producer as artist, but if we take the performer as the instrument and the producer as the player (she that produces the sound from the instrument), then the criticism cannot stand with regard to performance. We still have to deal with the erasing of the histories, though. With regards to themix, this is problematic, because we have disconnected from the map to produce the trace, and this trace, the final mix of the line, asks us to accept it as a count-as-one devoid of its 'true' composition (the multiplicity from which it is drawn) and thus to recognize this presentation as pure multiple without history (inconsistent) and as being composed by count-as-ones (consistent), but that this count is made of 'what is there' and that 'what is not there' (i.e., the takes that are left on the virtual cutting room floor) never existed. In real-world terms, the line can be analyzed with regards to its notes, its phrases, it syntax, etc., without taking into account the map from which the trace is made. Again this seems inauthentic, the idea that we are hiding the truth, the operation of the presentation relies on the muting (which is the erasure) of the parts of the takes that are not used. What then if we try to undo this monster? In real terms, we could leave all takes to sound together, at the moment of greatest contingency (available choices) we could reveal all lines of flight by playing all takes together. The result would be an audible map, though we may not be able to comprehend it as such if our brains cannot cope with that amount of material. This result might be pleasing to the modernist (and it is no surprise that *A Thousand Plateaus* begins with a musical quote from a complex score by Sylvano Bussoti. Let us not be mistaken into thinking that the postmodernist Deleuze was anything other than a modernist when it came to music), but it makes no sense in the world of pop in which we are producing. It would immediately be inauthentic pop music because it does not hold true to our beliefs (the rules of our game). The Beatles might have been proud to show off their engagement with modernism and Stockhausen in particular at times, but that is OK, the pop world includes the avant-garde in small doses. We might do the same, for a short time and for effect, but it seems that our paradigm is not happy with revealing the full extent of the map and relies on the trace for its authenticity (although we will explain later with regard to 'The Song of a Thousand Songs' how pop puts its traces back on to the map).

This trace, at presentation, has no history, and so we should not call it a trace at all but perhaps give it its more usual designation of the line or the track (in comping terms, the master track, but without its history it is no longer the master anymore but just a plain old track). The track as pure multiple is innocent (with no available history to condemn it); when we notice it as consistent with regard to its composition of ones, they are also innocent (they also have no available incriminating history either), themix is no longer available to us (themix is not) because if it is to be an authentic themix "the tracing should always be put back on the map" (ATP 13), but we have erased this so it cannot be viewed as trace. Themix then is guilty at this point because we know themix, if it is one, must have an available map to be put back on to. The bodies may be missing, but we can still prosecute (the 'no body, no murder' rule no longer applies) on the reasonable grounds that there must have been bodies in order to produce themix (there is enough circumstantial evidence to convict the comp), otherwise we cannot call it themix. Therefore, themix is guilty, whereas inconsistent and consistent mixes are innocent because they are perceived only at face value, as what is presented with no history assumed. Let us replace innocent with authentic and guilty with inauthentic, and then by step from authentic to real and inauthentic to not-real. These are not unreasonable steps to take, particularly when we note that the results can be read as this: "The mix is not real, it is fake, inauthentic" (themix); "the mix is real, true, it has all the elements I love" (consistent); "this track is *real*, I'm so lost in the moment right now" (inconsistent). Thus, we locate our truths according to the mode that dominates the presentation as it appears to us.

'The Song of a Thousand Songs' (Metaphysmix edition)

There would seem to be a cutoff point in the mix that enables easy separation. In our song process we have recorded and mixed, at which point we cease messing around with the parts because (having run out of time, money or patience) we have to send the track on to the mastering process, where it is given the audio version of steroids and made to conform to standards, so that when it is played on radio or streamed, etc., it will fit nicely. It will be fixed, normal and hence acceptable. We are not as cynical as we might sound about this; we are simply aware of this as a protocol rather than a deliberate removal of difference for any political reasons (although we are aware that this approach is also present in our industry). This protocol allows us to be admitted into the public domain, to become accessible, searchable, mixable, playable to and usable by the many—without it we could but be accessed by the few, maybe by those who wish only to fish in places where the protocol does not apply, a place where a particular notion of creativity has not been squeezed into a straightjacket. But that is not our philosophy—we have accepted the protocol so that we can be in the game and become the game.

If songs are separated largely by non-musical signifiers (the composers, the song titles, etc.) we can "nullify endings and beginnings" (ATP 25), for these starts and finishes are nothing more than segmentation caused by the effect of imposing non-audio signifiers onto audio. When we do this, we cease to operate within a representational system. There are no longer identifiable ones of songs—the removal of the artificial beginnings and ends has shown that they are actually all joined together. In fact, it is not correct to say that they are joined at all, once we have removed the sticky labels marked 'beginning' and 'end' we see underneath that there is nothing but continuity. 'The Song of a Thousand Songs' is the result of an asignifying rupture that is "against the oversignifying breaks separating structures or cutting across a single structure" (ATP 9). It is no longer possible to remark that 'one song is like another song' because in the being of the multiplicity there is no other, it is all 'The Song of a Thousand Songs'. Ones are not, multiplicity is. However, the pure multiple breaks down when we consider the consistent mix and themix; here we note that the multiplicity has localized calibrations. As we have said, the mix is lumpy; it has not set evenly and is somewhat congealed in places (which is not consistent with Plato's one, even a circle cannot be one because "the round is that of which all extreme points are equidistant from the centre" (P 59), hence the identification of center means it has parts and, although it may be a whole, it is not one. The lumps in themix similarly show it not to be one but many).

We should then nullify the mastering as signifying an end to the mix, for mixing goes on well beyond this point. The mastering is only the protocol required to connect to the map and enter into the becoming of the rhizome of pop music. Mastering of course has many other functions, but our concern here is only with its operation as an instrument for standardization. We need to explore the mix post-mastering, and some of these mixes have become art forms in their own right. The DJ seamlessly segues from one track to another by mixing the 'end' with the 'start' (by tearing off the 'sticky labels'). In this sense, the DJ is making a trace from the map, identifying and fixing a particular line of flight; although this is fleeting, he will do it again another night perhaps but it will follow a different line, again a fleeting moment that once gone reopens contingency. These are local and unstable segmentations, a highlighting of one possible way through the pop music rhizome. The mash up, the mixing of songs together, sometimes simultaneously although more frequently a DJ-like segueing of tracks, shows us another way of highlighting themix. Detractors might state that the mash up demonstrates the possibility of a reduction, the mere fact that songs can be played together proves they are shallow ornaments of an underlying truth such as a simple chord progression (this is the very stuff of Adorno's pseudo-individualism). But it does not have to be seen that way; to take the reductionist viewpoint is but a preference, not a proof. Again this is due to the imposition of separations caused by non-musical signifiers; the songs in themselves (if they are) hold within them many lines that accord to, for example, harmonic rules (a simple Schenkerian analysis may quickly reduce a whole song down to a I-V-I progression), so given the songs in the mash up are not separate at all, then they are no different to a melody and countermelody in a Beethoven symphony or the subject and countersubject of a Bach fugue. We are of course all DJs, all mixers, piecing together our playlists. These playlists were once fairly fixed; we made mix-tapes for ourselves and for each other, though we might record over them with a new set of tracks a few weeks later. Now our playlists constantly connect and disconnect from the map, new lines of flight taken, or old ones highlighted, bits kept, bits replaced. We have an immense available pool from which to draw upon courtesy of the mastering protocol that allowed them to be available, to become part of the map. We are all mixers of the music, regardless of how much time was spent on setting up the distance of instrument to microphone, the sweeping through of frequencies (and probably reducing things at 500 Hz), the panning of voices, etc.; once free of the studio we turn up the bass on our headphones, the treble on our car audio and the mix continues to be unfixed. We are well aware of this—we mix for little headphones, for big speakers, for cars, for radio—the continuation of the mix beyond our studio control is something we plan for, an attempt to be as usable as possible within the multiplicity of listening.

We have taken the inconsistent mix, the pure multiple, if it can be other than an idea, to be the state of *un*-analysis in the user, the inertia that delays the splitting of the multiple, the we have turned up the bass and, head nodding, we are singing along full volume. As odd as a form of a pure multiple might seem, it also seems to be the aim of the mixer, the producer. Do we want to make engines for engine analysts to analyze and then admire our greatness, to even notice that there are parts? Not really, that sounds like a game of hide-and-seek, just like a crossword puzzle, the answers known and then hidden for us to uncover. Not to put down crossword compilers and solvers, it is indeed an art form of its own, but our goal is not so that we might be analyzed and hence the pure mix, as we use the term, is our preferred state. With the consistent mix awareness, analysis, begins to seep in and cracks open the pure multiple. Themix is a more pragmatic, earthbound observation, it is how pop music works: one fabulous, continuous symphony. We realize that in our discussions we have crossed over the borders of pretentiousness and that we have played irresponsibly, footloose and fancy-free, with postmodernism and metaphysics in our quest to find out what is going on in the mix. So what, it's only rock 'n' roll, but we like it.

Note

1 'The Bins' was performed by The And Ensemble, a subgroup of The And and York St John University's leading Deleuzian pop ensemble. Our thanks to band members Abigail Hall, Joe Collins and Angus Williams for their fine vocal performances on the recording and their ebullient antics on the video.

Bibliography

Badiou, A. (2007). *Being and Event*, trans. Oliver Feltham. London and New York: Continuum.
Deleuze, G. and Guattari, F. (1987). *A Thousand Plateaus: Capitalism and Schizophrenia*, trans. Brian Massumi. Minneapolis, MN: University of Minnesota Press.
Johnson, C. (2016). *The Bins*. York: Q22 Records.
Jowett, B. (trans.) (1931). *The Dialogues of Plato*, Volume IV, 3rd edition. London: Oxford University Press.

13

Mixing Metaphors

Aesthetics, Mediation and the Rhetoric of Sound Mixing

Mark Marrington

Introduction

This chapter's purpose is to evaluate concepts of mixing that have been advanced through practice and writing on the subject, focusing on two aspects in particular. The first relates to the sound mixer's[1] conception of their role, entailing ideas about their position within the production process and their relationship to other individuals. The second is concerned with what sound mixers consider that they are doing with the recorded material when mixing takes place. In particular, I wish to draw attention to the rhetorical function of mixing, in other words, the role it plays in presenting its (musical) subject matter convincingly and persuasively to an imagined audience.[2] Sound mixers, in effect, speak on behalf of the artist, and in doing so take on a certain amount of responsibility for the realization of the artistic vision. In practice, this entails a degree of mediation, often involving mixing in reference to established 'codes' of record production (in rhetorical terms, stylistic figures which ornament musical discourse), as well as finding effective ways to articulate the material to maximize its impact. The notion of mixer idiolect, which refers to situations in which this mediation begins to incorporate aspects of the sound mixer's own rhetorical style, highlights an interesting question regarding the scope of the sound mixer's creative contribution to the mix and the role this potentially plays in determining the music's message.[3] Aside from these specific lines of inquiry, my general aim is to provide a useful frame of reference for interpreting the stated objectives of those who mix and the ideologies that inform their practice.

Locating Aesthetics Within Mixing Practice

Over the last two decades, theoretical perspectives on mixing have become consolidated in a substantial body of writing which includes practitioner-led 'how to' guides, interview literature and the academically couched discussions of production aesthetics contributed by the 'musicology of record production' camp.[4] From this literature can be discerned

particular ideological positions on mixing (and indeed on music production in general) which form the basis of mixing aesthetics. Before considering this area in more depth, I wish to make some general observations
regarding the sound mixer's place within the production process and their
relationship to the recorded material.

Mixing, in its most practical sense, is understood to be an autonomous
engineering specialty and largely the province of the mix or recording
engineer. This is reflected in the content of a number of the 'how to' type
of texts that focus on the principles of audio theory and offer extensive
commentaries on the various technical tools employed during the mixing
process (see, for example, Case's *Mix Smart* (2011), Gibson's *The Art of
Mixing* (1997/2005) and Izhaki's *Mixing Audio* (2008)). In such discussions, mixing is often painted as a series of problem-solving tasks, whose
function is to organize the recorded material to its best sonic effect—in
Stavrou's words, "You'll be assembling an intricate puzzle with dozens
of sounds. They have to fit together" (Stavrou, 2003: 161). Moylan offers
a concise summary of the activities that are typically part of the sound
mixer's remit:

> The mix of a piece of music/recording defines the relationships of individual
> sound sources to the overall texture. In the mixing process, the sound stage
> is crafted by giving all sound sources a distance location and an image size
> in stereo/surround location. Musical balance relationships are made during
> the mix, and relationships of musical balance with performance intensity are
> established. The sound quality of all of the sound sources is finalized at this
> stage also, as instruments receive any final signal processing to alter amplitude,
> time, and frequency elements to their timbre and environmental characteristics
> are added.
>
> (2007: 233)

It is also generally accepted that these kinds of tasks take place at a late
stage in the record production process, usually further to the realization and
recording of the music, with the implication that there is a certain degree of
separation between the sound mixer and the original musical conception.
Hepworth-Sawyer and Golding (2011: 178–193), for example, posit their
"Capture-Arrangement-Performance" (CAP) model to express what should
be done prior to the mixing stage. For these authors, the often maligned
situation of "fixing it in the mix" implies shortcomings in these earlier
activities and instead, "mixing should be something that flows naturally
from the intended arrangement as set out by the producer and artist" (2011:
230).[5] Izhaki (2008: 28) reiterates this point in his comment that "mixing is
largely dependent upon both the arrangement and the recordings". Sound
mixers themselves acknowledge the importance of their independence relative to the main production process. According to Owsinski (1999: 8),
the vast majority of mixers would "prefer to start the mix by themselves",
while acknowledging that ultimately they cannot ignore the ambitions
of the producer and artists for the mix outcome. Lipson comments that
"it's far more useful to have the artist hear something complete . . . than
for them to hear the whole process of me getting to that point" (Massey,

2009:197). Stavrou in more forceful terms suggests that the mixer, "turf the producer and band out of the control room" (2003: 151). Objectivity and neutrality are also valued by sound mixers when undertaking mixing tasks. Senior, for example, refers to the mix as "data" about which one needs to be able to "make objective decisions . . . irrespective of your own subjective preferences" (2011: 2).[6] To facilitate this objectivity, Stavrou, in an extreme gesture in keeping with his unique theory of mixing, has advocated playing the mix backwards because it "removes the language out of the music . . . leaving you only with shape, and harmonics galore", creating "a whole new psycho-acoustical playground to dive into" and forcing "an unbiased view of the sounds" (2003: 232). The idea that the mix constitutes a form of relatively neutral sonic data has had a particular currency with researchers in the area of automated mixing, involving the development of intelligent systems to facilitate the application of mixing techniques in relation to quantifiable aesthetic outcomes. De Man and Reiss (2013), who are at the forefront of this field, even go so far as to suggest that "mixing multi-channel audio comprises many expert but *non-artistic tasks* that, once accurately described, can be implemented in software or hardware".[7] It is apparent that the authors' notion of the mixing process in this sense is in part a response to the position held in the 'how to' type of texts that the practice can be broken down into a series of engineering specialties.[8]

Although mixing is commonly articulated in terms of its engineering techniques, it is rarely the case that sound mixers will remain completely detached from the musical material they are dealing with. Indeed, even in the purely engineering context, sound mixers will be likely at some point to engage with the sounds they are mixing in terms of their musical connotations, as illustrated by the common practice of relating musical pitch to frequency values when applying EQ to tracks. 'How to' type of texts directed at mix engineers and other audio professionals frequently advocate the usefulness of musical knowledge when decoding what is happening in the mix. Owsinski (1999: 12), for example, notes that the ability to balance the various instrumental timbres of the mix as it progresses is facilitated by "understanding the mechanics of a well written arrangement".[9] Moylan's comprehensive pedagogical treatise, *The Art of Recording*, throughout stresses the need for audio professionals (or recordists) to correlate their engineering knowledge to "sound's artistic elements (or the meanings or message of the sound)" (2007:4), in reference to pitch, loudness, duration, timbre and space. Case goes further, suggesting that the sound mixer develop a broad range of musicianship skills, advising the pursuit of "music studies, a music degree, and proficiency on more than one instrument", and suggesting that being a "gigging performer makes you an even better mixer" (2011: 226). For Case, the purpose is to enable the sound mixer to gain clearer insights into the song and therefore make mixing decisions that are "supportive of the music" (2011: 227). It is with such comments that we begin to move from a concern with overtly technical matters to a consideration of mixing aesthetics, a term which I use here to refer to the broader artistic objectives of the mixing process relative to the musical material.

From Objectivity to Mediation: The Development of Mixing Aesthetics

Mixing aesthetics to a certain extent have been shaped by the evolution of recording practice as well as technological developments that have expanded the possibilities for manipulating sound. The notion of mixing as a unique activity within the music production chain has its origins in the early era of record production, when the term referred to the work of an individual who 'balanced' the various instrumental parts during the process of capturing them.[10] One of the best known of these pioneering sound balancer mix engineers is Tom Dowd (1925–2002), a pivotal figure in the development of the Atlantic Records Rhythm and Blues sound during the 1950s and '60s. Dowd held the view, well into the multitrack era, that records should be cut in a single take with all performers present at once:

> During the late sixties and early seventies I did not want to use the overdubbing features of multi-track recording. If I had a five-piece group I wanted five guys to play simultaneously and if I had a ten-piece group I wanted ten to play simultaneously . . . Everything was done on the fly.
>
> (Buskin, 1999: 165)

Mixing in essence was both the start and the end-point of production, making its mark at the moment of recording itself. The remit of the mix engineer was also to a certain extent informed by the musical subject matter typically being recorded, which was either fully worked out prior to the sessions (arrangements of light music or classical music, for example) or dependent upon the live performance situation for its realization (for example, improvised jazz or popular music honed in a live band context).[11] The music thus possessed its own integrity and there was generally no expectation that its substance would be modified further once it had been recorded, other than to correct major errors. Certainly, the mix engineer would not have presumed to radically manipulate the material beyond the point of its live capture.[12]

What mixing tended to reflect in this early period was a particular ideology of record production that Lee Brown (2000: 361) has referred to in metaphorical terms as the "transparency perspective". This describes a situation in which "sound recording is understood on the model of a transparent windowpane through which we can see things undistorted", implying both accurate representation of the music and minimal mediation.[13] This attitude still remains current in the thinking of mix engineers today and is often expressed in terms of an aesthetic of 'responsibility' to the integrity of the music. For example, Armin Steiner, an engineer who possesses a background in pop, classical and film score recording and mixing, holds the view that "We are not interpreters. We are servants of the music, and all we're doing is taking down what the composer did, and hopefully putting it down in the perspective that he heard" (Droney, 2003: 127). Bruce Swedien, queried as to his "philosophy about mixing", has commented that

It comes from my early days in the studio with Duke Ellington and from there to Quincy [Jones]. I think the key word in that philosophy is what I would prefer to call responsibility . . . our first thought should be that our responsibility to the musical statement that we're going to make and to the individuals involved.

(Owsinski, 1999: 198)

Swedien also emphasizes the sound mixer's purposeful distance from the artistic conception when he says: "I firmly believe that what we do as engineers can never win out over the personality of the music itself" (Massey, 2009: 48).

As multitrack recording facilities improved from the 1960s onwards, bringing a re-assessment of the role of the studio in the creative process, this attitude towards the recorded material subsequently became modified. First, mixing decisions could now be postponed until much later in the production process and endlessly revisited. Second, there was scope for the mixer to begin to reflect upon the recorded material creatively and experiment with mix possibilities. This shift corresponded to the emergence of what have since been referred to as entrepreneur-type producers who began to make innovative use of the new possibilities of the multitrack studio to develop their own production aesthetics.[14] Having little technical know-how themselves, such producers were usually reliant upon their engineers to interpret their directions, as occurred, for example, with Phil Spector, who worked closely with engineer Larry Levine to develop his celebrated 'wall of sound' during the early 1960s.[15] This effectively gave license to the mix engineer to become a creative collaborator who could make subjective mixing decisions about the sonic quality of records, in some cases raising their status to a position equal to that of the producer.[16]

These developments naturally created a tension between old and new mixing traditions, with some engineers weaned on the transparency perspective increasingly finding themselves at odds with multitrack production strategies that apparently distorted the inherent artistic properties of the music and its performance. This is illustrated by Phill Brown's account of his experiences when working with the band Red Box in the mid-1980s. After a lengthy process of recording and mixing the band's projected single, 'Living in Domes', Brown took his completed mixes to Max Hole (A&R at Warner Brothers) for approval. Hole's response was to declare the mix of the track to be "too aggressive", which necessitated a return to the studio to revisit the material. However, once back in the studio Brown found it problematic to alter the mix because

the power (what Max had called the aggression) was in the playing and singing. I tried to tone it down using reverbs, echo and EQ. The chant vocals were placed back in the mix but I thought this just made it feel weak. Trying to stop the mix taking its natural course was frustrating.

(Brown, 2010: 268)

For Brown this was ultimately an irreconcilable situation: "once the record company gets a foot in the door on discussions about overall sound and mixes, it's all over" (2010: 269).

Speaking on Behalf of the Music: Mixing and Mediation

Brown's anecdote raises the important question of what constitutes mediation during the mixing process, and more particularly, where its limits lie. In this scenario, the musical recording has been deemed by the sound mixer to be speaking for itself; therefore, its message should be allowed to pass unimpeded. The conflict occurs when the record company demands a more active form of mediation to re-shape the material in accordance with its commercial remit. Making inappropriate adjustments to the mix to satisfy a commercial imperative might be regarded as an extreme form of mediation, but even when such factors are not in play, sound mixers will be still making a range of decisions that impact upon the substance of the music they are mixing. What ultimately determines the outcome here is the sound mixer's approach to interpreting the material they are dealing with. This of course implies a range of possible positions on the mix, which is why, in a bid to ground their mixing approach, sound mixers have frequently placed an emphasis on being able to accurately discern the artist's objectives. Stavrou, for example, remarks on the need for insight into "the purpose of the song—the distinctive thread behind its creation— or the heartbeat that inspired it" (2003: 175), while Gibson talks of the requirement for the sound mixer to be sensitive to the "essence of the song, or the primary message" (2005: 3) that is present, suggesting (in an echo of the transparency perspective) that the "art of mixing" is "the way in which the dynamics we create with the equipment in the studio interface with the dynamics apparent in music and songs" (2005: 147). This is where the aforementioned musical abilities and sympathies of the sound mixer come into play. In essence, the task of the sound mixer is to decode the inherent properties of the music, including melodic, harmonic and rhythmic elements, and articulate them in a way that that coheres with the artist's intentions. Or to put it in Moylan's words, sound mixers need to employ the parameters of sound in "support of any component of the musical idea" (2007: 67), with a view to "the communication of meaningful (musical) messages" (2007: 37). Mixing, in other words, necessitates being able to appreciate the rhetorical effect of the *music itself* and discover ways in which it can be enhanced by sonic treatment.

One metaphor that has been commonly used in the literature to suggest the interpretative aspect of the sound mixer's remit is 'performance', which implies the construction (or invention) of an ideal musical rendering of the multitrack from its constituent elements.[17] Performance here refers in particular to the articulation of the musical arrangement, which sound mixers will typically consider in terms of the spatial configuration of the mix and its evolution within the temporal domain.[18] The spatial rhetoric of the mix has been discussed and defined extensively in the literature, from Moore's "soundbox" (1993) to Moylan's "perceived performance environment" (2007: 263).[19] What is usually being referred to is a virtual three-dimensional space in which the elements of the mix are organized, comprising depth (the sound's apparent proximity to the listener), the horizontal axis (placement of sounds in the stereo field) and the vertical axis (individual frequency bands of sounds—sometimes expressed as height) to construct the virtual 'soundstage' for the performance.[20] A principal rhetorical use of spatial parameters is to mark significant musical elements of the mix for listener consciousness, the choice of which will depend upon the nature of the music being mixed and the sound mixer's appreciation of the relevant genre conventions (or codes).[21] In commercial pop song mixes, for example, EQ, panning and reverb may be used to foreground the voice relative to the other instruments to draw attention to the primary artist, the lead melody and the lyric.[22] The positioning of particular musical elements in space can often be critical in conveying the essential musical attributes of a given genre. In a house music track, for example, the common practice of placing the kick and snare drums centrally has the effect of maximizing the energy distribution of these sounds between the two speakers, ensuring that the beat, a key rhetorical figure in dance music, is communicated unambiguously.[23] The sound mixer's approach to space can also help to contribute to the authenticity of the listener's perception of a given musical genre, in certain cases entailing the construction of a spatial environment that did not exist during the recording process. In a rock context, for example, where liveness and authentic musicianship are valued, mix elements that were tracked separately might be staged in a way that suggests they were performed together in the same space at the same time, as well as treated sonically in a way that lends them immediacy.[24] As Moylan has observed, an important aspect of sound mixing is concerned with creating "illusions of space" (2007: 52), and such illusions comprise an important part of the rhetoric of recordings from the sound mixer's perspective. A final point to note is that genre-specific spatial configurations of mixes tend to become conventionalized (or overcoded) with use, as illustrated by Gibson's extensive taxonomy of genre-based mix types (2005) and Dockwray and Moore's recent survey of 'normative' stereo mixes created between 1965–72 (2010).[25] While this can provide useful consistency for the commercially inclined sound mixer, it also presents opportunities to play upon such expectations and employ a spatial rhetoric which is distinctive in its lack of observation of convention and becomes a creative gesture in its own right.[26]

In addition to spatial concepts, sound mixers are also concerned with the organization of the musical material in its own terms. This can entail making adjustments to the both the structure and content of the mix to improve the clarity and coherence of the musical 'oratory'. Many sound mixers, for example, consider that it is within their remit to contribute opinions on the musical arrangement, and they may even take it upon themselves to re-configure this during the mixing process.[27] The kinds of changes that a sound mixer might make to the recorded material can include structural reorganization (such as re-locating a verse or bridge section for example, or deleting a section) as well as the removal, replacement or addition of specific musical elements (such as a drum sound, for example). While from the mix engineer's perspective such activities may be regarded as being in the service of objective practicality—for example, to resolve conflicts between instruments or improve a particular instrument's timbre—their role in determining the music's character is nonetheless significant. Another way in which the sound mixer can potentially re-configure the message is by 'curating' the musical material—essentially making value judgments about particular details of the mix and their potential to improve the music's impact. Senior, for instance, has highlighted an interesting form of inadvertent creativity that results from discoveries made of overlooked aspects of the original material, such as a tastefully performed instrumental phrase or an imaginative ad-lib. He comments that

> several times I've been congratulated for "putting a new part" into a mix, when all I've done is dredge up one of the client's own buried sounds that had long since faded from their memory. The beauty of this trick when mixing other people's work is that it's low-risk, because anything in the original multitrack files implicitly bears the client's seal of approval, whereas there's nothing like that guarantee for any truly new parts you might add.
>
> (Senior, 2011: 85)[28]

The situation of the sound mixer as interpreter of the musical substance can become more problematic when confronted with mixes whose artistic vision is so specific and unique that there may be many possibilities for how the mix should be achieved. This is illustrated by the work of Scott Walker, who, since the 1980s, has developed an increasingly personal and uncompromising artistic language (see, for example, the songwriting and production on such albums as *Tilt* (1995) and *The Drift* (2006)). The ambiguity of Walker's song structures, coupled with the wide-ranging timbral palette of his arrangements, has presented particular challenges for the sound mixing process. As Walker's engineer Peter Walsh, has stated in reference to mixing the recent album, *Bish Bosch* (2012), "It's very difficult, 'cause where do you start? . . . It is very much each song for its own" (Doyle, 2013). Clearly, an objective approach is not an option here because there are no general standards by which the artistic vision can be appreciated—in other words, there is no established rhetorical category for mixing this music, nor an imagined audience. On the other hand, a subjective position would be liable to lead to a misinterpretation of the artist's intentions. In such instances, a close collaboration between sound mixer and artist is likely to yield the most satisfactory results—indeed Walsh has been Walker's long-time engineer since 1984, in effect qualifying himself to deal with the latter's artistic vision from the sound mixing perspective.

Regardless of the interpretative stance taken by the sound mixer towards the materials of the mix, all the decisions made during the mixing process inevitably modify the sonic properties of the music in one way or another, with potential consequences for its message. This applies even to the most minor tasks that mix engineers consider are simply corrective, such as making adjustments to the timing or tuning of audio material ('perfect' production is itself a rhetorical gesture), while with the use of staple tools such as the EQ and the compressor, the mixer is potentially moving into the territory of sound design.[29] Stavrou's position on mixing aesthetics (in his book, *Mixing with Your Mind*) is notable for its emphasis on the active role that studio devices can play in sculpting the musical vision. Indeed, the creative use of tools arguably constitutes his definition of the art of mixing. To this end, he devotes a great deal of discussion to re-thinking the employment of EQ, compression and reverb in terms of their musical possibilities. In regard to reverb, for example, Stavrou suggests that the effect's ability to alter time relationships in a track is more important than what it indicates about the space the mix is in: "Reverb isn't so much about being true to an acoustic space, it's more about the effect it has on the groove". He also suggests that in certain cases "the only purpose of some instrument is to become a key input for an interesting reverb effect" (2003: 182).[30] In other words, the mix engineer ought to be actively seeking out material in the multitrack that would provide an opportunity to use a particular tool to embellish the music.

Idiolect in Mixing: Mediation as Transformation

One final aspect of the rhetoric of mixing that I wish to consider, which has already been touched upon in the preceding discussion, is the notion of mixing idiolect. Idiolect is a term derived from semiotic theory, which is usually employed to account for the specific identifying fingerprints of individuals within areas of broadly consistent practice.[31] In effect, this implies the assertion of the sound mixer's creative personality upon the material of the mix and might thus be regarded as the antithesis of the transparency perspective. Certainly in some quarters, this has been regarded as crossing the line. Bill Bottrell, for example, has expressed a particularly negative view of engineers who allow their own creativity to take precedence during the mixing process:

> It all started to go wrong in the late '70s when engineers and producers started being allowed to impose their frustrations as musicians on the records. And that should never have happened. Because the artists know what to do, if they are really an artist, and the producer should just set up a situation where the artist feels free to do what they do. The engineer should just record it and get out of the way.
>
> (Droney, 2003: 17)

Bottrell even goes so far as to suggest that artists have been "brainwashed" to think that it is an engineer's prerogative to impose ideas, and that "there's a whole generation of artists who think that's how records are made, and they don't question it" (2003: 17). It is clear that Bottrell's view is not widely shared, however, as evidenced by the many sound mixers who place an importance on the development of a personal mixing style. Often the notion of idiolect in sound mixing is tantalizingly hinted at in the mixing literature without being specified. Senior, for example, who makes much of the notion of the mixer's objectivity, acknowledges that "entirely subjective decisions do have a place in the mixing process" and that these "distinguish one mix engineer's personality from another" and are "essential for the development of new and exciting music". However, he gives no examples, stating that "I can't help you with the truly subjective decisions" (Senior, 2011: 57).

To provide an illustration of what idiolect might entail in this context, I wish to consider Martin Hannett's mixing approach with post-punk band, Joy Division (and their later incarnation as New Order).[32] Hannett (1948–1991) is an interesting figure for a number of reasons. He possessed a scientific background as well as extensive knowledge of music technology, including expertise in synthesis. He was also a performing musician with an appreciation for avant-garde electronic music and dub, both of which arguably had a significant influence on his attitude to mixing—the former because it foregrounded sound design and signal processing, the latter because it encouraged mixing on the fly, the use of effects and the treatment of the multitrack with a certain amount of malleability.[33] We thus have an individual within whom were combined an accomplished engineering skillset and a distinctive repertoire of musical influences.

In the studio Hannett clearly saw himself as moving freely between producer and mixing engineer roles, taking responsibility for the organization of the recording process as well as mixing the material 'hands on' at the desk.[34] One of the most notable aspects of Hannett's approach in his producer guise was his method of garnering performances from the band in a form that would allow him a great deal of flexibility during the mixing process. To this end, he usually insisted upon the separation of the band members during tracking, taking this to extremes with drummer Steven Morris who was required on more than one occasion to dismantle his kit and record the various drums onto individual channels.[35] In traditional fashion, Hannett undertook the mixing after the recordings had been completed and the band had left the studio. His approach during this process was to re-contextualize the raw material within the spatial domain using the various studio tools in his possession, and in particular the AMS DMX 15–80 digital delay, which he employed repeatedly on the band's recordings. For Hannett, delay, chorusing and reverb were "gifts to the imagination" because they made it possible to add "little attention-grabbing things, into the ambient environment, just in case interest was flagging in the music" (Savage, 1992). This is redolent of the mixing aesthetic advocated by Stavrou, in which the creative possibilities of technological tools are foregrounded in the mixing process. Interestingly, Hannett's motivation for using such effects appears to have been to compensate for the sparseness of the raw material. In a 1989 interview, he stated that, "there used to be a lot of room in the music, and they were a gift to a producer, 'cos they didn't have a clue" (Savage, 1992). In other words, he took what he perceived as the band's underdeveloped artistic vision as a cue to elaborate their material. The effect of Hannett's experiments was to substantially alter the dynamic relationship between the band members, particularly where the strength of the guitar was concerned, giving priority to Ian Curtis's vocal and thereby underpinning the latter's idiosyncratic poetry. Journalist Chris Ott has provided an apt summation of Hannett's re-configuration of the guitar part on the album *Unknown Pleasures* (1979):

> Hannett's equalization cuts the brunt of Sumner's fuller live sound down to an echoing squeal. In search of vocal clarity and space for delay and reverb to ring out, Hannett relegates the guitar to hard-panned stereo placement in later tracks and thins the robust double-humbucker sound of Sumner's Gibson SG.
>
> (Ott, 2010: 64)

While the result was a sonically groundbreaking record which cemented Joy Division's reputation, the response of some of the band members to what Hannett had done was decidedly hostile, as guitarist Bernard Sumner commented:

> We played the album live loud and heavy. We felt that Martin toned it down, especially the guitars. The production inflicted his dark, doomy mood over the album; we'd drawn this picture in black and white, and Martin coloured it in for us. We resented it, but Rob [Gretton] loved it, [Tony] Wilson loved it, the press loved it, and the public loved it. We were just the poor, stupid musicians who wrote it.
>
> (quoted in Savage, 1994)

Tankel (1990), in reference to the practice of remixing, has used the word "recoding" to describe what takes place when remixers re-imagine the materials of an existing studio multitrack to discover new perspectives on the original recording. 'Recoding' would also seem to be an apt word for what took place in Hannett's case, especially given the subsequent conflict between the band's sense of its 'live' identity and what was created in the studio.[36] Remixing, for the purposes of the present discussion, can usefully be regarded as a subbranch of mixing practice in which the sound mixer is given *carte blanche* to be creative. In other words, there is no question here that idiolect should be sought. What the practice also highlights is the fact that the mix is not a fixed entity, or a definitive statement, and that there can be many legitimate opinions on what the best approach should be. In this sense, it is a thought-provoking question as to whether the record would have been as successful had the basic recorded material been mixed in a way that simply clarified its essence ("in black and white") rather than elaborated it ("in colour"). A final point worth adding is that in the wider context of record production, Hannett is often cited as an architect of the 1980s' Manchester post-punk sound.[37] This suggests that his idiolect was potentially more far-reaching than his re-molding of Joy Division, serving to define (or re-define) a domain of practice within that scene more generally.[38]

Conclusion

This chapter has considered some of the ways in which sound mixers can be implicated in the communication of the message of the recordings they mix. I have suggested that although on one level mixing can be regarded as a technically oriented engineering activity whose purpose is to put the best sonic face on a given musical recording, it cannot avoid having a bearing upon the articulation of the artistic elements therein. This is because sound mixers are caught up in particular ideologies of record production that define their mixing aesthetics and determine how they mediate the material. Sound mixers must make a variety of decisions in accordance with these aesthetics, whether in the service of achieving a transparent rendering of the music, responding to a commercial remit or searching for the best means of expressing a unique artistic vision. The sound mixer also has a rhetorical function relative to the recorded material, because they essentially speak on behalf of it. This requires an ability to interpret the material in reference to the specific rhetorical devices expected by the audience in a particular musical context.[39] As Gibson has commented, "the mix should be *appropriate*. Appropriate for the style of music, appropriate for the song and all of its details, and appropriate for the people involved" (2005: 48). To this end, conversance with a wide range of record production approaches is advocated by sound mixers, as Case (2011: 226) comments, "We need to know our history—enjoy and study the most important recordings in our chosen styles of music . . . avidly seek out and analyze the most popular contemporary recordings". Indeed, it is this conversance with a broad range of record production strategies that potentially determines the scope of what a sound mixer's creative contribution might be to a mix. Thus, those sound mixers who adhere closely to specific genre contexts might produce conservative but nonetheless commercially viable mixes, while those who are open to drawing intertextually from across different genres are more likely to be innovative.[40] Ultimately, it is an eclectic approach that enables the 'art' of mixing to progress, because it admits of elements that are likely to re-configure the parameters of a given mixing domain.[41] This latter point I have illustrated in terms of idiolect, whereby the sound mixer may place a stamp of individual identity on the musical material that has little to do with the artist's intentions or commercial imperative. Finally, it must not be forgotten that the intended target of this rhetorical activity is the audience, whose interpretation of the message is not necessarily predictable. Thus, sound mixers must also take pains to anticipate the general effect of their work here, as illustrated by the common practice of gauging the impact of completed mixes through playback via typical consumer audio systems, such as car stereos, mobile phones and laptop speakers. If the sound mixer can secure the audience's accurate perception of the mix in acoustic terms, they may stand a good chance of ensuring the accurate reception of its message in artistic terms.

Notes

1 I use the term 'sound mixer', derived from Kealy (1979), throughout this discussion to refer to an individual who mixes the multitrack at a point *post* the initial production stage. This is not to discount the fact that mixing can also take place during the production process itself. However, as will become apparent, my discussion is founded on the idea that sound mixing takes place in reference to an already defined vision of the production.

2 In addition to the general discussions of rhetoric as persuasive speech that can be found in classic modern treatises, such as Corbett (1998), my thinking on rhetoric in relation to record production and popular music studies is also informed by discussions in Brackett (2000), Zak (2001) and Toft (2010).

3 My use of the words 'code', 'idiolect' and 'message' also implies semiotic theory, of which rhetoric is considered to be an adjunct. See, for example, Eco (1976: 276–288) and Tarasti (2012: 271–300).

4 Approaches to discussing the subject in the practitioner literature are also reflected in the columns of the trade periodicals (*Sound on Sound*, *Computer Music*, *Tape Op*, *Mix*, etc.). Supplementing the practitioner guides are also a number of interview compilations—Schwartz and Stone (1992), Buskin (1999/2012), Droney (2003), Massey Vols. 1 and 2 (2000/2009)—which frequently draw out references to mixing practice, as well as some useful autobiographical accounts by mixing engineers and producers including Phil Ramone's *Making Records* (2007) and Phill Brown's *Are We Still Rolling?* (2010). The other major source of contribution to the discussion has come from academics working in the "musicology of production" field, including Moore (1993/2001), Zak (2001), Doyle (2005) and Zagorski-Thomas (2014). Typically, writing in this area has been concerned with notions of 'staging' and the classification of mixing strategies relative to era and particular musical genre.

5 See the discussion of fixing it in the mix in Hepworth-Sawyer and Golding (2011: 227–28). 'Fixing it in the mix' is an ideological position that is becoming more acceptable, particularly in the context of the DAW where compositional and arrangement decisions can be postponed well into the mixing stage.

6 Indeed, "objectivity" is a word which recurs frequently during Senior's discussions of mixing.

7 Italics are my emphasis.

8 De Man and Reiss's provocative conjecture that mixing involves "non-artistic" tasks is particularly debatable given that even the most minor adjustments to the sound may have the potential to impact upon the music's message.

9 Gibson (1997: 4) notes that "The recording engineer is quite commonly the most knowledgeable person in the studio when it comes to being aware of all the types of musical instruments and sounds available."

10 Indeed, many mix engineers still refer to mixing in these terms. The subheading of Stavrou's book, *Mixing with Your Mind* (2003), for example, states that the author is offering "Closely guarded secrets of sound balance engineering". See Hodgson (2010: 149–157) and Owsinski (1999: 2–6) for further discussion of the historical evolution of mixing and the mix engineer.

11 For a detailed discussion of the ontology of recorded music, see Gracyk (1996).

12 This is not to imply that the recorded material was not re-edited extensively during the production process in this era, as occurred even in the case of classical music. The purpose, in most instances, was nonetheless to achieve a close correlation to the live rendition.

13 See also Moorefield's (2010) notion of the "illusion of reality".

14 See Kealy's classic essay 'From Craft to Art' (1979) for a full assessment of the developments that occurred this period. Moorefield's *The Producer as Composer* (2010) also contains relevant discussion.

15 It is worth noting that much of Levine's contribution to the mixing process occurred during the recording stage (i.e., in the live capture mode).

16 Indeed, it is from this point that production and mixing tend to become more synonymous as producers explored mixing, and sound mixers explored production. This perspective is accommodated in mixing theory. Moylan's use of the word "recordist", for example, describes the conflation of recording producer and engineer—in other words, an individual who mixes music that they have also played a key role in shaping during the recording process. Kealy (1979) has used the term "artist-mixer".

17 Zak (2001: 141), for example, states that the object of mixing is to create "a composite image of an apparently unitary musical performance". Moylan (2007: 319) suggests that "the actual process of executing the mix is very similar to performing". Golding and Hepworth-Sawyer (2012: 182) have notably compared the role of the modern mix engineer to that of an orchestral conductor, presumably because sound mixers, like conductors, are essentially dealing with a type of blueprint for a potential musical performance.

18 Musicologists have used words such as "montage" (Zak, 2001: 141) and "sonic narrative" (Liu-Rosenbaum, 2012) to account for the ways in which sound mixers may be thinking when considering the mix in terms of its temporal evolution.

19 Albin Zak (2001:141–160) has also outlined a spatial theory of mixing based upon George Massenburg's "four dimensional space model".

20 I borrow this term from Moylan (2007: 328).

21 It is apt, where rhetorical ideas are concerned, that Hodgson (2010) refers to mixing as "the space of communications".

22 There is an interesting discussion in Senior (2011: 122) concerning what is sacrificed or reduced in the service of foregrounding particular elements that suit the mix style.

23 For a thoughtful account of the rhetoric of dance music, see Hawkins (2003). Snoman (2014) also provides many useful insights into dance music mixing.

24 As Moylan (2007: 195) notes, "the characteristics of this envisioned performance environment will greatly influence the conceptual setting for the artistic message of the work."

25 For a particularly interesting discussion of the spatial strategies of a particular era of record production, see also Doyle (2005).

26 For suggestions as to what this flouting of convention might entail, see Dockwray and Moore (2010) and Liu-Rosenbaum (2012).

27 Gibson (1997: 1) notes that "even when there is a producer, he or she will rely heavily on the values and critiques of the engineer. In fact, groups often go to major studios solely because of the production assistance they get from professional recording engineers."

28 Again, the transparency perspective is implicit in the sense that mixing here is about locating some essence of the artistic vision that needs to be accurately perceived and faithfully transmitted. Stavrou (2003: 175–176) also echoes Senior's idea when he recommends "searching the tracks for the most inspiring player".

29 An EQ, for example, is potentially acting like a synthesizer filter and can certainly remove a sound a considerable distance from its natural state, while the compressor can be used to substantially alter a sound's envelope and rhythmic characteristics. For further discussion of sound shaping aspects of recording practice, see Moylan (2007: 46–47).

30 Stavrou's discussion of reverb (2003: 181–189) contains many interesting observations regarding its creative use.

31 For further contextualization of "idiolect" and examples of its use in analysis see Eco (1976), Middleton (1990: 174) and Ibrahim and Moore (2004: 139–158).

32 This includes their seminal album, *Unknown Pleasures* (1979) and its follow-up, *Closer* (1980), as well as their early work under the name of New Order on the album *Movement* (1981).

33 Discussion of Hannett's background can be found in Sharp (2007).

34 While Hannett was assisted by Chris Nagle (engineer at Strawberry Studios, Stockport) during a number of Joy Division sessions, it is clear that Hannett regarded himself as the one doing the mixing.

35 For accounts of Hannett's recording approach by members of Joy Division, see Reynolds (2009: 229–243) and Hook (2013: 150–159).

36 To hear the difference, it is instructive to listen to the band's early recordings under the name of Warsaw, particularly their 1978 version of 'Transmission,' which Hannett re-recorded and mixed in 1979.

37 See, for example, the sleeve notes for the retrospective single collection, *Zero: A Martin Hannett Story 1977–1991*.

38 Reynolds (2005: 187) notes, with great insight, that "Hannett believed punk was sonically conservative precisely because of its refusal to exploit the recording studio's capacity to create space." For discussion of Hannett's contribution in the general context of post-punk, see also Witts (2009).

39 Hepworth-Sawyer and Golding's concept of "Production Plotting" (2011: 177) is also of interest in this regard.

40 Albin Zak's discussion of "resonance" in *The Poetics of Rock* (2001: 184–197) provides much food for thought here.

41 I use "domain" here in the Csikszentmihalyi sense of the term. Summarized broadly, Csikszentmihalyi (1988: 325–339) considers creativity relative to the particular environment within which the individual operates. He uses the term "domain" to refer to an existing context of practice from which one assimilates patterns of creative approach (the rules of the game as it were) and "field" to refer to the social factors (namely people and institutions) which determine those creative contributions that are most likely to be accepted into the domain.

Bibliography

Bonds, M.E. (1991). *Wordless Rhetoric: Musical Form and the Metaphor of the Oration*. Cambridge, MA: Harvard University Press.

Brackett, D. (2000). *Interpreting Popular Music*. Berkeley and Los Angeles, CA: University of California Press.

Brown, L.B. (2000). Phonography, Rock Records, and the Ontology of Recorded Music. *The Journal of Aesthetics and Art Criticism* 58 (4): 361–372.

Brown, P. (2010). *Are We Still Rolling?: Studios, Drugs and Rock 'n' Roll—One Man's Journey Recording Classic Albums*. n.p.: Tape Op Books.

Buskin, R. (1999). *Inside Tracks*. New York: Avon.

Buskin, R. (2012). *Classic Tracks: The Real Stories Behind 68 Seminal Recordings*. London: Sample Magic.

Case, A.U. (2011). *Mix Smart: Pro Audio Tips for Your Multitrack Mix*. Oxford: Focal Press.

Corbett, E.P.J. and Connors, R.J. (1998). *Classical Rhetoric for the Modern Student*. Oxford: Oxford University Press.

Csikszentmihalyi, M. (1988). 'Society, Culture, and Person: A Systems View of Creativity.' In *The Nature of Creativity*, ed. R.J. Sternberg. New York: Cambridge University Press, pp. 325–339.

Day, T. (2002). *A Century of Recorded Music: Listening to Musical History*. New Haven, CT: Yale University Press.

De Man, B. and Reiss, J.D. (2013). A Semantic Approach to Autonomous Mixing. *Journal on the Art of Record Production* 8. [Online]. Available at: http://arpjournal.com/a-semantic-approach-to-autonomous-mixing/ [Accessed: 5 January 2016].

Dockwray, R. and Moore, A.F. (2008). The Establishment of the Virtual Performance Space in Rock. *Twentieth-Century Music* 5 (2): 219–241.

Dockwray, R. and Moore, A.F. (2010). Configuring the Sound-Box 1965–1972. *Popular Music* 29 (2): 181–197.

Doyle, P. (2005). *Echo and Reverb: Fabricating Space in Popular Music Recording, 1900–1960*. Middletown, CT: Wesleyan University Press.

Doyle, T. (2013). Peter Walsh and Scott Walker: Producing Bish Bosch. *Sound on Sound*. [Online]. Available at: http://www.soundonsound.com/sos/jan13/articles/scott-walker.htm.

Droney, M. (2003). *Mix Masters: Platinum Engineers Reveal Their Secrets for Success*. Boston, MA: Berklee Press.

Eco, E. (1976). *A Theory of Semiotics*. Bloomington, IN: Indiana University Press.

Eisenberg, E. (2005). *The Recording Angel: Music, Records and Culture from Aristotle to Zappa*. New Haven, CT, and London: Yale University Press.

Gibson, D. (1997). *The Art of Mixing: A Visual Guide to Recording, Engineering, and Production*. Vallejo, CA: MixBooks.

Gibson, D. (2005). *The Art of Mixing: A Visual Guide to Recording, Engineering, and Production*. Boston, MA: Thomson Course Technology.

Gracyk, T. (1996). *Rhythm and Noise: An Aesthetics of Rock*. Durham, NC: Duke University Press.

Hawkins, S. (2003). 'Feel the Beat Come Down: House Music as Rhetoric.' In *Analyzing Popular Music*, ed. A.F. Moore. Cambridge: Cambridge University Press, pp. 80–102.

Hepworth-Sawyer, R. and Golding, C. (2011). *What Is Music Production? Professional Techniques to Make a Good Recording Great*. Oxford: Focal Press.

Hodgson, J. (2010). *Understanding Records: A Field Guide to Recording Practice*. New York: Continuum Books.

Hook, P. (2013). *Unknown Pleasures: Inside Joy Division*. London: Simon and Schuster.

Ibrahim, A. and Moore, A.F. (2004). 'Sounds Like Teen Spirit: Identifying Radiohead's Idiolect.' In *The Music and Art of Radiohead*, ed. J. Tate. Farnham: Ashgate, pp. 139–158.

Izhaki, R. (2008). *Mixing Audio: Concepts, Practices and Tools*. Oxford: Focal Press.

Jones, S. (1992). *Rock Formation: Music, Technology, and Mass Communication*. London: Sage Publications.

Kealy, E.R. (1979). From Craft to Art: The Case of Sound Mixers and Popular Music. *Work and Occupations* 6 (1): 3–29.

Liu-Rosenbaum, A. (2012). The Meaning in the Mix: Tracing a Sonic Narrative in 'When the Levee Breaks'. *Journal on the Art of Record Production* 7. [Online]. Available at: http://arpjournal.com/the-meaning-in-the-mix-tracing-a-sonic-narrative-in-'when-the-levee-breaks' [Accessed: 5 January 2016].

Massey, H. (2000). *Behind the Glass Vol. 1: Top Record Producers Tell how They Craft the Hits*. San Francisco, CA: Backbeat Books.

Massey, H. (2009). *Behind the Glass Vol. 2: Top Record Producers Tell how They Craft the Hits*. Milwaukee, WI: Backbeat Books.

Middleton, R. (1990). *Studying Popular Music*. Buckingham: Open University Press.

Moore, A.F. (1993). *Rock: The Primary Text*. Milton Keynes: Open University Press.

Moore, A.F. (2001). *Rock: The Primary Text*. Aldershot: Ashgate.

Moorefield, V. (2010). *The Producer as Composer: Shaping the Sounds of Popular Music*. Cambridge, MA: MIT Press.

Moylan, W. (2007). *Understanding and Crafting the Mix: The Art of Recording*, Second edition. Oxford: Focal Press.

Ott, C. (2010). *Joy Division's Unknown Pleasures*. New York: Continuum Books.

Owsinski, B. (1999). *The Mixing Engineer's Handbook*. Vallejo, CA: Mix Books.

Ramone, P. (2007). *Making Records: The Scenes Behind the Music*. New York: Hyperion.

Reynolds, S. (2005). *Rip it Up and Start Again: Postpunk 1978–1984*. London: Faber & Faber.

Reynolds, S. (2009). *Totally Wired: Postpunk Interviews and Overviews*. London: Faber & Faber.

Savage, J. (1992). An Interview with Martin Hannett, 29th May 1989. *Touch-Vagabond*. [Online]. Available at: http://www.rocksbackpages.com/Library/Article/an-interview-with-martin-hannett-29th-may-1989 [Accessed: 20 November 2015].

Savage, J. (1994). Joy Division: Someone Take These Dreams Away. *Mojo*. [Online]. Available from: http://www.rocksbackpages.com/Library/Article/joy-division-someone-take-these-dreams-away [Accessed: 20 November 2015].

Schwartz, D. and Stone, T., eds. (1992). Music Producers: Conversations and Interviews with Some of Today's Top Record Makers. London: Omnibus Press.

Senior, M. (2011). *Mixing Secrets for the Small Studio*. Oxford: Focal Press.

Sharp, C. (2007). *Who Killed Martin Hannett?: The Story of Factory Records' Musical Magician*. London: Aurum.

Snoman, R. (2014). *Dance Music Manual: Tools, Toys, and Techniques*. Burlington, MA: Focal Press.

Stavrou, M.P. (2003). *Mixing with Your Mind: Closely Guarded Secrets of Sound Balance Engineering*. Mosman, New South Wales: Flux Research.

Tankel, J.D. (1990). The Practice of Recording Music: Remixing as Recoding. *Journal of Communication* 40 (3): 34–46.

Tarasti, E. (2012). *Semiotics of Classical Music: How Mozart, Brahms and Wagner Talk to Us*. New York: Walter de Gruyter.

Toft, R. (2010). *Hits and Misses: Crafting Top 40 Singles, 1963–1971*. New York: Continuum Books.

Witts, R. (2009). 'Records and Recordings in Post-Punk England, 1978–80.' In *The Cambridge Companion to Recorded Music*, eds. E. Clarke, N. Cook, D. Leech-Wilkinson and J. Rink. Cambridge: Cambridge University Press, pp. 80–83.

Zagorski-Thomas, S. (2014). *The Musicology of Record Production*. Cambridge: Cambridge University Press.

Zak, A.J. (2001). *The Poetics of Rock: Cutting Tracks, Making Records*. Berkeley and Los Angeles, CA: University of California Press.

Zak, A.J. (2005). 'Edition-ing' Rock. *American Music* 23 (1): 95–107.

Zak, A.J. (2009). 'Getting Sounds: The Art of Sound Engineering.' In *The Cambridge Companion to Recorded Music*, eds. E. Clarke, N. Cook, D. Leech-Wilkinson and J. Rink. Cambridge: Cambridge University Press, pp. 63–76.

14

Mix as Auditory Response

Dr. Jay Hodgson

Research on record production almost always reproduces a crucial mis-understanding about the material nature of recorded musical communi-cations, namely, that when we listen to records we hear distinct acoustic phenomena—like kick drums, electric guitars, sequenced claps, singing, et cetera—rather than a single acoustic phenomenon, produced by speaker and headphone technology, designed to trick the human hearing appara-tus into believing it detects the presence of distinct acoustic phenomena. Hearing Led Zeppelin's 'When The Levee Breaks' (1971), for instance, one might think they hear a kick drum, a snare drum, hi-hats, cymbals, rack and floor toms, an electric guitar or two, electric bass, some blues harp and Robert Plant's plaintive vocal wails. But what they actually hear is a single acoustic phenomenon, a single sound, produced by speakers and headphones, designed to trick their auditory apparatus into believing it detects the presence of numerous distinct acoustic phenomena (like kick drums, and snare drums, and a blues harp, and so on). And the same goes for less figurative records, of course. Listening to Deepchild's 'Neukoln Burning' (2012), for instance, we probably think we hear a sequenced kick, design-intensive synths, slowed vocals and numerous other 'non-veridic' sounds.[1] What we actually hear, though, is a single sound that *portrays* those sounds. In overlooking—or, perhaps, in simply not realizing—this basic fact, analysts mistake the *subject* of a recorded musical communi-cation (i.e., the performances and broader musical contents recordists use sound to portray) for communication itself. In turn, the artistry of record-ing practice—the myriad musical things that recordists do—are kept from the scholarly record, except perhaps as a mere 'technical support' for the 'true arts' of performance and composition. It's as though an exciting new field has emerged that wants to study, say, fashion photography, but only in terms of how models pose.

If analysts are to speak about records with any sort of clarity—if we are to truly grasp what recordists and listeners actually say, and hear, when they communicate by record—a complete critical realignment of music production studies at large is required. At present, this academic terrain is largely dominated by cultural studies and musicological work, which fol-low a similar methodological paradigm, both of which tend to see record production only as an indication of broader social and historic trends.[2] But something more permanent happens each time someone makes and hears a recorded musical communication, regardless of their social situations and historical contexts. Indeed, a 'universal poetic' undergirds every recorded musical communication, and analysts of every disciplinary stripe, with any level of musical expertise, can see and hear this 'universal poetic' at work, if only they would don the right analytic lens. Thus do I offer this 'theory of record production', which considers recorded musical communications as exchanges of *auditory* rather than *acoustic* information, that is, which hears recorded musical communications as *auditory representations* of sounds rather than as sounds per se. This theory is unique, insofar as its broader analytic perspective on record production doesn't exist anywhere else in print. This said, some of its components derive from cited sources, and an expanded and significantly altered version of what follows was used as the text for an electronic photo essay that I composed with the Canadian visual artist Steve MacLeod.[3] In keeping with the broader theme of this volume, though, I emphasize the crucial role mixing plays in recorded musical communications at large in this essay. In fact, I will even go so far as to argue that the most meaningful way to consider the aesthetic content of any record is *as* a mix, that is, as an auditory response to sounds—a past-tense, precisely repeatable auditory narrative that explains how some sounds were once (ideally) heard—rather than as sounds per se. It is my humble hope that doing this provides analysts with a useful clarification of the communications paradigm they study, and that it spurs recordists, and recording musicians in general, to reconsider certain fundamental aspects of their creativity.

Past-Tense Auditory Narratives

Aesthetically speaking, two things are heard simultaneously whenever a record plays: (i) some sounds in particular and (ii) a certain spatial arrangement of those sounds (i.e., a mix). The sounds one hears when they listen to a record simply cannot exist except in such an arrangement. Thus, one only ever hears a mix when they listen to a record.

This is the case even when the record features only solo performances. Consider, for instance, Nick Drake's 'Horn' (1972/2003).[4] The fifth track on Drake's classic *Pink Moon* (Island: 1972) LP, 'Horn' is an instrumental segue, played by Drake on an acoustic guitar, which connects 'Which Will' (1972) to 'Things Behind the Sun' (1972), tracks four and six respectively on the album. 'Horn' consists of a simple improvised melody in C# minor, plucked over a tonic drone. However, the guitar is situated somewhat right of center, and it fades out during the last four seconds. This happens every time the record plays; there is no getting around this auditory point of view. As such, 'Horn' is not simply a sequence of pitches and silences performed on an acoustic guitar. It is, rather, an *auditory response* to a sequence of pitches and silences, a retelling of how an acoustic guitar performance was once (ideally) heard.

In hearing, each mix spatially situates listeners in relation to what they hear (which is, ultimately, what the mix 'heard' first). The mix for The Beatles' 'Can't Buy Me Love (2015 Version)' (1964/2015), to name an arbitrary example, creates a listening position which is before, and facing, the band.[5] The mix 'hears' Paul McCartney's double-tracked vocals furthest in front, slightly ahead of George Harrison's and John Lennon's electric and acoustic guitars, McCartney's bass track, and Ringo Starr's drums. In so doing, the mix for 'Can't Buy Me Love' hears The Beatles perform such that Paul McCartney's vocals are always prioritized. The reception paradigm, which this mix models, is that which obtains given the separation of performer from audience in the concert-hall division of stage (performer's space) from seating (audience's space).

Of course, this division is obviously idealized on the record. Every listener is situated exactly the same by the mix, regardless of where or how they listen to it. Listeners *must* hear the band perform from a front-row and center perspective, in other words. Thus, they must always hear McCartney's double-tracked vocals front and center and panned slightly to the left and right. Ringo's drums must always be heard farthest back. And Lennon's and Harrison's guitars must always be heard somewhere in between those two extremes. The record plays in the past tense, in other words. What we hear, when we listen to it, is a sequence of *already heard* sounds, that is, *auditory* rather than *acoustic* information.

Parameters of Audition: The Soundbox

Western notation cannot objectify the auditory response a mix construes. Notation is, after all, a prescriptive technology developed for the express purpose of reasoning and objectifying sonic phenomena as such, made under the auspices of sound production. A mix is anything but a linear sequence of sounds per se. Obviously, then, Western notation can only haphazardly conceptualize it. And, of course, it doesn't claim to do otherwise.

In light of this, to grasp the parameters of a mix's auditory response—to delineate its auditory capabilities, as it were—I consider it best to deploy an analytic tool used by most mix engineers when they first learn the craft, namely, a 'soundbox'.[6] This tool affords burgeoning engineers with a means of conceptually objectifying the aural perspective each mix construes, so they may begin to grasp the 'anchor points' and other spatial formulae which structure most modern records. Moreover, the soundbox is used in many beginner classrooms, including my own, and thus should be easy enough to understand for analysts with any level of musical expertise.

There are six components of a soundbox, as I teach it. I call these components (i) The Auditory Horizon, (ii) The Horizontal Plane, (iii) The Horizontal Span, (iv) The Proximity Plane, (v) The Vertical Plane and (vi) The Vertical Span. Each mix can be described according to these six components. To explain them, I will use the mix for Pink Floyd's 'Speak to Me' (1973/2011), which strikes me as ideally suited for this task.

Composed by drummer Nick Mason, 'Speak to Me' constitutes roughly the first minute of a key artifact of recording practice, namely, Pink Floyd's *Dark Side of the Moon* (Capitol, 1973/2011) LP.[7] The album's opening track, 'Speak To Me' is divisible into three sections. Section One runs to thirty-eight seconds. Section Two runs from thirty-eight seconds to one minute and eleven seconds. Section Three comprises the last five seconds of the track.

'Speak To Me' begins with silence (eleven seconds on digital formats; more or less eleven seconds on vinyl LP or audio cassette). A heartbeat then fades to audibility. After another twenty seconds, the ticking hands of stopwatches fade in. These are followed shortly by the pendulum swings of a grandfather clock. Band roadie Pete Watts then confesses, "I've been mad for fucking years, absolutely years man, over the edge working with bands". In the meantime, a looped cash register opens and slams shut at an obsessive rate across the stereo spectrum. Jerry Driscoll, the doorman at Abbey Road Studios where Pink Floyd recorded *Dark Side of the Moon*, then says, "I've always been mad, I know I've been mad, like most are. . . . Very hard to explain why you're mad, even if you're not mad", and a loop of nervous, even deranged, laughter becomes audible. Something like a pneumatic drill fades in, coupled with an electronic drone. These two tracks are then pumped to an increasingly higher volume such that they overtake all but Clare Torry's melodic screams, and a cymbal roll, both of which sound, for the remaining five seconds, over and above the growing din. After one minute and sixteen seconds, 'Speak To Me' cross-fades into 'Breathe' (1973/2011), which is the second track on *Dark Side of the Moon*.

As a collection of mostly 'found sounds', 'Speak To Me' is usually explained, at least in published accounts, as the first scene of a narrative which all told comprises *Dark Side of the Moon*.[8] The remaining tracks on the record are said to recount, via a series of flashbacks, the protagonist's encounters with what Roger Waters called "anti-life forces" (in order: authority, paranoia, time, money and war), each of which is alleged, throughout the album, to exact an universally deadening toll upon the psyche.

In this respect, and given a familiarity with *Dark Side of the Moon* as a whole, 'Speak To Me' constitutes a sonic analogy for the album's protagonist *in medias mental collapse*, as it were. The sounds comprising the track amble randomly about the stereo spectrum, analogizing the protagonist's sudden incapacity to reason, or situate, one sound in relation to another according to their inherent symbolic connotations. The remainder of tracks on the album elucidate this collapse and, in so doing, polemicize the capitalist mode of production as a hijacker of desire and, eventually, of sanity. During 'Money' (1973/2011), for instance, the fifth track on *Dark Side of the Moon*, the album's protagonist emerges as something like Herbert Marcuse's 'One-Dimensional Man' in the extreme.[9] On the album's penultimate track, 'The Great Gig in the Sky' (1973/2011), the protagonist contemplates suicide, having realized that he's achieved nothing but great wealth (a cause for celebration, to my mind!). By the time of 'The Lunatic'

(1973/2011), the concluding track on *Dark Side of the Moon*, the protagonist has lost his mind completely.

Regardless of how one interprets the track—I've always heard a childbirth from the child's perspective, in fact—'Speak To Me' remains, in the first instance, a finely detailed mix. As such, it provides us with a valuable tool for elucidating the aural parameters of a mix's auditory narrative, using soundbox analysis. I will thus turn my attention now to elucidating the track's sonic details, knowing that it is only through these details that the jarring narrative interpretations I mention above can in the first instance emerge.

As already noted, the first component of a soundbox is what I call 'The Auditory Horizon'.[10] In 'Speak To Me', the Auditory Horizon is established during the first eleven seconds of the track, when the heartbeat fades in and reaches its maximum volume. Behind that Horizon is silence. The Auditory Horizon thus constitutes the total reach of a mix's 'earshot'. If a track fades in—as with the heartbeat in 'Speak To Me', for instance—it begins its trek towards the Auditory Horizon from behind it, which is to say, from a place 'too far away' to be heard. Conversely, if a track fades out—as with, for example, the heartbeat heard during 'The Dark Side of the Moon' (1973/2011), the album's final track—it ends its trek past the Auditory Horizon, beyond 'earshot'.

A mix's 'Horizontal Plane', the next component of a soundbox, describes where a sound is heard in relation to center, and we call the total horizontal expanse of a mix its 'Horizontal Span'. What I call the 'pneumatic drill' and 'electronic drone', which sound at fifty-five seconds into 'Speak To Me', are panned variously throughout their brief twenty-one second existences. They oscillate between, rather than leap from, left to right positions along the Horizontal Plane. The track that sounds farthest to the right, however, is the loop of ticking clocks, first heard at thirty-one seconds in. Sixteen seconds later, a male voice confesses that he's "always been mad" at the farthest position left along the Horizontal Plane. Together, these two tracks create the Horizontal Span of the mix for 'Speak To Me'.

Perhaps the most significant component of a soundbox is its Proximity Plane. This component describes the position of sounds in a mix vis-à-vis its Auditory Horizon. As such, the Proximity Plane represents a mix's ability to hear in depth, with the Auditory Horizon comprising its far limit. In 'Speak To Me', as what I call 'the pneumatic drill' pans along the Horizontal Plane, it is also faded to an ever higher volume and, thereby, moved ever 'closer' along the Proximity Plane. In this way, the track is made to sound as though it is ever increasing in 'proximity' and, thus, ever encroaching upon the listener.

Alongside an Auditory Horizon, Horizontal Plane, Horizontal Span and Proximity Plane, every soundbox should also feature a Vertical Plane and Vertical Span. Just as the Horizontal Plane and Horizontal Span together describe a mix's total horizontal earshot, the Vertical Plane and Vertical Span describes its capacity to hear vertically. The loop of ticking clocks heard at a hard-right position in 'Speak To Me', for instance, also sounds 'over and above' everything else in the mix, given that it features a plethora of high-frequency content. As such, it sounds 'highest' of all the tracks in the mix. Conversely, 'lowest' of all the tracks in 'Speak To Me' is the maniacal laughter, with its throaty bass energy. Together, these two tracks comprise the total Vertical Span of the mix.

The six components of a soundbox that I elucidate above, which are the six components of a mix, together comprise an *auditory response* to sounds, rather than sounds per se. That is, they present an auditory narrative that explains how a sequence of sounds was once (ideally) heard, not the sequence of sounds per se. There is simply no other way to hear the sounds a record portrays except through the broader psychoacoustic point of view its mix construes. Each sound is thus mixed to create that point of view, and that point of view remains always, then, most fundamentally the actual aesthetic content of any recorded musical communication.

Psychoacoustic Profiles

Indeed, every sound on a record has been 'heard before', and from a clearly specified vantage. As noted, the Auditory Horizon, Horizontal Plane, Horizontal Span, Proximity Plane, Vertical Plane and Vertical Span together describe an *auditory response* to particular sounds, not the particular sounds as such. And we know this because what I call 'psychoacoustic profiles' are fixed on record, even as they are ceaselessly dynamic in concert. Crafting and fine-tuning these profiles, then, must comprise the aesthetic craft of mixing *in toto*.

The concept of a psychoacoustic profile defies any simple or narrow definition. For now, though, it will suffice to conceptualize a psychoacoustic profile as the sum of modifications exacted on a soundwave by acoustic and psychoacoustic interferences. Even if recordists use randomizing algorithms or modular analog processes, virtual synthesis and digital sequencers, to generate truly aleatoric (random) timbres, pitches and pauses, every 'sound' on the resulting record would nonetheless bear a fixed and precisely repeatable psychoacoustic profile, including completely dry (conspicuously nonreverberant) tracks. This means that sequenced synthesis tracks without any signal processing applied bear a psychoacoustic profile which is just as fixed as, say, Robert Plant's vocal track on Led Zeppelin's 'No Quarter' (1973), or Philip Glass's synthesizer track on the Nonesuch re-release of 'Two Pages' (1994), or any of the vocal tracks on Noah Pred's *Third Culture* (2013) LP. The conspicuous absence of reverberation is a reverberation profile, after all!

To be clear, psychoacoustic profiles don't *modify* the sounds a mix hears. They are an integral property, a holistic part and parcel, of those sounds. Listeners cannot disentangle a recorded snare drum hit from its reverberations, for instance, whether those reverberations were captured during a live tracking session, synthesized or applied during mixdown using signal processing. We might move to the left or right of our speakers, or even walk between and through them, but the snare drum remains fixed wherever the mix engineer positioned it. Likewise, we might don a pair of headphones and run a city block, but the snare drum nevertheless stays forever put wherever it is mixed to be. And we certainly can't move around the singer, to garner a less obstructed perspective on the snare, as we might at a concert (security allowing).

Single-point perspectival painting provides an obvious analog here. Every shade of darkness and light, every geometric distortion, in a perspectival painting ultimately combines to form a broader visual perspective for the painting at large. It is from this perspective, from this precisely specified spatiotemporal point of view, that every painted object is conveyed to viewers, making it a kind of narrator for the painting. People don't just see a vase when they look at a single-point perspectival painting of flowers on a table, for instance. Rather, they see a vase geometrically shaped and shaded to triangulate with the painting's vanishing point. Whatever can be seen in perspectival paintings has already been seen, then, and from a clearly specified place in time and space. And this is true whether the vista is abstract (as in Cézanne) or irrational (as in Picasso) or hyperrealist (as in Charles Bell). It is possible, even, to consider this point of view the painting's primary subject.

As with perspectival painting, the broader auditory perspective a record construes—its mix—simply cannot be moved, modified or superseded. Listeners can move about the listening environment and alter their perspective on a mix vis-à-vis room acoustics, to be sure, and they can equalize the output however they deem fit. But they don't change the psychoacoustic profiles they hear in so doing, nor do they in any manner alter the broader mix those profiles combine to construe. What they do change is their perspective on those profiles, their perspective on the auditory perspective a mix already construes. Returning to our case study of 'Speak To Me' above, listeners might choose to hear the record from inside a shower two rooms away, with the bass boosted to an obscene measure, and though they would likely hear only frequencies under about 500 Hz as a consequence, not a single psychoacoustic profile would change—the heartbeat, pneumatic drill, electronic drone, maniacal laughter, confessions of madness, melodic screams, and every other sound the mix hears would all stay forever put wherever in the soundbox they happen to be. And this can only be the case if 'Speak To Me' comprises an idealized auditory response to sounds, not sounds per se.

Records as Models of Hearing

Ultimately, records don't *reproduce* sounds. They model hearing.

I mean this literally. Records present listeners with a single acoustic phenomenon, and mixes comprise the aesthetic content of that phenomenon. Mixes are the 'conjuring trick' engineers use to generate the auditory delusion of musical activity within a single acoustic phenomenon, in other words. In fact, mixes—and, thus, records in general—inscribe particular tympanic motions, simply reversed and fed through electronic circuitry, and stored as data of whichever kind the release format demands. This data, if used correctly, moves speakers and headphones at the precise rates, and in roughly inverse directions and amplitudes, that human tympana should move to register particular auditory phenomena. Mix engineers thus ultimately paint on the canvas of human psychophysiology. They orchestrate psychophysiological reactions to disturbances in air pressure, each of which conjures a unique auditory delusion.

Tweaking the reverb and equalization on an acoustic guitar track, for instance, mix engineers can orchestrate a psychophysiological reaction in listeners that *compels* them to localize the now-dulled instrument behind a less reverberant and brighter lead-vocal track. Even then, though, listeners don't actually hear a guitar or a voice when they play the record. What they hear instead is a single acoustic phenomenon, designed by mix engineers to provoke a particular auditory response in listeners, specifically, to an acoustic guitar and singer performing in tandem. This process, that is, programming particular auditory responses via speaker and headphone technology, comprises the substantive basis of all mixing. As such, it comprises the substantive basis of all recorded musical communications.

Conclusion

So what does this all mean for recording practice in general? To answer this question, I'll need to quickly consider the role that sample rates play in facilitating the auditory responses I note above. Indeed, few realize that recording practice comprises a kind of stop-motion auditory animation. Every technical phenotype of storage medium, from wax cylinders to the Voice Memo app on your iPhone, stores discontiguous packets of data called samples. Film provides a useful analogy here. Video cameras—analog *and* digital phenotypes—generate discrete snapshots quickly enough that when they are displayed in sequence and at the same rate they were encoded, the images seem to animate, taking on an illusory life beyond empirical two-dimensionality. The animated images don't actually move, of course, just as soundwaves don't actually undulate on records.

Mixes, that is, recorded musical communications, are comprised of audio samples which portray soundwaves in various states of propagation and decay. Once we record you plucking the second-lowest string of your acoustic guitar, for instance, we can always scrub past the first three seconds to hear the exact same guitar timbre in precisely the same state of decay, and according to the same precisely fixed acoustic and psychoacoustic variables (we can always skip the first three seconds' worth of samples, in other words). We might even replace a few samples we don't like with other samples, using the 'sample replace' function on whichever DAW we use, as is so often done by mix engineers working on drums nowadays. We don't hear the plucking of an acoustic guitar when we play the record, after all, and we don't hear an A below middle-C propagate and decay. What we hear are discrete samples played at a very precise rate (44.1 kHz, most often), each of which depicts a different *moment* of propagation and decay. *Thus does the map cover the terrain . . .*

Recording practice depends on sample rates completely. Its communications paradigm ultimately boils down to this single measure, in fact. Get the sample rate wrong and you can forget about hearing a recorded musical communication. What you'll hear instead is a stubborn silence, or the hum and chatter of failing machinery. Tracking and mixing are software programming, in the end—coding samples of data meant to vibrate speakers and headphones at robotically precise rates and amplitudes. These processes transform 'live' performance from a communicative act in and of itself into a *subject* of communications, something *represented* (rather than achieved) through the communicative activity of recording practice.

Simply put, recordists *portray* rather than *reproduce* sounds, and then music, by provoking particular auditory responses in listeners to a single acoustic phenomenon. Mixing is just one of the many things they do to achieve this auditory sleight-of-hand. Musicians perform, sequence, sample, and so on, for microphones and direct-injection technology when they record, to be sure, just as models pose for cameras and lighting arrays in the photographer's studio. But when recordists track and mix records, they do the sonic equivalent of a fashion photographer framing a model's pose and editing, say, the picture's color and contrast. In other words, creative agency is shared in recording practice among recordists and the artists whose musical work they portray. And again, in neglecting this basic fact about recording practice—i.e., that its communications paradigm ultimately amounts to stop-motion auditory portraiture—we mistake the subject of recorded musical communications for communication itself. In so doing, we relegate record production in general, of which mixing is but one of many component procedures, to a mere act of conveyance rather than as musical practice per se. It is my hope that this essay helps analysts hear this directly for themselves, so they may engage record production, and its component procedures, with intensified clarity.

Notes

1 For more on 'veridic' and 'non-veridic' sounds in modern recording practice, see John Andrew Fisher's 'Rock and Recording: The Ontological Complexity of Rock Music' in *Musical Worlds: New Directions in the Philosophy of Music*, ed. P. Alperson (Pennsylvania: University of Pennsylvania Press, 1998), 109–123. I complicate Fisher's taxonomy in my own book, *Understanding Records: A Field Guide to Recording Practice* (New York: Bloomsbury, 2010), 71–72.

2 To be clear, this is not a criticism of so-called 'interdisciplinary' studies of modern recording practices. I simply agree with Louise Meintjes's observation that "sociology and media studies have led the way in generating discussions about studio-based creativity", in Louise Meintjes, *Sound of Africa! Making Music Zulu in a South African Studio* (Washington, DC: Duke University Press, 2003), 27.

3 See Jay Hodgson with Steve MacLeod, *Representing Sound: Notes on the Ontology of Recorded Musical Communications* (Kitchener/Waterloo: Wilfrid Laurier University Press, 2013).

4 The mix I consider here appears as track five on the 2003 remaster of Nick Drake's *Pink Moon* (Island: B000025XKM, 1972).

5 The mix I consider here appears on The Beatles, 'Can't Buy Me Love (2015 Stereo Version)', track five on *The Beatles, 1* (2015 Version) (Capitol Records: B01576X99U, 2015).

6 For an edifying academic discussion of the 'soundbox', see Ruth Dockwray and Allan
   Moore, 'Configuring the Soundbox 1965–1972' (2010) in *Popular Music* 29 (2): 181–
   197. To see how it is typically used in audio engineering pedagogy, see David Gibson, *The
   Art of Mixing: A Visual Guide to Recording, Engineering and Production, Second Edition*
   (Boston, MA: Course Technology PTR, 2005).
7 Specifically, the 2011 remastered version of Pink Floyd's *Dark Side of the Moon* (Capitol:
   B004ZN9RWK, 1973/2011).
8 See Nicholas Schaffner, *Saucerful of Secrets: The Pink Floyd Odyssey* (Philadelphia, PA:
   Delta Books, 1992).
9 See Herbert Marcuse, *One Dimensional Man: Studies in the Ideology of Advanced Indus-
   trial Society, Second Edition* (Boston, MA: Beacon Press, 1991).
10 For more on the practical mixing tools and techniques recordists use to *create* these components of a 'soundbox' see, for instance, Roey Izhaki, *Mixing Audio: Concepts, Practices and Tools, Second Edition* (Boston, MA: Focal Press, 2011); Mike Senior, *Mixing
   Secrets in the Small Studio* (Boston, MA: Focal Press, 2011); David Gibson, *The Art
   of Mixing: A Visual Guide to Recording, Engineering and Production, Second Edition*
   (Boston, MA: Course Technology PTR, 2005); Alexander Case, *Sound FX: Unlocking
   the Creative Potential of Recording Studio Effects* (Boston, MA: Focal Press, 2007); William Moylan, *Understanding and Crafting the Mix: The Art of Recording, Second Edition*
   (Boston, MA: Focal Press, 2007); Bobby Owsinski, *The Mixing Engineer's Handbook,
   Second Edition* (New York: Cengage, 2006).

Bibliography

Case, Alexander (2007). *Sound FX: Unlocking the Creative Potential of Recording Studio
    Effects*. Boston, MA: Focal Press.
Dockwray, Ruth and Moore, Allan (2010). Configuring the Soundbox 1965–1972. *Popular
    Music* 29 (2): 181–197.
Fisher, John Andrew (1998). 'Rock and Recording: The Ontological Complexity of Rock
    Music.' In *Musical Worlds: New Directions in the Philosophy of Music*, ed. P. Alperson.
    Pittsburgh, PA: University of Pennsylvania Press, pp. 109–123.
Gibson, David (2005). *The Art of Mixing: A Visual Guide to Recording, Engineering and
    Production*, Second edition. Boston, MA: Course Technology PTR.
Hodgson, Jay (2010). *Understanding Records: A Field Guide to Recording Practice*. New
    York: Bloomsbury.
Hodgson, Jay with MacLeod, Steve (2013). *Representing Sound: Notes on the Ontology of
    Recorded Musical Communications*. Kitchener/Waterloo: Wilfrid Laurier University Press.
Izhaki, Roey (2011). *Mixing Audio: Concepts, Practices and Tools, Second edition*. Boston,
    MA: Focal Press.
Marcuse, Herbert (1991). *One Dimensional Man: Studies in the Ideology of Advanced Indus-
    trial Society*, Second edition. Boston, MA: Beacon Press.
Meintjes, Louise (2003). *Sound of Africa! Making Music Zulu in a South African Studio*.
    Washington, DC: Duke University Press.
Moylan, William (2007). *Understanding and Crafting the Mix: The Art of Recording*, Second
    edition. Boston, MA: Focal Press.
Owsinski, Bobby (2006). *The Mixing Engineer's Handbook*, Second edition. New York: Cengage.
Schaffner, Nicholas (1992). *Saucerful of Secrets: The Pink Floyd Odyssey*. Philadelphia, PA:
    Delta Books.
Senior, Mike. *Mixing Secrets in the Small Studio*. Boston, MA: Focal Press, 2011.

Discography

The Beatles, *1 (2015 Version)* (Capitol Records: B01576X99U, 2015).
Nick Drake, *Pink Moon* (Island: B000025XKM, 1972).
Pink Floyd, *Dark Side of the Moon* (Capitol: B004ZN9RWK, 1973/2011).

15

An Intelligent Systems Approach to Mixing Multitrack Audio

Joshua D. Reiss

Introduction

Although audio production tasks are challenging and technical, much of the initial work follows established rules and best practices. Yet multitrack audio content is still often manipulated 'by hand', using no computerized signal analysis. This is a time-consuming process, and prone to errors. Only if time and resources permit does the sound engineer refine his or her choices to produce an aesthetically pleasing mix which best captures the intended sound.

In order to address this challenge, a new form of multitrack audio signal processing has emerged. Intelligent tools have been devised that analyze the relationships between all channels in order to automate the mixing of multitrack audio content. By 'intelligent', we mean that these tools are expert systems that perceive, reason, learn and act intelligently. This implies that they must analyze the signals upon which they act, dynamically adapt to audio inputs and sound scene, automatically configure parameter settings, and exploit best practices in sound engineering to modify the signals appropriately. They derive the parameters in the editing of recordings or live audio based on analysis of the audio content and on objective and perceptual criteria. In parallel, intelligent audio production interfaces have arisen that guide the user, learn his or her preferences and present intuitive, perceptually relevant controls.

An assumption that is often, but not always, made about mixing is that it is an iterative process (Pestana, 2013). There is no fixed order in the sequence of steps applied, and an iterative, coarse-to-fine approach is applied (Figure 15.1) whereby mixing is treated as an optimization problem, with targets and criteria set for the final mix. Such a view lends itself well to an intelligent systems approach, whereby the steps can be sequenced and diverse optimization or adaptive techniques can be applied in order to achieve given objectives.

For progress towards intelligent systems in this domain, significant problems must be overcome that have not yet been tackled by the research community. First, multitrack audio editing tools demand manual intervention. Although audio editors are capable of saving a set of static scenes

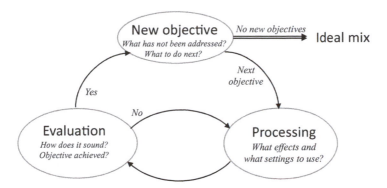

Figure 15.1   The iterative approach to mixing multitrack audio

for later use, they lack the ability to take intelligent decisions, such as adapting to different acoustic environments or different set of inputs. Second, most state-of-the-art audio signal processing techniques focus on single-channel signals. Yet multichannel or multitrack signals are pervasive, and the interaction and dependency between channels plays a critical role in audio production quality. This issue has been addressed in the context of audio source separation research, but the challenge in source separation is generally dependent on how the sources were mixed, not on the respective content of each source. New, multi-input multi-output audio signal processing methods are required, which can analyze the content of all sources in order to improve the quality of capturing, editing and combining multitrack audio. Finally, advances in machine learning must be tailored towards problems and practical applications in the domain of audio production. This chapter presents an overview of recent advances in this area.

Enabling concepts

The idea of automating the audio production process, although relatively unexplored, is not new. In *Automation for the People* (White, 2008), the editor of *Sound on Sound* magazine wrote, "There's no reason why a band recording using reasonably conventional instrumentation shouldn't be EQ'd and balanced automatically by advanced DAW software". He also wrote that mixing tools can "come with a 'gain learn' mode . . . DAWs could optimise their own mixer and plug-in gain structure while preserving the same mix balance". This would address the needs of the musician who doesn't have the time, expertise or inclination to perform all the audio engineering required. Similarly, Moorer (2000) introduced the concept of an Intelligent Assistant, incorporating psychoacoustic models of loudness and audibility, intended to "take over the mundane aspects of music production, leaving the creative side to the professionals, where it belongs".

Automatic mixing research has received a lot of attention in recent years. The state of the art was described in Reiss (2011), but since then the field has grown rapidly. This section describes the key concepts in automatic mixing.

Intelligent and Adaptive Digital Audio Effects

Rather than have sound engineers manually apply many audio effects to all audio inputs and determine their appropriate parameter settings, intelligent, adaptive digital audio effects may be applied instead (Verfaille et al., 2006). The parameter settings of adaptive effects are determined by analysis of the audio content, where the analysis is achieved by a feature extraction component built into the effect. Intelligent audio effects also analyze or 'listen' to the audio signal, but are furthermore imbued with knowledge of their intended use and control their own operation in a manner similar to manual operation by a trained engineer. The knowledge of their use may be derived from established best practices in sound engineering, psychoacoustic studies that provide understanding of human preference for audio editing techniques or machine learning from training data based on previous use. Thus, an intelligent audio effect may be used to set the appropriate equalization, automate the parameters on dynamics processors and adjust stereo recordings to more effectively distinguish the sources.

A block diagram of an intelligent audio effect is given in Figure 15.2. Any additional processing is performed in a separate section so that the audio signal flow is unaffected. This side chain is essential for low latency, real-time signal flow. The side chain is comprised of a feature extraction section and an analysis section.

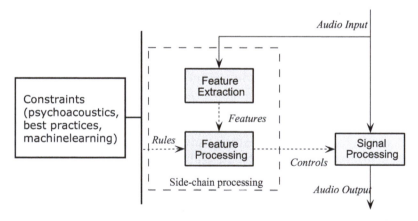

Figure 15.2 Block diagram of an intelligent audio effect. Features are extracted by analysis of the audio signal. These features are then processed based on a set of rules intended to mimic the behavior of a trained engineer. A set of controls are produced which are used to modify the audio signal

The feature extraction is in charge of extracting a series of features from the input channel. Accumulative averaging, described in a later section, is used to ensure real-time signal processing operations, even when the feature extraction process is non-real time. The analysis section outputs control signals to the signal processing side in order to trigger the desired parameter control change command.

Reiss (2011) described several intelligent, adaptive effects for use with single-channel audio, which automate many parameters and enable a higher level of audio editing and manipulation. This included adaptive effects that control the panning of a sound source between two user-defined points, depending on the sound level or frequency content of the source, and noise gates with parameters which are automatically derived from the signal content.

Cross-Adaptive Digital Audio Effects

When editing multitrack audio, one performs signal processing changes on a given signal source not only because of the source content but also because there is a simultaneous need to blend it with the content of other sources, so that a high-quality mix is achieved. The relationship between all the sources involved in the audio mix must be taken into account. Thus, a cross-adaptive effect processing architecture is ideal for automatic mixing.

In a cross-adaptive effect, also known as inter-channel dependent or MIMO (multi-input/multi-output) effect, the signal processing of an individual source is the result of the relationships between all involved sources. That is, these effects analyze the signal content of several input channels in order to produce several output channels. This generalizes the single-channel adaptive signal processing mentioned above.

In an intelligent multitrack audio editing system, as shown in Figure 15.3, the side chain will consist of a feature extraction section for each channel and a single analysis section that processes the features extracted from many channels. The cross-adaptive processing section of an intelligent multitrack audio editing system exploits the interdependence of the input features in order to output the appropriate control data. This data controls the parameters in the signal processing of the multitrack content. The cross-adaptive feature processing can be implemented by a set of constrained rules that consider the interdependence between channels.

In principle, cross-adaptive digital audio effects have been in use since the development of the microphone mixer. However, such systems are only concerned with automatic gain handling and require a significant amount of human interaction during setup to ensure a stable operation.

Intelligent, Multitrack Digital Audio Effects

In Reiss (2011), and references therein, several cross-adaptive digital audio effects were described that explored the possibility of reproducing

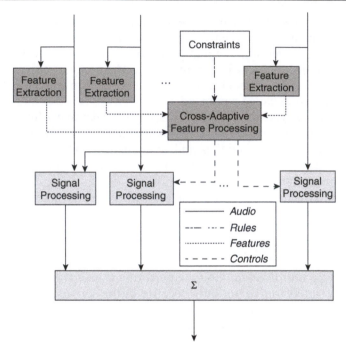

Figure 15.3 Block diagram of an intelligent, cross-adaptive mixing system. Extracted features from all channels are sent to the same feature-processing block, where controls are produced. The output channels are summed to produce a mix that depends on the relationships between all input channels

the mixing decisions of a skilled audio engineer with minimal or no human interaction. Each of these effects produces a set of mixes where each output may be given by the following equation;

$$mix_l[n] = \sum_{m=0}^{M-1}\sum_{k=0}^{K-1} c_{k,m,l}[n] * x_m[n], \tag{1}$$

where there are M input tracks and L channels in the output mix. K is the length of the control vector c and x is the multitrack input. Thus, the resultant mixed signal at time n is a sum over all input channels, of a control vectors convolved with the input signal.

Any cross-adaptive digital audio effect that employs linear filters may be described in this manner. For automatic faders and source enhancement, the control vectors are simple scalars, and hence the convolution operation becomes multiplication. For polarity correction, a binary valued scalar, ±1, is used. For automatic panners, two mixes are created, where panning is also determined with a scalar multiplication (typically, the sine-cosine panning law). For delay correction, the control vectors become a single delay operation. This applies even when different delay estimation methods are used, or when there are multiple active sources. If multitrack convolutional reverb is applied, then c represents direct application of a finite room impulse response. And automatic equalization employs impulse responses for the control vectors based on transfer functions representing each equalization curve applied to each channel. And though dynamic range compression is a nonlinear effect due to its level dependence, the application of feedforward compression is still as a simple gain function. So multitrack dynamic range compression would be based on a time-varying gain for each control vector.

Real-Time, Multitrack Intelligent Audio Signal Processing

The standard approach adopted by the research community for real-time audio signal processing is to perform a direct translation of a computationally efficient off-line routine into one that operates on a window-by-window basis. However, effective use in live sound or interactive audio requires not only that the methods be real-time, but also that there is no perceptible latency. The minimal latency requirement is necessary because there should be no perceptible delay between when a sound is produced and when the modified sound is heard by the listener. Thus, many common real-time technologies, such as look-ahead and the use of long windows, are not possible. The windowed approach produces an inherent delay (the length of a window) that renders such techniques impractical for many applications. Nor can one assume time invariance; sources move and content changes during performance. To surmount these barriers, perceptually relevant features must be found which can be quickly extracted in the time domain, analysis must rapidly adapt to varying conditions and constraints, and effects must be produced in advance of a change in signal content.

In this section, we look at some of the main enabling technologies that are used.

Reference Signals and Adaptive Thresholds

An important consideration to be taken into account during analysis of an audio signal is the presence of noise. The existence of interference, cross talk and ambient noise will influence the ability to derive information about the source. For many tasks, the signal analysis should only be based on signal content when the source is active, and the presence of significant noise can make this difficult to identify.

One of the most common methods used for ensuring that an intelligent tool can operate with widely varying input data is adaptive gating, where a gating threshold adapts according to the existing noise. A reference microphone placed far from the source signal may be used to capture an estimation of ambient noise. This microphone signal can then be used to derive the adaptive threshold. Although automatic gating is typically applied to gate an audio signal, it can also be used to gate whether the extracted features will be processed.

The most straightforward way to implement this is to apply a gate that ensures that the control vector is only updated when the signal level of the mth channel is larger than the level of the reference, as given in the following equation;

$$c_m[n+1] = \begin{cases} c_m[n] & x^2_{m,RMS}[n] \leq r^2_{RMS}[n] \\ \alpha c_m'[n+1] + (1-\alpha)c_m[n] & otherwise \end{cases} \quad (2)$$

Where c' represents an instantaneous estimation of the control vector. Thus, the current control vector is a weighted sum of the previous control vector and some function of the extracted features. Initially, computation of RMS level of a signal x is given by

$$x^2_{rms}[n] = \frac{1}{M} \sum_0^{M-1} x^2[n-m] \quad (3)$$

And later values may either be given by a sliding window, which reduces to

$$x^2_{RMS}(n+1) = x^2(n+1)/M + x^2_{RMS}(n) - x^2(n+1-M)/M, \quad (4)$$

or a low-pass one pole filter (also known as an exponential moving average filter),

$$x^2_{RMS}(n+1) = \beta x^2(n+1) + (1-\beta)x^2_{RMS}(n). \quad (5)$$

α and β and represent time constants of IIR filters and allow for the control vector and RMS estimation, respectively, to smoothly change with varying conditions. Eq. (4) represents a form of dynamic real-time extraction of a feature (in this case, RMS), and Eq. (5) represents an accumulative form.

Incorporating Best Practices Into Constrained Control Rules

In order to develop intelligent software tools, it is essential to formalize and analyze audio production methods and techniques. This will establish required functionality of such tools. Furthermore, analysis of the mixing and mastering process will identify techniques that facilitate the mixing of multitracks, and repetitive tasks which can be automated. By establishing methodologies of audio production used by professional sound engineers, features and constraints can be specified that will enable automation.

Many of the best practices in sound engineering are well known and have been described in the literature (Pestana et al., 2014b). In live sound, for instance, the maximum acoustic gain of the lead vocalist, if present, tends to be the reference to which the rest of the channels are mixed, and this maximum acoustic gain is constrained by the level at which acoustic feedback occurs. Furthermore, resonances and background hum should be removed from individual sources before mixing, all active sources should be heard, delays should be set so as to prevent comb filtering, dynamic range compression should reduce drastic changes in loudness of one source as compared to the rest of the mix, panning should be balanced, spectral and psychoacoustic masking of sources must be minimized, and so on.

Similarly, many aspects of sound spatialization obey standard rules. For instance, a stereo mix should be balanced and hard panning avoided. When spatial audio is rendered with height, low-frequency sound sources are typically placed near the ground, and high-frequency sources are placed above, in accordance with human auditory preference. Sources with similar frequency content should be placed far apart, in order to prevent spatial masking and improve the intelligibility of content. Interestingly, Wakefield et al. (2015) showed that this avoidance of spatial masking may be a far more effective way to address general masking issues in a mix than alternative approaches using equalizers, compressors and level balancing.

These best practices and common approaches translate directly into constraints that are built into intelligent software tools. For example, De Man et al. (2013a, 2013b) described autonomous systems that were built entirely on best practices found in the literature. Also, many parameters on digital audio effects can be set based on an understanding of best practices and analysis of signal content, e.g., attack and release on dynamics processors are kept short for percussive sounds.

Psychoacoustic Studies

Important questions arise concerning the psychoacoustics of mixing multitrack content. For instance, little has been formally established concerning user preference for relative amounts of dynamic range compression used on each track. Admittedly, such choices are often artistic decisions, but there are many technical tasks in the production process for which listening tests have not yet been performed to even establish whether a listener preference exists.

Listening tests must be performed to ascertain the extent to which listeners can detect undesired artifacts that commonly occur in the audio production process. Important work in this area has addressed issues such as level balance preference (King et al., 2010, 2012), reverberation level preference (Leonard et al., 2012, 2013), 'punch' (Fenton et al., 2015), perceived loudness and dynamic range compression (Wilson et al., 2016), as well as the design and interpretation of such listening tests.

Before they are ready for practical use, intelligent software tools need to be evaluated by both amateurs and professional sound engineers to assess their effectiveness and compare different approaches. In contrast to separation of sources in multitrack content, there has been little published work on subjective evaluation of the intelligent tools for mixing multitrack audio. Where possible, prototypes should also be tested with engineers from the live sound and post-production communities in order to assess the user experience and compare performance and parameter settings with manual operation. This research would both identify preferred sound engineering approaches and allow automatic mixing criteria derived from best practices to be replaced with more rigorous criteria based on psychoacoustic studies.

Recent developments

Table 15.1 provides an overview of intelligent mixing systems since the early ones described in Reiss (2011). These technologies are classified in terms of their overall goal, whether they are multitrack or single track, whether or not they are intended for real-time use and how their rules are found.

Many of the tools deal with masking in some form. Lopez et al. (2010), Aichinger et al. (2011) and Ma et al. (2014) all propose measures of masking in multitrack mixes, but do not contain intelligent approaches to masking reduction.

Faders

The most common form of multitrack automatic mixing system is based around simple level adjustments on each track. In almost all cases, it begins with the assumption that each track is meant to be heard at roughly equal loudness levels.

Mansbridge et al. (2012b) provided a real-time system, using ITU 1770 as the loudness model. The off-line system described in Ward et al. (2012) attempted to control faders with auditory models of loudness and partial loudness. In theory, this approach should be more aligned with perception and take into account masking, at the expense of computational efficiency. But Wichern et al. (2015) showed that the use of an auditory model offered little improvement over simple single-band, energy-based approaches. Interestingly, the evaluation in Mansbridge et al. (2012b) showed that autonomous faders could compete with manual approaches by professionals, and test subjects gave the autonomous system highly consistent ratings, regardless of the song (and its genre and instrumentation) used for testing. This suggests that the equal loudness rule is broadly applicable, whereas preference for decisions in manual mixes differs widely dependent on content.

Equalization

The rules and best practices for equalization typically fall into two categories: artifact correction, such as hum removal (Brandt et al., 2014) and the equalization of salient frequencies (Bitzer et al., 2008), or creative equalization (which may still follow rules and best practices), where equalizers are applied in order to achieve a certain overall spectrum (Pestana et al., 2013; Deruty et al., 2014).

Table 15.1 Classification of intelligent audio production tools since those described in Reiss (2011)

Single or multitrack	Audio effect	Reference	Real-time?	Rules
Single track	Equalization	Ma et al., 2013	Yes	Mix analysis
		Sabin et al., 2008, 2009a, 2009b; Pardo et al., 2012a, 2012b; Cartwright et al., 2013	No	Machine learning
	Compression	Giannoulis et al., 2013; Mason et al., 2015	Yes	Psychoacoustics; best practices
	Reverberation	Rafii et al., 2009; Chourdakis et al., 2016a	No	Machine learning
	Distortion	De Man et al., 2014b		Best practices
Multitrack	Faders and gains	Mansbridge et al., 2012b	Yes	Best practices
		Scott et al., 2011; Ward et al., 2012; Wichern et al., 2015	No	Best practices
	Equalization	Hafezi et al., 2015	Yes	Best practices
	Compression	Maddams et al., 2012; Ma et al., 2015	Yes	Psychoacoustics; best practices
	Stereo panning	Mansbridge et al., 2012a; Pestana et al., 2014a	Yes	Best practices
	Delay and polarity	Clifford et al., 2010, 2011a, 2013; Jillings et al., 2013	Yes	Acoustics
	Interference reduction	Clifford et al., 2011b; Kokkinis et al., 2011	No	Acoustics
	Exploration	Cartwright et al., 2014	Yes	Machine learning
	Knowledge engineered mix	De Man et al., 2013a, 2013b	No	Best practices

Ma et al. (2013) described an intelligent equalization tool that, in real time, equalized an incoming audio stream towards a target frequency spectrum. The target spectrum was derived from analysis of fifty years of commercially successful recordings (Pestana et al., 2013). Since the input signal to be equalized is continually changing, the desired magnitude response of the target filter is also changing (though the target output spectrum remains the same). Thus, smoothing was applied from frame to frame on the desired magnitude response and on the applied filter. Targeting was achieved using the Yule-Walker method, which can be used to design an IIR filter with a desired magnitude response.

Hafezi et al. (2015) created a multitrack intelligent equalizer that used a measure of masking and rules based on best practices from the literature to apply, in real time, different multiband equalization curves to each track. Results of objective and subjective evaluation were mixed and showed lots of room for improvement, but they indicated that masking was reduced and the resultant mixes were preferred over amateur, manual mixes.

Stereo Positioning

The premise of Mansbridge et al. (2012a) is that one of the primary goals of stereo panning is to 'fill out' the stereo field and reduce masking. It set target criteria of source balancing (equal numbering and symmetric positioning of sources on either side of the stereo field), spatial balancing (uniform distribution of levels) and spectral balancing (uniform distribution of content within each frequency band). It further assumes that the higher the frequency content of a source, the more it will be panned, and that no hard panning will be applied. Finally, it used a multitude of techniques to position the sources; amplitude panning, timing differences and double tracking.

Pestana et al. (2014a) took a different approach, where different frequency bands of each multitrack are assigned different spatial positions in the mix. This approach is unique among the intelligent multitrack mixing tools since it does not emulate, even approximately, what might be performed by a practitioner. That is, practitioners aim for a single position (albeit sometimes diffuse) of each source. However, it captures the spirit of many practical approaches since it greatly reduces masking and makes effective use of the entire stereo field. In fact, Matz et al. (2015) showed that dynamic spectral panning had a larger effect in the overall improvement provided by automatic mixing than any of the other tools they considered (intelligent distortion, autonomous faders and multitrack EQ).

Dynamic Range Compression

Automating dynamic range compression is much more challenging than other effects for several reasons. It is a nonlinear effect with feedback, there are complicated relationships between its parameters and its use is less understood than other effects. Nevertheless, Giannoulis et al. (2013) automated most of the parameters of a compressor such that a single parameter determines the overall amount of compression and all other parameters are optimized to the signal. This was taken one step further by Mason et al. (2015), where the amount of dynamic range compression applied is determined based on a measurement of the background noise level in the environment.

A first attempt at multitrack dynamic range compression was provided by Maddams et al. (2012). Results of evaluation were mixed, and it was difficult to identify a preference between an automatic mix, a manual mix and no compression applied at all. Furthermore, it wasn't possible to tell whether this was due to a genuine lack of preference or due to limitations in the experimental design (e.g., poor stimuli, untrained test subjects).

A more rigorous approach was taken in Ma et al. (2015). The challenge was to formalize and quantify the relevant best practices described in Pestana et al. (2014b). First, a method of adjustment test was performed to establish preferred parameter settings for a wide variety of content. Then least squares regression was used to identify the best combination of candidate features that map to parameter settings. Thus, a rule such as 'more compression is applied to percussive tracks' translates to 'the ratio setting of the compressor is a particular function of a certain measure of percussivity in the input audio track'. Perceptual evaluation then showed a clear preference for automatic dynamic range compression over amateur application and over no compression, and sometimes performed close to professionals.

Studies have also investigated the dynamic range (or loudness range) of commercial content (Deruty et al., 2014; Kirchberger et al., 2016). Though the relationship between this range and the settings of dynamic range compressors is a complicated one, this direction of research may lead the way towards automatic dynamic range compression based on matching the dynamics of popular recordings, similar to the approach taken in Ma et al. (2013) for equalization.

Delay, Polarity and Interference

Delay and interference reduction are actually well-established signal processing techniques, more generally known as time alignment and source separation, but in Clifford et al. (2010, 2011a, 2011b, 2013) and Jillings et al. (2013) they are used and customized for mixing applications. That is, they deal with optimizing parameter settings for real-world scenarios, such as microphone placement around a drum kit, moving sources on stage and interference reduction under the constraint that no additional artifacts may be introduced.

Reverb

Of all the standard audio effects found on a mixing console or as built-in algorithms in a digital audio workstation, there has perhaps been the least effort on intelligent systems design for reverberation. Chourdakis et al. (2016a, 2016b) proposed an adaptive digital audio effect for artificial reverberation that allows it to learn from the user in a supervised way. They first perform feature selection and dimensionality reduction on features extracted from a training data set. Then, a user provides examples of reverberation parameters for the training data. Finally, a set of classifiers is trained, and they are compared using 10-fold cross validation to compare classification success ratios and mean squared errors. Tracks from the Open Multitrack Testbed (De Man et al., 2014a) were used in order to train and test the models.

Adaptive and Intuitive Mixing Interfaces

In this section, we provide an overview of the state of the art concerning interfaces for intelligent or adaptive mixing, with an emphasis on perceptual adaptive and intuitive controls. Various approaches for learning a listener's preferences for an equalization curve with a small number of frequency bands have been applied to research in the setting of hearing aids (Neuman et al., 1987; Durant et al., 2004) and cochlear implants (Wakefield et al., 2005), and the modified simplex procedure (Kuk et al., 1992; Stelmachowicz et al., 1994) is now an established approach for selecting hearing aid frequency responses. However, many recent innovations have emerged in the field of music production.

Dewey et al. (2013) and Mycroft et al. (2013) looked at the effect of the complexity of the interface for an equalizer, and suggested that simplified interfaces may encourage the user to focus on the aural properties of the signal, rather than the interpretation of visual information. Loviscach (2008) presented an interface for a five-band parametric equalizer, where the user simply freehand draws the desired transfer function and an evolutionary optimization strategy (chosen for real-time interaction) finds the closest match. Informal testing suggested that this interface reduced the set-up time for a parametric equalizer compared to more traditional interfaces. Building on this, Heise et al. (2010) proposed a procedure to achieve equalization and other effects using a black-box genetic optimization strategy. Users are confronted with a series of comparisons of two differently processed sound examples. Parameter settings are optimized by learning from the users' choices. Though these interfaces are novel and easy to use by the nonexpert, they make no use of semantics or descriptors.

Considerable research has aimed at the development of technologies that let musicians or sound engineers perform equalization using perceptually relevant or intuitive terms, e.g., brightness, warmth, presence. Reed (2000) presented an assistive sound equalization expert system. Inductive learning based on nearest neighbor pattern recognition was used to acquire expert skills. These are then applied to adjust the timbral qualities of sound in a context-dependent fashion. They emphasized that the system must be context dependent; that is, the equalization depends on the input signal system and hence operates as an adaptive audio effect. In Mecklenburg et al. (2006), a self-organizing map was trained to represent common equalizer settings in a two-dimensional space organized by similarity. The space was hand-labeled with descriptors that the researchers considered intuitive. However, informal subjective evaluation suggested that users would like to choose their own descriptors.

The work of Bryan Pardo and his collaborators has focused on new, intelligent and adaptive interfaces for equalization tasks. They address the challenge that complex interfaces for equalizers can prevent novices from achieving their desired modifications. Sabin et al. (2008, 2009b, 2011) described and evaluated an algorithm to rapidly learn a listener's desired equalization curve. Listeners were asked to indicate how well an equalized sound could be described by a perceptual term. After rating, weightings for each frequency band were found by correlating the gain at each frequency band with listener responses, thus providing a mapping from the descriptors to audio processing parameters. Listeners reported that the resultant sounds captured their intended meanings of descriptors, and machine ratings generated by computing the similarity of a given curve to the weighting function were highly correlated to listener responses. This allows automated construction of a simple and intuitive audio equalizer interface. In Pardo et al. (2012a), active and transfer learning techniques were applied to exploit knowledge from prior concepts taught to the system from prior users, greatly enhancing the performance of the equalization learning algorithm.

The early work on intelligent equalization based on intuitive descriptors was hampered by a limited set of descriptors with a limited set of training data to map those descriptors to equalizer settings. Cartwright et al. (2013) addressed this with SocialEQ, a web-based crowd-sourcing application aimed at learning the vocabulary of audio equalization descriptors. To date, 633 participants have participated in a total of 1,102 training sessions (one session per learned word), of which 731 sessions were deemed reliable in the sense that users were self-consistent in their answers (Pardo, 2015). This resulted in 324 distinct terms, and data on these terms is made available for download.

Building on the mappings from descriptors to equalization curves, Sabin et al. (2009a) described a simple equalizer where the entire set of curves were represented in a two-dimensional space (similar to Mecklenburg et al., 2006), thus assigning spatial locations to each descriptor. Equalization is performed by the user dragging a single dot around the interface, which simultaneously manipulates forty bands of a graphic equalizer. This approach was extended to multitrack equalization in Cartwright et al. (2014), which provided an interface that, by varying simple graphic equalizers applied to each track in a multitrack, allowed the user to intuitively explore a diverse set of mixes.

The concepts of perceptual control, learned from crowdsourcing, intuitive interface design and mapping of a high-dimensional parameter space to a lower dimensional representation were all employed in Stasis et al. (2015). This approach scaled equalizer parameters to spectral features of the input signal, then mapped the equalizer's thirteen controls to a 2D space. The system was trained with a large set of parameter space data representing warmth and brightness, measured across a range of musical instrument samples, allowing users to perform equalization using a perceptually and semantically relevant, simple interface. A similar approach, also incorporating gestural control, was applied to dynamic range compression in Wilson et al. (2015).

Current and Future Research Directions

Open Multitrack Testbed

The availability multitrack audio is of vital importance to research in this field, but existence of such tracks alone is not sufficient. The content should be highly diverse in terms of genre, instrumentation and quality, so that sufficient data is available for most applications. Where training on large datasets is needed, such as with machine learning applications, a large number of audio samples is especially critical.

Data that can be shared without limits, because of a Creative Commons or similar license, facilitates collaboration, reproducibility and demonstration of research and even allows it to be used in commercial settings, making the testbed appealing to a larger audience.

Moreover, reliable metadata can serve as a ground truth that is necessary for applications such as instrument identification, where the algorithm's output needs to be compared to the 'actual' instrument. Providing this data makes the testbed an attractive resource for training or testing such algorithms, as it obviates the need for manual annotation of the audio, which can be particularly tedious if the number of files becomes large. Similarly, for the testbed to be highly usable, it is mandatory that the desired type of data can be easily retrieved by filtering or searches pertaining to this metadata.

Existing online resources of multitrack audio content have a relatively low number of songs, show little variation in content, contain content of which the use is restricted due to copyright, provide little to no metadata, rarely have mixed versions including the parameter settings, and/or do not come with facilities to search the content for specific criteria. However, two initiatives (Bittner et al., 2014; De Man et al. 2014a) have tried to address this problem. MedleyDB is an annotated, royalty-free dataset of multitrack recordings, initially developed to support research on melody extraction, but generally applicable to a wide range of multitrack research problems. The Open Multitrack Testbed (which also links to the MedleyDB content) was designed for broad and diverse use by researchers, educators and enthusiasts. Such initiatives are a strong indicator that research in this field will continue to grow.

Mix Evaluation

One of the chief distinguishing characteristics between the early work on intelligent mixing systems and those described herein is that very few of the early systems had any form of subjective evaluation, whereas now this is standard practice. A popular form of evaluation for such systems has become multistimulus rating, similar to that used in MUSHRA.

Mansbridge et al. (2012b) compared their proposed autonomous faders technique with a manual mix, an earlier implementation, a simple sum of sources and a semi-autonomous version. Mansbridge et al. (2012a) compared an autonomous panning technique with a monaural mix and panning configurations set manually by three different engineers. Both showed that fully autonomous mixing systems can compete with manual mixes.

Similar listening tests for the multitrack dynamic range compression system described in Maddams et al. (2012) were inconclusive, however, since the range of responses was too large for statistically significant differences between means and since no dynamic range compression was often preferred, even over the settings made by a professional sound engineer. However, a more rigorous listening test was performed in Ma et al. (2015), where it was shown that compression applied by an amateur was on a par with no compression at all, and an advanced implementation of intelligent multitrack dynamic range compression was on a par with the settings chosen by a professional.

In Wichern et al. (2015), the authors first examined human mixes from a multitrack dataset to determine instrument-dependent target loudness templates. Three automatic level balancing approaches were then compared to human mixes. Results of a listening test showed that subjects preferred the automatic mixes created from the simple energy-based model, indicating that the complex psychoacoustic model may not be necessary in an automated level setting application.

One of the most exciting and interesting developments has been perceptual evaluation of complete automatic mixing systems. In Matz et al. (2015), various implementations of an automatic mixing system are compared, where different combinations of autonomous multitrack audio effects were applied, so that one could see the relative importance of each individual tool. Although no comparison was made with manual mixes, it is clear that the application of these tools provides an improvement over the original recording, and that the combination of all tools results in a dramatic improvement.

Conclusions

In this chapter, we described how mixing of multitrack audio could be made simpler and more efficient through the use of intelligent software tools. Ideally, intelligent systems for mixing multitrack audio should be able to pass a Turing test. That is, they should be able to produce music indistinguishable from that which could be handcrafted by a professional human engineer. This would require the systems to be able to make artistic as well as technical decisions, and achieve this with almost arbitrary audio content. However, considerable progress is still needed in order for systems to even be able to 'understand' the musician's intent. But, in the near term, such software tools may result in two types of systems. The first would be a set of tools for the sound engineer that automate repetitive tasks. This would allow professional audio engineers to focus on the creative aspects of their craft, and help inexperienced users create high-quality mixes. The other type of system would be a 'black box' for the musician that allows decent live sound without an engineer. This would be most beneficial for the small band or small venue that doesn't have or can't afford a sound engineer, or for recording practice sessions where a sound engineer is not typically available.

There are major concerns with such an approach. Much of what a sound engineer does is creative and based on artistic decisions. It is doubtful that such decisions could be effectively reproduced by a machine. But if the automation is successful, then machines may replace sound engineers. However, it is important to note that these tools are not intended to remove the creativity from audio production. Nor do they require software to reproduce artistic decisions, although this would be an interesting direction for future research. Rather, the tools rely on the fact that many of the challenges are technical engineering tasks, some of which are perceived as creative decisions because there is a wide range of approaches without a clear understanding of listener preferences. By automating those engineering aspects of record production, it will allow the musicians to concentrate on the music and allow the audio engineers to concentrate on the more interesting, creative challenges.

Bibliography

Aichinger, P., et al. (2011). 'Describing the Transparency of Mixdowns: The Masked-to-Unmasked Ratio.' In *130th Audio Eng. Soc Convention*.

Bittner, R., et al. (2014). 'MedleyDB: A Multitrack Dataset for Annotation-Intensive MIR Research.' In *15th International Society for Music Information Retrieval Conference (ISMIR)*.

Bitzer, J., et al. (2008). 'Evaluating Perception of Salient Frequencies: Do Mixing Engineers Hear the Same Thing?' In *124th Audio Engineering Society Convention*, Amsterdam.

Brandt, M., et al. (2014). Automatic Detection of Hum in Audio Signals. *Journal of Audio Engineering Society* 62 (9): 584–595.

Cartwright, M., et al. (2013). 'Social-EQ: Crowdsourcing an Equalization Descriptor Map.' In *14th International Society for Music Information Retrieval*, Curitiba, Brazil.

Cartwright, M., et al. (2014). 'Mixploration: Rethinking the Audio Mixer Interface.' In *19th International Conference on Intelligent User Interfaces Proceedings (IUI14)*, Haifa, Israel.

Chourdakis, E.T., et al. (2016a). '*Automatic Control of a Digital Reverberation Effect using Hybrid Models.*' In *AES 60th International Conference Leuven*, Belgium.

Chourdakis, E.T., et al. (2016b). A Machine Learning Approach to Design and Evaluation of Intelligent Artificial Reverberation. *Journal of Audio Engineering Society (to appear)*.

Clifford, A., et al. (2010). 'Calculating Time Delays of Multiple Active Sources in Live Sound.' In *129th AES Convention*, San Francisco.

Clifford, A., et al. (2011a). Reducing Comb Filtering on Different Musical Instruments Using Time Delay Estimation. *Journal on the Art of Record Production* 5: 1–13.

Clifford, A., et al. (2011b). 'Microphone Interference Reduction in Live Sound.' In *14th Int. Conference on Digital Audio Effects (DAFx-11)*, Paris.

Clifford, A., et al. (2013). Using Delay Estimation to Reduce Comb Filtering of Arbitrary Musical Sources. *Journal of the Audio Engineering Society* 61 (11): 917–927.

De Man, B., et al. (2013a). 'A Knowledge-Engineered Autonomous Mixing System.' In *135th AES Convention*, New York.

De Man, B., et al. (2013b). A Semantic Approach to Autonomous Mixing. *Journal on the Art of Record Production (JARP)* 8: 1–23.

De Man, B., et al. (2014a). 'The Open Multitrack Testbed.' In *137th AES Convention*, Los Angeles.

De Man, B., et al. (2014b). 'An Intelligent Multiband Distortion Effect.' In *AES 53rd International Conference on Semantic Audio*, London, UK.

Deruty, E., et al. (2014). Human–Made Rock Mixes Feature Tight Relations between Spectrum and Loudness. *Journal of Audio Engineering Society* 62 (10): 643–653.

Dewey, C., et al. (2013). 'Novel Designs for the Parametric Peaking EQ User Interface.' In *134th AES Convention*, Rome.

Durant, E.A., et al. (2004). Efficient Perceptual Tuning of Hearing Aids with Genetic Algorithms. *IEEE Transactions on Speech and Audio Processing* 12 (2): 144–155.

Fenton, S., et al. (2015). 'Towards a Perceptual Model of "Punch" in Musical Signals.' In *139th AES Convention*, New York.

Giannoulis, D., et al. (2013). Parameter Automation in a Dynamic Range Compressor. *Journal of Audio Engineering Society* 61 (10): 716–726.

Hafezi, S., et al. (2015). Autonomous Multitrack Equalisation Based on Masking Reduction. *Journal of the Audio Engineering Society* 63 (5): 312–323.

Heise, S., et al. (2010). 'A Computer-Aided Audio Effect Setup Procedure for Untrained Users.' In *128th Audio Engineering Society Convention*, London.

Jillings, N., et al. (2013). 'Performance Optimization of GCC-PHAT for Delay and Polarity Correction under Real World Conditions.' In *134th AES Convention*, Rome.

King, R., et al. (2010). 'Variance in Level Preference of Balance Engineers: A Study of Mixing Preference and Variance Over Time.' In *129th Audio Engineering Society Convention*, San Francisco.

King, R., et al. (2012). 'Consistency of Balance Preferences in Three Musical Genres.' In *133rd AES Convention*.

Kirchberger, M., et al. (2016). Dynamic Range Across Music Genres and the Perception of Dynamic Compression in Hearing-Impaired Listeners. *Trends in Hearing* 20: 1–16.

Kokkinis, E., et al. (2011). 'Detection of 'Solo Intervals' in Multiple Microphone Multiple Source Audio Applications.' In *130th AES Convention*.

Kuk, F.K., et al. (1992). The Reliability of a Modified Simplex Procedure in Hearing Aid Frequency Response Selection. *Journal of Speech and Hearing Research* 35 (2): 418–429.

Leonard, B., et al. (2012). 'The Effect of Acoustical Environment on Reverberation Level Preference.' In *133rd AES Convention*, San Francisco.

Leonard, B., et al. (2013). 'The Effect of Playback System on Reverberation Level Preference.' In *134th Audio Engineering Society Convention*, Rome.

Lopez, S.V., et al. (2010). Quantifying Masking in Multi-Track Recordings. *Sound and Music Computing* 1–8.

Loviscach, J. (2008). 'Graphical Control of a Parametric Equalizer.' In *Audio Engineering Society Convention* 124.

Ma, Z., et al. (2013). 'Implementation of an Intelligent Equalization Tool Using Yule-Walker for Music Mixing and Mastering.' In *134th AES Convention*, Rome.

Ma, Z., et al. (2014). 'Partial Loudness in Multitrack Mixing.' In *AES 53rd International Conference on Semantic Audio*, London.

Ma, Z., et al. (2015). Intelligent Multitrack Dynamic Range Compression. *Journal of Audio Engineering Society* 63 (6): 412–426.

Maddams, J., et al. (2012). 'An Autonomous Method for Multi-Track Dynamic Range Compression.' In *Digital Audio Effects (DAFx)*, York, 1–8.

Mansbridge, S., et al. (2012a). 'An Autonomous System for Multi-track Stereo Pan Positioning.' In *133rd AES Convention*, San Francisco.

Mansbridge, S., et al. (2012b). 'Implementation and Evaluation of Autonomous Multi-Track Fader Control.' In *132nd Audio Engineering Society Convention*, Budapest, 1–8.

Mason, A., et al. (2015). 'Adaptive Audio Reproduction Using Personalised Compression.' In *AES 57th International Conference*, Hollywood, CA.

Matz, D., et al. (2015). 'New Sonorities for Early Jazz Recordings Using Sound Source Separation and Automatic Mixing Tools.' In *ISMIR*, Malage.

Mecklenburg, S., et al. (2006). 'subjEQt: Controlling an Equalizer through Subjective Terms.' In *Computer-Human Interaction EA '06 Extended Abstracts on Human Factors in Computing*, Montreal.

Moorer, J.A. (2000). Audio in the New Millennium. *Journal of the Audio Engineering Society* 48 (5): 490–498.

Mycroft, J., et al. (2013). 'The Influence of Graphical User Interface Design on Critical Listening Skills.' In *Sound and Music Computing (SMC)*, Stockholm.

Neuman, A.C., et al. (1987). An Evaluation of Three Adaptive Hearing Aid Selection Strategies. *The Journal of Acoustical Society of America* 82 (6): 1967–1976.

Pardo, B. (2015). *Data on SocialEQ*. J.D. Reiss.

Pardo, B., et al. (2012a). 'Building a Personalized Audio Equalizer Interface with Transfer Learning and Active Learning.' In *2nd International ACM Workshop on Music Information Retrieval with User-Centered and Multimodal Strategies (MIRUM)*, Nara, Japan.

Pardo, B., et al. (2012b). 'Towards Speeding Audio EQ Interface Building with Transfer Learning.' In *New Interfaces for Musical Expression*, Ann Arbor, MI.

Pestana, P. (2013). *Automatic Mixing Systems Using Adaptive Audio Effects*. PhD, Universidade Catolica Portuguesa.

Pestana, P.D., et al. (2013). 'Spectral Characteristics of Popular Commercial Recordings 1950–2010.' In *135th AES Convention*, New York.

Pestana, P.D., et al. (2014a). 'A Cross-Adaptive Dynamic Spectral Panning Technique.' In *17th Int. Conference on Digital Audio Effects (DAFx-14)*, Erlangen, Germany.

Pestana, P.D., et al. (2014b). 'Intelligent Audio Production Strategies Informed by Best Practices.' In *AES 53rd International Conference on Semantic Audio*, London, UK.

Rafii, Z., et al. (2009). 'Learning to Control a Reverberator using Subjective Perceptual Descriptors.' In *10th Int. Conf. Music Inf. Retrieval (ISMIR)*, Kobe, Japan.

Reed, D. (2000). 'A Perceptual Assistant to Do Sound Equalization.' In *5th International Conference on Intelligent User Interfaces*, New Orleans.

Reiss, J.D. (2011). 'Intelligent Systems for Mixing Multichannel Audio.' In *17th International Conference on Digital Signal Processing (DSP2011)*, Corfu, Greece, 1–6.

Sabin, A., et al. (2008). 'Rapid Learning of Subjective Preference in Equalization.' In *125th Audio Engineering Society Convention*, San Francisco.

Sabin, A., et al. (2009a). '2DEQ: An Intuitive Audio Equalizer.' In *ACM Creativity and Cognition*, Berkeley, CA.

Sabin, A., et al. (2009b). 'A Method for Rapid Personalization of Audio Equalization Parameters.' In *ACM Multimedia*, Beijing, China.

Sabin, A.T., et al. (2011). Weighted-Function-Based Rapid Mapping of Descriptors to Audio Processing Parameters. *Journal of Audio Engineering Society* 59 (6): 419–430.

Scott, J., et al. (2011). 'Automatic Multi-Track Mixing Using Linear Dynamical Systems.' In *8th Sound and Music Computing Conference (SMC)*, Padova.

Stasis, S., et al. (2015). 'A Model for Adaptive Reduced-Dimensionality Equalisation (Best Paper, 2nd Prize).' In *18th Int. Conference on Digital Audio Effects (DAFx-15)*, Trondheim, Norway.

Stelmachowicz, P.G., et al. (1994). Preferred Hearing-Aid Frequency Responses in Simulated Listening Environments. *Journal of Speech and Hearing Research* 37 (3): 712–719.

Verfaille, V., et al. (2006). Adaptive Digital Audio Effects (A-DAFx): A New Class of Sound Transformations. *IEEE Transactions on Audio, Speech and Language Processing* 14 (5): 1817–1831.

Wakefield, G.H., et al. (2005). Genetic Algorithms for Adaptive Psychophysical Procedures: Recipient-Directed Design of Speech-Processor MAPs. *Ear Hear* 26 (4): 57S–72S.

Wakefield, J., et al. (2015). 'An Investigation into the Efficacy of Methods Commonly Employed by Mix Engineers to Reduce Frequency Masking in the Mixing of Multitrack Musical Recordings'. In *138th AES Convention*.

Ward, D., et al. (2012). 'Multi-Track Mixing Using a Model of Loudness and Partial Loudness.' In *Audio Engineering Society (AES) 133rd Convention*, San Francisco.

White, P. (2008). Automation for the People. *Sound on Sound* 23 (12).

Wichern, G., et al. (2015). 'Comparison of Loudness Features for Automatic Level Adjustment in Mixing.' In 139th AES Convention, New York.

Wilson, T., et al. (2015). 'A Semantically Motivated Gestural Interface for the Control of a Dynamic Range Compressor.' In *138th AES Convention*.

Wilson, A., et al. (2016). Perception of Audio Quality in Productions of Popular Music. *Journal of Audio Engineering Society* 64 (1/2): 23–34.

16

How Can Academic Practice Inform Mix-Craft?

Gary Bromham

The democratization of music technology has led to a stark change in the way we approach music production and specifically the art of mixing. There is now a far greater need for an understanding of the processes and technical knowledge that mix engineers, both amateur and professional, need to execute their work practices. Where interns once learned their craft from assisting professionals in the studio, it is now more common-place for them to learn from a book, an instruction video or within a col-lege or university environment. There are, as Paul Théberge suggests, a "lack of apprenticeship placements" (Théberge, 2012). Unlike the 1960s, '70s and '80s, these days anyone can make a record, quite often work-ing remotely and increasingly using headphones rather than speakers, and in many cases, using spaces that weren't designed to be studios. We use sound libraries, plugin presets and templates and in doing so create the illusion that an understanding of how we arrive at the finished sound isn't necessary. Technology gives us infinite options and possibilities, but often stifles creativity and inhibits decision-making. There is a feeling that we are using the technology without understanding it.

I would propose that it is therefore more important than ever that the student and practitioner alike learn how to approach mixing with some grounded theory and scientific knowledge. The DAW and indeed the proj-ect studio have become ubiquitous, yet they are metaphors for the tradi-tional recording studio. Academic research practices such as ethnographic studies, scientific evaluation and auditory analysis (acoustic and psycho-acoustic) can all greatly assist the understanding of the mix engineer's craft. How can we learn from aesthetical, musicological, historical and scientific studies and apply them to a contemporary studio workflow?

Introduction

There is a tendency when talking about music production, or more specifi-cally about mixing, to ask how academics and to an extent non-professional mix engineers might benefit from the wisdom of practitioners. While this is an extremely valid question and provides an essential framework for analyzing workflows, aesthetics and creativity in the mix room, it does not necessarily give precedence to the science and physics behind the mix process. For this to happen, we need to make a clear distinction between the critical or technical aspects of mixing and the more artistic and ana- lytical decisions that inform mix-craft. A scientific approach to mixing is concerned predominantly with physics, whereas an alternative approach taken by a musicologist might be more focused on aesthetics or artistic expression. In attempting to answer this question, a number of interviews were conducted with both academics and practitioners, some who bridge the gap between the two fields. The discussions that took place helped to affirm and also inform some of the deductions made during my own career as a producer and engineer.

According to Dr. Andrew Bourbon,[1] Senior Lecturer in Audio Technol- ogy at The London College of Music, there is a myth among non-academics that you cannot have technical knowledge and still be creative in mixing. Sometimes it is difficult to practice the two simultaneously, but they can co-exist. Artist Peter Gabriel practices what he refers to as his Alpha and Zen approach to producing music in the studio, in which he does not attempt to be in the creative space while performing mundane editing or time-consuming post-production tasks. Mixing music offers a similar challenge when decid- ing how much we should allow scientific practice to influence creativity in the mix workspace. His view contrasts with that of legendary engineer and producer Tony Platt,[2] who feels that we often get academic practice confused with technical know-how and intuition in the studio: "Academics need to find a reason for the way things are done. If I'm academic about mixing it won't work. I mess around and suddenly it works and I don't know why!" Producer and engineer, Dennis Weinreich[3] echoes this sentiment when talking about one of his mentors, Ken Scott:

> He followed his nose. He knew how to do stuff but didn't know why. He was determined not to allow himself to become constrained by knowledge. He saw no barriers. He was as Tom Magliozzi aptly put it "Unencumbered by the thought process!"

It should be pointed out that Weinreich subsequently made it very clear that this is the view of Ken Scott and that he in fact grew up with a scien- tific background and still believes that physics form a very important part of sound engineering and mix-craft.

Dr. George Fazekas,[4] lecturer at Queen Mary University of London, makes the pertinent point that the study of engineering and mixing is now far more of an interdisciplinary area than ever before, when old skills need to be studied in the context of new ones to inform future work practices. You cannot look at music production from just one angle; it has to be viewed from an academic as well as a practitioner's viewpoint. This has not, of course, always been the case. The men in 'white coats' at Abbey Road Studios were hired for their technical or electrical skills, not their creative skills, but the field has changed dramatically in the last ten years. Fazekas also makes the point that mixing and production were not really viewed as fields of research in academia before this point in time: "Generally researchers don't tend to have much practical experience. We need collaboration between academics and practitioners as this leads to new methodologies, better work practices".

The gradual reduction in the number of large recording studio facilities has meant that there are far fewer opportunities for the aspiring intern. This has in turn led to a plethora of recording schools, colleges and universities now offering an alternative to the conventional studio apprenticeship. Many factors have led to this situation, but the shrinking size of the music industry, the decline of record sales and most importantly the accessibility of mainstream music technology have all had a pivotal role to play in this process. For too long, the music business has aligned itself to an analog business model while operating in a digital age. Digital distribution and marketing models are still relatively new, in many ways in their infancy, but the production chain is now changing faster than ever and requires an updated methodology. The speed at which an artist can record, mix, master and distribute a new song has had a huge impact on the way we use technology to achieve results. The mix engineer must be able to instantly recall work and be able to track changes made at all stages of the music production process. From editing, tuning and altering timing to drum replacement, re-amping and even mastering, they are expected to interpret and anticipate the journey taken from concept to consumer. The luxury of spending ten years acquiring the necessary skills to mix music in a studio environment is now sadly lacking. The main route to achieving this goal is now predominantly through academia.

Apprenticeship

Many people might argue that an apprenticeship or internship in a more traditional analog based recording studio might provide a better grounding in recording techniques, production aesthetics and audio manipulation. There can however be problems with this hypothesis. Dr. Francis Rumsey makes the point that traditional recording practice requires learning from a mentor, such as an experienced recording or mix engineer, but that this process might be flawed if the apprentice cannot discern what is good or bad. Academia in a sense arbitrates this process by assuming a certain level of training and accomplishment on the part of the lecturer. He says,

> If that person is a bullshitter then you don't really stand a chance. That system can be very arbitrary. Academic practice is however more likely to be reliable. In academia from a scientific point of view at least you're more likely to find someone who knows what they are talking about. In studios you might be more likely to find out more about what constitutes creativity. It's a bit of a lottery in many ways.[5]

With the adoption of digital recording and the ubiquitous nature of hard-disk-based recording systems, the DAW has become a metaphor for the recording studio in its conventional form. There is a greater need than ever for an understanding of how established, often analog-based, practices translate in the digital domain. Without knowing the context of how and why original tools were learned by an apprentice, some of this knowledge could easily be lost on the novice engineer. Where you once had to learn and understand how a patchbay worked or how one part of a signal chain might impact another, the process has in a sense been covered up when using plugins. There doesn't seem to be as much control over how they interact with each other, and in many ways we are taught that this doesn't really matter. Some of the most creative moments in recording have come from accidents where devices have been plugged up incorrectly and technology misappropriated; a knowledge of how this interaction takes place in the analog domain can certainly help to inform how this might be used creatively in the digital domain.

Andrew Bourbon makes the salient point about how an apprenticeship in a recording studio has always created an academic environment. Assistant engineers would always form arguments of their own leading to new ideas being formed.

I don't see the demise of the intern in the traditional recording studio as necessarily being a negative thing but rather an excuse for a new breed of engineer to emerge, one with a solid grounded knowledge gained on an audio technology course and applied to a modern, quite possibly digital, workflow. It is a changing of the guard.

Work Practices—The Limitation of the Medium

Mixing begins in the first hour of recording!

Teaching someone where to place a microphone can do more to help their mixes than can any amount of post-production trickery. The reason so many guitar amplifiers have a Shure SM57 placed in front of them at the recording stage is because the natural EQ curve of the microphone sits well in the balance of a track. With correct placement, very little processing is needed. The microphone choice and placement can act like the ultimate equalizer or the perfect natural environment to place a sound within. Capturing the sound of a real space can add more depth than any artificial reverberation algorithm or impulse response used in a convolution reverb. The challenge here is that the student needs to be shown this and to understand why the phenomenon will work. There is a myth that we can 'fix it in the mix' because we have unlimited access to plugins and editing software and that this will lead to some kind of technical nirvana (Emmerson, 2000). The reality is that unlimited choices and options can often lead to confusion and cause the novice to lose sight of their original aims and objectives. The interesting thing here, however, is that the inherent limitation of a dynamic microphone, such as the now ubiquitous Shure SM57, in fact becomes its very strength.

There is a commonly held belief among professional mix engineers that balance is the foundation of mixing and that this process should begin during the recording stage. When Roger Nichols was mixing Steely Dan, he mixed everything as it was going down to tape. According to producer and engineer Dennis Weinreich,

> People have forgotten the process and function of balance; engineers got that right at the recording stage. Big differences existed between the live source and the medium which it was being played back from, a lot of engineering in the early days was overcoming the limitations of that medium. Tape definitely had shortcomings![6]

A great deal of this statement stems from necessity, and it wasn't really until the mid-1980s with the advent of 48-track sessions that the mix started to become a post-production process. From the late 1980s onwards, and with the emergence of digital recording technologies, the demographic would change again! The digital revolution in a sense changed mixing forever. No longer were there concerns about degradation of audio quality or about restrictions imposed by having a limited track count. Instead, the possibilities became endless. This changed the role of the mix engineer into an interpreter who would have to imagine what the artists' intentions were and act as someone who was expected to have vision and creativity when shaping the sound of a mix. Thus was the 'superstar' mix engineer conceived. Probably no one personified this more than Bob Clearmountain, a house engineer at the Power Station in New York, who started to be hired for his creative skills as a mix engineer. The important thing is to acknowledge his pedigree as a recording engineer in the first instance. Andrew Bourbon says, "You need to see recording as part of mixing, they are not a separate entity. Look at all the great mix engineers and they are all good recording engineers".

In a sense, the function of the studio has come full circle with the use of the laptop as mobile recording and production medium. In his chapter 'The end of the world as we know it' in *The Art of Record Production*, Paul Théberge (2012) refers to the studio as a mobile entity and talks about how this has impacted record production and mix practice. The bedroom studio is now ubiquitous, and the use of headphones for monitoring is now commonplace. The typical home studio is entirely unsuitable for mixing records, so there is a greater need than ever to grasp how acoustics will impact our environment and how to work around these inherent shortcomings.

Listening Skills

Developing technical and critical listening skills is of paramount importance to the sound engineer when listening to and evaluating sound. The ability to clearly identify a problematic area in a mix, to hear a certain thing and act upon, it is arguably one of the biggest challenges. Learning to describe sounds in technical terms will generally provide a much faster route towards getting the results you want in a studio, and by using a structured learning process you can arguably get results far quicker than gradually picking things up as you go along. Many instructors now insist on critical listening being a key part of their curriculum. The fact that many of us work in less-than-perfect recording spaces has meant that a grounded knowledge in both technical listening skills and acoustics is not a luxury anymore. It is not uncommon to see a mix engineer walking around the control room in a studio when evaluating bass response, for example. When used in combination with a basic technical understanding of room modes, room shape, reflections and diffusion, for example, this knowledge can help to inform creative and aesthetic decisions that impact the sonic quality of the final mix. Positioning and choice of speakers are very much determined by the sound of the room they are placed in. We cannot ignore the fact that most popular music is now consumed via headphones, and it is therefore remiss of the sound engineer to ignore these facts when mixing a song. Some type of formal academic training in the area of listening skills and physics is now almost certainly essential to making critical decisions when mixing sound. In Jason Corey's book *Audio Production and Critical Listening—Technical Ear Training* (2010), "Technical ear training is a type of perceptual learning focused on timbral, dynamic, and spatial attributes of sound as they relate to audio recording and production". A concept of spatial evaluation is also necessary when learning to listen to sound. An ability to assess the perceived width and depth of the sound stage is important when creating a space to mix and record in. Nowhere will this have more impact than in our choice of placement of speakers, as has already been mentioned.

It is also important to make a distinction between critical and analytical listening. In *Understanding and Crafting the Mix* (2006), William Moylan talks about technical and scientific listening skills vs. artistic or creative expression. He says that understanding sound is perceiving sound for its inherent qualities. Francis Rumsey and Tim McCormick discuss a similar phenomenon in their book *Sound and Recording* (2009) when talking about auditory perception. Though not essential to becoming an accomplished mixing engineer, a basic understanding of how psychoacoustics work can be helpful. The word 'perception' is used several times in this chapter, particularly in the context of using vintage technologies and their inherent sonic imprints. Psychoacoustics is essentially the scientific study of sound perception. More specifically, it is the measurement of psychological and physiological responses to sound. Quite often how we perceive something to sound is invariably not how it actually sounds; it is influenced by many factors outside of the auditory experience. The GUI (graphic user interface) in a DAW, an audio plugin or an iOS app, depending on their visual reference point, will influence the outcome of processing a sound. This becomes an important factor for the engineer to consider, as employing a certain degree of scientific study will tell them that they are not always hearing what they think they are hearing. It is also particularly relevant in an age when there are so many visual representations in the studio and we are often told that we should listen to sound and not look at it. This challenge is more pronounced today than it was thirty years ago when engineers relied far less on visual feedback. Most music today is composed and produced on a computer.

Sonic Trends—The Branding of Mix Engineers

Since 2002, the software company Waves Audio has released a string of collections of plugins endorsed by famous mix engineers and producers. Known as its Signature Series and including such luminaries as Eddie Kramer, Chris Lord-Alge and Jack Joseph Puig, the plugin bundles provide the user with the opportunity to use, albeit virtually, a software re-creation of the analog hardware used by the aforementioned professionals in their studios. Lists of presets created by these same famous mix engineers for kick drum, rhythm guitar and lead vocal, among others, provide an insight and a glimpse into the engineers' technical and aesthetic worlds. Slate Digital and Universal Audio have created a similar business model feeding off the same retro obsession. They draw upon the phenomena of technostalgia to sell us a tiny part of a golden age of studio technology used to make classic recordings that only a privileged few were lucky enough to have been acquainted with. Accompanying this trend is a myth that you, the amateur, or professional amateur, can mix records to sound like one of your idols. An interesting observation might concern the way such practices impact creativity. Creating an original sonic stamp or signature often comes from the misuse or misappropriation of technology. It could be argued that the use of 'famous mix engineer' presets doesn't encourage the use of digital plugin technology in an enquiring way but more likely encourages its use in an unquestioning manner.

The practice of taking some of the sonic signature from mix engineers and producers without necessarily understanding why you might use them promotes a sort of sonic tourism without a map. Without an ability to interrogate these technologies and their accompanying workflows, the use of endless lists of presets and templates to mix a song merely serves as a shortcut without providing an understanding of the underlying tools used to execute a mix. Paul Théberge has discussed the history behind the use and purpose of presets in *Any Sound You Can Imagine* (Théberge, 1997). The original concept was to provide a keyboard player with a means of taking the sounds they had created in the studio into a live context, but more and more this idea has become an excuse for manufacturers to fill a new machine with hundreds of presets to market new technologies. The role and function of presets for mixing has largely been ignored or swept under the carpet; this is for good reason because of the negative stigma attached to such practices. It could be argued that the use of presets can also limit creativity. A paint-by-numbers approach to using presets in mixing can easily produce similar generic results in people's mixes. It has been suggested that many tracks produced in current popular music sound the same today and that they have similar sonic imprints. This might largely be because the same plugins with the same presets are being used. A 'one size fits all' approach to mixing without any understanding of the tools being used will probably not yield a unique-sounding mix. Nowhere is this seen more forcibly today than in mastering, where it is not uncommon to see a preset chain used to correct a less than satisfactory mix.

Academic practice and study can help us understand some of the techniques used by well-known studio practitioners. A student should aim to reflect on the practices and aesthetics used by producers and engineers and not merely try to emulate them. Academic practice encourages us to interrogate and understand why some of these digital recreations are being used. For many young producers and engineers, or those without a conventional internship, the context of the original equipment could easily be lost on them.

Another interesting aspect is the obvious marketing benefits gained by selling products based upon the legacy of producers and engineers and the nostalgic value attached to their equipment. There is an assumption with limited validity that there might be a shortcut to you sounding just like them. The new Stephen Slate, Chris Lord-Alge mixing course found at www.audiolegends.com takes this concept to a new level. A full session in a DAW of choice is sold and subsequently downloaded with all the plugins needed to copy the mixing workflow of Chris Lord-Alge. It is in many ways an ingenious marketing strategy! It might arguably be more important to ask ourselves what the impact of copying the techniques and technologies of others might be. Can it have a negative impact on our ability to produce original work? Misappropriating technology can often produce more interesting results, and using some of our favorite mix engineers' techniques out of context might be of more interest when considering their impact on creativity: "Sometimes, not knowing the theoretical operation of a tool can result in more interesting results by thinking outside the box" (Cascone, 2004).

Emulation Not Innovation!

Why innovate when you can emulate?[7]
It is not uncommon for plugins to use references points from hardware equivalents used in the past. Equally, it is not that unusual to see manufacturers producing much cheaper recreations of classic studio equipment than the original incarnations. The real issue is the cost to innovation by an insistence on embracing retro aesthetics in design and manufacturing. The perception that current record production and mixing techniques are somehow better as a result of these retro technologies can lead to confusion. There is, as Andrew Bourbon proposes, a whole list of problems as a result what he calls "emulation culture": "The tendency is to emulate not innovate. To innovate means that you then have to educate the market".

Familiarity plays a defining role in this process. Users feel a sense of security both culturally and scientifically when they connect with technology that has traditionally been used to shape work practices of an engineer. As Brian Eno says,

The trouble begins with a design philosophy that equates 'more options' with 'greater freedom'. Designers struggle endlessly with a problem that is almost non-existent for users. How do we pack the maximum number of options into the minimum space and price? In my experience, the instruments and tools that endure (because they are loved by their users) have limited options.

(Eno, 1999)

Eno goes on to argue that this is why we resort to retro technologies, to the familiar, because we are stifled creatively by having too many possibilities. "Indeed, familiarity breeds content. When you use familiar tools, you draw on a long cultural conversation—a whole shared history of usage—as your backdrop, as the canvas to juxtapose your work" (Eno, 1999). An interesting parallel can be drawn here with the work of Andrew McPherson and Victor Zappi with their D-Box musical instrument. D-Box is an instrument that can be hacked or modified by the user and in doing so the meaning or appropriation can be changed in unconventional ways. The key to the creative aspect of the D-Box is its simplicity and limited controls.[8]

Recording technology didn't have the same significance as it does now—it wasn't held in such high esteem. There is an air of iconicity in today's home studio where virtual recreations are ubiquitous; they assume an almost untouchable and unquestioning status. It is okay for commercial software and hardware companies to sell products based on nostalgia, but there is a need to interrogate these technologies and understand the context that they are used in. It is useful for contemporary mix engineers and producers to have a notion of the implications of such technologies, as this can have far-reaching consequences on creativity in the studio. When we look at a software recreation of a classic piece of analog equipment, such as a Universal Audio 1176 compressor, we are instantly transported to a place of nostalgic significance. It takes on a whole new context when semantic terms such as 'warm', 'punchy' or 'classic' are ascribed to it. As Alan Williams (2015) says in his paper 'Technology and the Cry of the Lonely Recordist',

> Software emulations are not inherently nostalgic, though much of the marketing surrounding them capitalizes on the desire to harness the past. Since digital audio processes are distinctly different from analog electronic and acoustic ones, these products present a functionality that masks the actual technology involved.

Creativity can come from misappropriating technology, and a traditional mix engineer might use the equipment in ways it wasn't intended. This is less likely, though certainly not impossible, to happen in the digital domain, as there is a sense of pre-defined purpose and intention when using a series of plugins chained together, in contrast to patching pieces of outboard in series. There is arguably less left to chance when making a virtual connection than there is when making a physical one. The debate about 'analog versus digital' and 'in the box' versus 'outside the box' is one that will surely continue for some time, but it is only through academic study that we can start to make an informed judgment about the true implications.

The Language of Mixing

Surprisingly, one of the most underresearched areas of the mix process and one which so often leads to much misunderstanding is the use of language in the studio. A language does not really exist for describing sound quality. How do we define 'warmth' when the term could equally mean 'muddiness' to another set of ears? The use of semantics in the recording studio, where different meanings or descriptors are used to describe the same thing, is common. When doing an acoustical analysis of a mix, there will inevitably be differences in the terminology used when describing the attributes. At Queen Mary University of London, a great deal of research has been done in this area. The EPSRC[9] funded FAST[10] project is focused on ways that semantic (web) technologies can assist in informing music consumption. This process starts in the recording studio at the production stage and aims to define how the use of language can help to define better work practices. Another important technology developed in collaboration between Birmingham City University and Queen Mary University is the SAFE[11] project. A series of audio processing plugins have been designed in which the user can add their own semantic descriptors and in doing so help to describe sounds using more accessible terms. Processes such as data mining, semantic analysis and machine learning can all be applied to the descriptors, which are collated in such projects. These could help to offer a potential solution to resolving the subjectivity of understanding the language used in production and mixing practices. This in turn helps inform the formulation of best practices used in the creation of new tools, which can be used by engineers.

New Technologies to Assist Users in Mixing

Interacting with the tools we use is important in many areas of performance, music production and mixing. It can be argued that most DAWs have restrictions and limitations of some kind, and in a sense we become locked into a systematic way of working determined by the software we use (Lanier, 2011). It could also be argued that the DAW we choose dictates the workflow. Logic Pro, Cubase or Ableton Live, for example, are good tools for being creative, whereas Avid Pro Tools, Sony Soundforge or SADiE subjectively perform better as post-production tools. Regardless of their feature-sets or attributes, they all work predominantly through the medium of mouse and computer. This in itself presents challenges when evaluating new ways of controlling or manipulating some of the technologies that can assist mix engineers. Dr. Mathieu Barthet,[12] Lecturer in Digital Media at Queen Mary University London, says,

> The ability to use gestural control or vocalization devices to manipulate sounds in the DAW has interesting possibilities. Through methods used in distributed cognition we can begin to understand the correlation between a physical object and a virtual one. The interaction between humans and technologies is an area of great interest to researchers.

Perceptual experiments are another great way to help us evaluate the mood created when listening to music, and the ability to measure similarities and differences between mixes can assist in informing future technologies for engineers. This can help us to identify attributes and patterns between mixes. Research at Queen Mary University of London using technologies found at www.isophonics.net, such as Sonic Visualiser, Soundbite and Mood Player, can help to inform mix practices by analyzing sound attributes and assessing their impact on workflow. A great deal can be learned from conducting mood-based studies and analyzing their effects upon attitudes to a mix.

Conclusion

The word 'academia' can send mixed messages when mentioned in creative circles. Some would argue, as Tony Platt has suggested in our discussion, that academics have a need to 'pigeonhole' work processes or put everything into neat boxes to explain what is taking place in the studio. He finds it frustrating that academics think there is a golden key, which will somehow unlock the door to knowledge. Francis Rumsey conversely suggests the possibility that maybe academic training is a substitute for the fact that recording studios aren't providing a conventional learning environment anymore.

Maybe Academic Training Is the ONLY Option for Many?

Andrew Bourbon and George Fazekas believe there is a greater need than ever for an interdisciplinary approach to mixing practices. There is a need for knowledge about old skills, which can in turn help to inform and investigate new technologies and practices. Bourbon continually makes the point that we should interrogate some of the technologies we so often use without questioning. More importantly, we need to learn to ask questions about some of the techniques used by mix engineers and reflect upon them. The pathway to knowledge is through understanding how we arrive at a finished mix and we should be asking, what makes something sound like a record?

Whatever our conclusion, I would propose that because of the way music is now produced and, more pertinently, mixed, it is more important than ever to conduct some form of scientific evaluation alongside an aesthetic reflection of different workflows of mix engineers to help inform future practices. A good way to do this is by educating via an academic institution or medium. There will never be a substitute for creativity, but there may be an alternative route to gaining scientific knowledge for the intern, which could take years to assimilate under a conventional recording and increasingly less common studio model.

Notes

1 Dr. Andrew Bourbon. Interview with author 2015.
2 Tony Platt. Interview with author 2015.
3 Dennis Weinreich. Interview with author 2015.

4 Dr. George Fazekas. Interview with author 2015.
5 Dr. Francis Rumsey. Interview with author 2015.
6 Dennis Weinreich. Interview with author 2015.
7 Andrew Bourbon. Interview with author 2015.
8 D-Box. A hackable musical instrument which is the collaborative work of Victor Zappi from the University of British Columbia in Canada and Andrew McPherson from Queen Mary University of London.
9 EPSRC. Engineering and Physical Sciences Research Council. https://www.epsrc.ac.uk
10 FAST IMPACt. Fusing Audio and Semantic Technologies for Intelligent Music Production and Consumption. http://www.semanticaudio.ac.uk
11 SAFE. Semantic Audio Feature Extraction. http://www.semanticaudio.co.uk
12 Dr. Mathieu Barthet. Interview with author 2015.

Bibliography

Bech, Søren and Zacharov, Nick (2006). *Perceptual Audio Evaluation*. Chichester: Wiley.
Cascone, Kim (2004). 'Audio Culture: Readings in Modern Music.' In *The Aesthetics of Failure: "Post-Digital" Tendencies in Contemporary Computer Music*. London: The Continuum International Publishing Group Inc., pg. 392.
Corey, Jason (2010). *Audio Production and Critical Listening: Technical Ear Training*. Oxford: Focal Press.
Czikszentmihalyi, Mihalyi (1996). *The Psychology of Discovery and Invention*. New York: Harper Collins.
De Man, Brecht and Reiss, Joshua (2015). Analysis of Peer Review in Music Production. *Journal on the Art of Record Production* (Vol. 10).
Emmerson, Simon (2000). *Music, Electronic Media and Culture*. Farnham: Ashgate Publishing Ltd.
Eno, Brian (1999). The Revenge of the Intuitive. *Wired Magazine* (7).
Fazekas, G., Raimond, Y., Jacobson, K. and Sandler, M. (2010). An Overview of Web activities in the OMRAS2 Project. *Journal of New Music Research on Music Informatics and the OMRAS2 Project*.
Lanier, Jaron (2011). *You Are Not a Gadget*. London: Penguin Books.
Moylan, William (2006). *Understanding and Crafting the Mix: The Art of Recording*, Second edition. Oxford: Focal Press.
Rumsey, Francis and McCormick, Tim (2009). *Sound and Recording*, Sixth edition. Oxford: Focal Press.
Stables, R., Enderby, S., De Man, B., Fazekas, G. and Reiss, J.D. (2014). 'SAFE: A System for the Extraction and Retrieval of Semantic Audio Descriptors.' In *15th International Society for Music Information Retrieval Conference (ISMIR 2014)*.
Taylor, Timothy (2011). *Strange Sounds: Music, Technology and Culture*. Abingdon: Routledge Publishing.
Théberge, Paul (1997). *Any Sound You Can Imagine: Making Music/Consuming Technology*. Middletown, CT: Wesleyan University Press.
Théberge, Paul (2012). 'The End of the World as We Know It: The Changing Role of the Studio in the Age of the Internet.' In *The Art of Record Production: An Introductory Reader for a New Academic Field*, eds. Simon Frith and Simon Zagorski-Thomas. Farnham, Surrey: Ashgate.
Williams, Alan (2015). Technostalgia and the Cry of the Lonely Recordist. *Journal on the Art of Record Production* (Vol. 9).
Wilmering, T., Fazekas, G. and Sandler, Mark (2012). 'High-Level Semantic Metadata for the Control of Multitrack Adaptive Digital Audio Effects.' In *133rd AES Convention*, San Francisco.
Zappi, Victor and McPherson, Andrew (2015). 'The D-Box: How to Rethink a Digital Musical Instrument.' In *ISEA 2015: Proceedings of the 21st International Symposium on Electronic Art*, Vancouver, Canada.

17

The Dreaded Mix Sign-Off

Handing Over to Mastering

Rob Toulson

Introduction

The final stage of mixing, and indeed the final responsibility of the mix engineer, is usually the handover to mastering, which brings a number of creative and technical considerations. In the traditional approach to music production, mastering is conducted by a specialist audio engineer once the final mixes have been consolidated to a stereo format. In the early days, mixes would be recorded to a physical two-track analog tape that would then be shipped to the mastering engineer. Nowadays it is more common for the stereo mixes to be sent as lossless audio files through an Internet file transfer. The final signing-off of the mix is an intrepid point in the process, requiring the artist, music producer and mix engineer to agree that they have completed the mixes, which can reveal any uncertainties or insecurities that they may bear in relation to the project. The mix sign-off and handover to mastering is therefore seen as a critical and crucial point in the music production process.

Approaches to, and technologies for, mixing and mastering have evolved, as have all aspects of music production. New methods and approaches bring opportunities to simplify and reduce the cost of production, although with the potential for practitioners to inadvertently cut corners and underperform in both creative and technical contexts. Modern processing tools enable mix engineers to also master their own music, and there are a number of arguments for and against the use of mastering techniques at the mixing stage. For example, it can be argued that mix engineers need to take a greater responsibility towards technical attributes such as dynamics and noise cancellation. Whereas, in contrast, the use of mix-bus limiting when generating draft listening copies can confuse and falsify the sign-off process. Furthermore, it may be seen that mastering engineers prefer or are requested to work from mix stems (i.e., a number of consolidated audio tracks that collectively make up the mix), but does that mean they are effectively mixing as well as mastering the songs?

This chapter discusses the critical point of completing the mix and moving towards mastering, that is, it considers the crucial process of 'signing off' a mix and reaching agreement between stakeholders that a song is ready for mastering. The discussion draws on the experience and expertise of a number of award-winning mix and mastering engineers through direct discussion and interview, particularly with respect to methods and contemporary practices that are common at the mix-completion stage. The mix and mastering engineers contributing to this chapter are George Massenburg, Mandy Parnell, Ronald Prent, Darcy Proper and Michael Romanowski, whose professional insights give a firsthand reflection on best practice for finalizing the mix and handing over to mastering.

Decision-Making and Signing-Off the Mix

The final signing-off of the mix is a hugely difficult task, as it is the critical point in the music production process where the most significant creative and technical decisions are committed for the final time. Signing off can expose any uncertainties or insecurities that the artist, producer or mix engineer may bear in relation to the project. Historically, mixes would be bounced down through an analog mixing desk to a physical two-track analog mix-master tape. The tape would then be shipped to the mastering engineer, who would then manipulate the audio where necessary and cut the master disc that would be used to manufacture vinyl records. Therefore, the sign-off process was final—it was far too time-consuming and costly to make any changes after final mixes were committed to tape. In many ways, the modern process makes things more flexible. Total recall of 'in-the-box' mixes means that it is very easy to make small changes at any time, and the possibility to quickly share files over the Internet means that, on the surface, there is little additional cost or time constraint with making such changes. However, it appears that the modern process, though much more flexible and reversible, often makes it harder for artists and engineers to sign off a mix, which can be considerably disruptive if the mastering has already started.

Mandy Parnell, of Black Saloon Studios in London, is a Grammy Award–winning mastering engineer who has worked with the likes of Bjork, Annie Lennox and Aphex Twin, among many others. Mandy explains that artists and mix engineers now regularly find it hard to agree on the final mix:

> I've noticed over time that when clients can hear their music on professional loudspeakers, they realize there are details they hadn't heard before, so one of the trends is that the number of remixes that are presented to mastering has increased over the years.

Mandy explains that the time and cost implications of this 'back and forth' can be substantial:

> People just want to tweak. It's a big problem. We have to take notes on everything; in mastering we are still using a lot of analog tools, so to recall a whole album is a lot of work if the client has gone back in and tweaked the mix.

Ronald Prent, a recording and mix engineer at Wisseloord Studios in The Netherlands who has worked with many successful artists, including Elton

John, Def Leppard and Rammstein, concurs with Parnell's assessment. Ronald explains:

> We live in a world of instant recall and I think that's a reason why people want to change things after they've signed off. But if you strive for a little bit more quality, so you don't mix in the box, and instead you use an analog console or you use outboard equipment, then it's not as simple as pressing a recall button, it takes two hours.
>
> I find, because I work that way, artists get the time to listen and sign it off, and until I finish the whole album they have plenty of time to reflect and request changes. But they know that two days after I've finished the project, and I'm on another album, if they want to change anything then that costs money. That really stops people from doing the stupid "can you increase the guitar 1 or 2 dB?" I choose the analog approach because it makes the client listen more carefully and listen more to the essence of the song. Is the song really going to be better if I put the guitar up by 2 dB, or does it really matter?

George Massenburg, whose career has covered many aspects of music production and technology development, earning him Grammy Awards and production credits with artists Earth, Wind & Fire, Billy Joel and Lyle Lovett, agrees on the benefits of working with an analog mix setup:

> Working on a desk means you are learning the song as it goes by. You mix live as the song goes by and you're making a commitment to the levels.

Ronald Prent emphasizes further that mixing in an analog environment helps with decision-making and reflecting on the context of the whole song as opposed to just short passages at a time:

> When you work with an analog tape machine, you press play and you listen to and mix the whole song, and then when it's done you rewind and press play again. Nowadays people loop a 10-second section for half an hour, then they loop the next 10 seconds for half an hour too. I insist that the students who I teach listen to the whole song. I tell them not to loop, to play if from the beginning, so when they make a change they know whether it works or not for the whole song. The rewind time is important too, because you think and reflect while the tape is turning back. I play from the beginning to the end and I mix the song. If you hear things 20 times in a loop you get bored and you lose judgement of its context with the whole song.

George Massenburg explains a wise approach for agreeing and signing off aspects of the mix as it is in process:

> In the very earliest steps of the mix, especially a big complicated mix, the thing you have to get right is the vocal. To get that right you need the artist to sign off on a vocal sound before you build the mix around it.

Darcy Proper is a multiple Grammy Award–winning mastering engineer from Wisseloord Studios, who identifies that an artist's confidence in their work contributes significantly to the challenge of signing off a mix:

Part of the reason why people are putting off decision-making until the last possible second is because they are working in environments that make it difficult to make those final decisions with any confidence. One of the downsides to people working at home in an uncontrolled environment is that it's very difficult to have the same level of confidence as there was years ago when all the processes needed to be conducted in a professional studio. I think insecurity is part of the equation, it's not just a lack of knowledge, but, because of the circumstances in which people are working—and in many cases within the budget which they are working—it's very difficult for artists and mix engineers to say "ok, yeah now I'm sure that it is done".

Darcy adds that sometimes being too close, or too emotionally attached, to the music can cloud the sign-off process, and means that moments of accidental beauty might be discarded if someone has an unattainable vision of perfection. It is usually therefore beneficial for artists and songwriters to work in teams with specialist mix and mastering engineers, rather than alone:

There could be a beautiful natural performance where there is a fantastic break in the artist's voice, just at the moment where it is reaching its emotional peak of poignancy, and as a first-time listener you hear that and it just makes your heart break and you think, "that's perfect because it's not perfect". That fresh perspective when you hear it for the first time [as the mastering engineer] is the same perspective that listeners will have when they hear it for the first time. I think that's an important part in the decision-making process. If you leave those decisions to the people who have been on that journey the whole time, their tendency might be to fix things that aren't broken and thereby take the beauty and the joy out of the nuances and the beautiful flaws.

Ronald Prent also emphasizes that the reversible processes that have been allowed by digital audio workstations (DAWs) actually encourage practice that is counterproductive with regards to decision-making:

I think in an education context we should ask Avid to make a Pro Tools version that doesn't have an undo button and only allows destructive recording. I'm serious, because that's how we learn to make decisions, but not only technical decisions, very much musical decisions too. If you add something new and it doesn't work, then it's because what you have recorded is good already and there's no space for hundreds of keyboard overdubs and percussion samples and programming, because you've already determined musically what is right. If you do destructive recording, once you hit record the other content is gone, so you have to make musical decisions based on technical knowledge—the undo button killed music as far as I'm concerned.

Critical decision-making is certainly one of the most important skills required for creating a good record, throughout the complete recording, mixing and mastering process. As Mandy Parnell simply states:

Mixers need to own their mix.

Michael Romanowski, a Grammy-nominated engineer from Coast Mastering in San Francisco with several thousand credits to his name, echoes this sentiment, noting the crucial importance of the artist taking responsibility for their music:

> An important trait for artists to adopt is having courage, owning your own music, having intent—why are you making it sound like this?

It is clear, therefore, that if artists and mix engineers are brave and focused, and have reasons for their artistic choices, then this positivity can drive the record production forward in a successful and productive manner. However, one further issue at the mix sign-off stage is with artists being unsure how their mix will translate into the real world of music consumption, with its unlimited number of playback systems and formats. Ronald Prent explains:

> Signing off the mix has changed in many ways. More and more often, the artist is not there when you are mixing, and they want an mp3 to listen to after the session, because that is how it will sound on iTunes or an iPhone. This is the quality they are used to judging; if they are in the studio listening through a pair of real speakers, it scares the shit out of them sometimes, because they hear the detail that we are used to hearing, but they don't know how to deal with it.

Ronald's solution is a novel one:

> I stream to my client when I mix live and then they can choose whichever mobile device, earbuds or computer speakers they want to listen on. I stream MP4 live the whole day and they can listen in whenever they want; if they are well equipped I can stream 96/24 lossless too. Artists are starting to demand that and it really works. The comments I get back are good comments—they can Internet chat with me in shorthand and I get far better comments, they are coherent and easy to translate.

Comparison Is the Route of All Discontent

In signing off a mix, artists and mix engineers are often drawn to compare their work with that of other practitioners. This can be a dangerous element to throw into the critical and delicate process of mix sign-off. George Massenburg recalls:

> There is an old aphorism that says "comparison is the route of all discontent".

Michael Romanowski equally dispels the theory that comparison is necessary for sign-off, and states that comparison and competition has no valid place in music production:

> I don't understand or agree with the idea of competition in music production. This is art and it shouldn't be 'this painting is more blue than that one, therefore this one wins'—we don't do that. Own your mixes rather than compare them to somebody else.

Darcy Proper also agrees that making music in comparison to some other art is essentially a contradictory process:

> It's a funny message that we get from our clients; their music is completely unique and impossible to characterize and they are very proud of that fact, but they also want it to sound exactly like whatever is on the radio right now! Artists should celebrate their music for the fact that it is unique and let every part of the production process celebrate that uniqueness, rather than trying to turn it into a song that sounds like something that is already there in the Top 40.

There is clearly a fine line between using reference material as a source of inspiration, which is deemed positive, and as a source for comparison or competition, which is regularly counterproductive. While professionals agree that good music is something that should be celebrated for its uniqueness, using examples of high-quality recordings for inspiration can clearly help the focus and direction of the mix and mastering processes, as George Massenburg states:

> It's good advice for the mix engineer to get reference material from the artist at the very beginning of the process. Play me a CD or whatever, give me some idea where you want to go with this and we can at least tell you if it's not possible! If necessary you can tell the artist "you can't get there from here", because that's not what you have on the recording. But identifying the artist's irrationalities early on is helpful!

Equally, Mandy Parnell describes the importance of reference material and creative insights at the mastering stage, in enabling her to giving the mix its necessary final polish:

> If artists can give us direction that helps. Some of my clients describe things in pictures and abstract art, they come in with laptops, show me pictures, play me music, we talk a lot. I think there has to be sufficient communication; you can't just send the mix without the context or a reference point.

Ronald Prent also highlights misconceptions on the level of manipulation and improvement that can be achieved at the mix stage:

> If you just mix a project, it's two or three songs of total guesswork and sometimes you find out after mixing three songs you've just been enhancing what they've recorded and carried on with that vision. But it's totally not what they want, they want a completely different record, like they've recorded acoustic guitars and nice piano but they want an AC/DC sound. And sometimes I have to sit there and say, sorry you didn't record that.

In particular, Ronald explains his go-to records for defining a high-quality final mix:

> One of the best reference points for me is *Songs in the Key of Life* by Stevie Wonder, the song 'Isn't She Lovely', it's very low volume. It's better than many modern references; they often sound crappy and distorted.

Another great reference record is *Joshua Judges Ruth* by Lyle Lovett, which was mixed by George Massenburg. That record changed my life. It has the right footprint sonically, dynamically, because it's made with passion for music and audio and you can put it on in any studio in any place and if you know how that record is supposed to sound, there you have the best reference for mixing. I grew up with these records and if you get anywhere close then you're doing a great thing.

Mandy Parnell highlights the challenge of finding valuable reference material for mix engineers, agreeing with Ronald Prent that older records are more realistic comparisons, given the high levels of loudness that are prevalent on modern recordings:

I find especially with young mix engineers and producers, they don't have a suitable reference point. The modern references they have are already mastered, a lot of them to a ridiculously loud level, so they are chasing those productions and mixing to compete with something that isn't technically correct. I'm trying to educate people to listen back to classic recordings. But even some of the classic albums have been remastered recently, and they are not a great reference point sonically, in comparison to the records I grew up with.

To Stem or Not to Stem?

A modern approach to mastering uses a number of instrument stems, allowing the mastering engineer to manipulate individual aspects of the mix, rather than just the entire mix. The stems are created by the engineer, on completion of the mixing, by soloing groups of instruments that have contributed to the final mix. Each stem will usually be a stereo file and there may be, for example, an individual stem for each of drums, bass, keyboards, guitar and vocals, depending on the type of music. Usually effects, reverbs, delays and parallel processing will be present in each stem too.

At first, mixing to stems appears to be advantageous for allowing the mastering engineer more options and flexibility. Many mastering engineers, however, report that if a producer desires the mastering to be conducted from stems, then it indicates a failing earlier in the mix or recording process, or an unwillingness to decide on a final balance for the mix. Mandy Parnell explains:

People are asking me to master from stems, but basically they are just failing to make decisions or sign off.

Furthermore, the use of stems at the mastering stage means that the evaluation of balance—the relative volumes of instruments—becomes the ultimate responsibility of the mastering engineer, which Mandy points out requires different listening skills:

Working with stems requires a different headspace. Mastering engineers listen differently to mixers. Stems take us out of our comfort zone and put us in a mix zone, which isn't our expertise and usually the results are not as good. Mastering engineers tend to work on emotion, whereas mixing is all about balance.

Darcy Proper also emphasizes that a different mindset is required for evaluating balance versus the sonic attributes of the song as a whole once complete:

> The reason I'm not a mix engineer is because mixing is a different mindset. When you are rebalancing things, it's a very different focus than the fine-tuning of EQ and compression that's involved in mastering.

Michael Romanowski feels even more strongly, altogether refusing to master music from mix stems:

> I don't do stems; that's not my job as a mastering engineer, that's mixing. I feel strongly that if you can't make the decision then you are not ready for mastering. It's lazy; if you're hired as a mixing engineer to do the mixes, then mix it, don't bring it to me to do your job!

> With mastering, there are just two channels. You are asking "how does the content present itself when it's finished; dynamically; EQ-wise; fades; spaces; order?" Mixing is balance. If I'm asked to take stems that means I'm not being completely objective with the presentation of the music. If I can sit back and listen to the presentation, then I can develop an opinion on how it sounds and I can make an appropriate judgement and decide the actions to take from there.

When the mix engineer creates mix stems, an important aspect is to ensure that the individual stems, when summed together, are identical to the stereo output that was used to monitor and sign off the mix. Unfortunately, this is very often not the case, as Ronald Prent explains:

> If you give the stems for mastering—no matter how well you set up your routing on your console and how well you manage your gain structure—if you add the stems up, it's not the same mix, because it doesn't use the processing that you use at the end to make your mix.

Darcy Proper agrees; although she is willing to work with stems if necessary, the best results come when a mix has been properly signed off:

> The projects that have turned out the best in my career, they have all come from stereo mixes, not stems. Of course stems have an important place in film and game music, but, most of the time, the sum of the stems doesn't equal 'the mix'—it's never quite as simple as summing up the stems and then playing those all off at zero.

The argument that stems allow different aspects of the mix to be mastered separately is not one that simplifies the process, indeed it makes mastering much more complicated, as Darcy explains:

> If they haven't submitted a complete mix, then what's often needed is far more complicated than just adjusting the individual stems by a certain level for the whole piece. I may likely need to look across the whole song, raising vocals during the chorus and dropping them maybe in the verse. It really is mixing and it's very different to what we should be doing at the mastering stage.

George Massenburg emphasizes this point even further:

> Does anybody like stems? I hate having to send stems for mastering. It's not as simple as people think. For example, if the A&R department gets back to you and says "well, you got to bring up the bass", it's a lot more than that; you have to add the bass and a little more kick and whatever else is in the low end, the low end of the piano, the low end of the voice, everything changes. These things do not exist independently, and the fiction of stems is that you can allow the record company to make what is in their mind an improvement, and they are just idiots, it's not an improvement.

Massenburg goes on to emphasize that the use of stems is indeed a method of avoiding key decision-making, and allows record companies and A&R representatives to take more of an active role in the music production process:

> Working with stems is basically a means for A&R to get their hands on your record and to control something about the record and, most importantly for them, to own a part of it. It allows insecure A&R men to take ownership of something.

Ronald Prent also highlights somewhat underhanded processes that can take place following the delivery of stems to the record label:

> We are asked to print stems just in case. It's usually the client, artist, producer or in some cases record companies that demand stems for remixing purposes. But they don't want to pay any more, or they want to give your stems to someone else to remix, which I find very underhanded.

Mixing and Mastering in a Single Process

The widespread availability of consumer digital audio workstations and advanced audio processing plug-ins means that it is now possible for anyone to record, mix and master their own music at relatively little cost. Given the challenges with raising budgets for music production projects, it is therefore possible to cut costs by working autonomously. In particular, it is now possible for an engineer to attempt to mix and master their own music, sometimes in a single process. Michael Romanowski suggests that mixing and mastering should rarely be conducted at the same time, or by the same person, because the objectives for each process is very different:

> There are cases where with the right person who knows what they are doing, with the right perspective and intent can take both roles. But it is decidedly two hats, because your objectives for mixing and mastering are completely different. It's not simple to do—putting something on the output bus of a mix and saying that is mastering—that's not mastering, that's something completely different, that's finishing the mix.

Michael also emphasizes the importance of critical and objective listening at the mastering stage, and highlights the challenges with moving between creative and technical processes in music production in general:

> This is art and art is expression. We have a left brain and a right brain, and I think of it like a pendulum. The brain has a creative side and a technical side and if you're the artist and you're going to try to record and mix yourself, you think "I've got a great idea for a song, now I'm gonna open up a preset on a plug-in and connect this thing in here, wait a minute, what was that lick again?" I feel like the pendulum never swings nearly as far as it should on either side because it's constantly going back and forth and you're not getting your full artistic expression, nor your best technical level of capturing that. So consider hiring professions throughout the chain to be able to do the best they can to represent that art.

Darcy Proper also emphasizes the difficulties with taking the responsibility for both mixing and mastering a record:

> As the artist or mix engineer, by the time you get to the mastering process you are in one of two states: Either you feel that it's absolutely your best work and you may be tempted to believe that it's perfect and doesn't need further mastering, while a few minor adjustments could really make it shine. Very often the opposite is true too, in that it doesn't excite you anymore, because you're just tired of it. Then you start fixing things that aren't broken, and when you start fixing things that aren't broken you are actually doing harm and damage. That's one of the advantages of having another person involved who doesn't bring any baggage to the session.

There can be a temptation to consider mastering as simply raising the loudness of mixes to a commercial standard, and, as such, it is possible to add loudness by using a limiter plug-in on the mix output channel. However, most professional mastering and mix engineers agree that this is not good practice for a number of reasons. Michael Romanowski explains:

> I've seen people say "I'm gonna start a song and I'm gonna put a couple of EQs and a compressor on the master bus and I'm gonna mix the song". That to me is like saying "I'm gonna start with a cup of salt and I'm gonna now make me some soup!" Do either of those things make any sense to anybody?

> When someone says to me they are going to master a track by adding some EQ to the master bus I always go back to the question 'why?' And if they say it's because the guitars are not bright enough—well fix the guitars, work on that. That's not mastering, that's just fixing a mix and for us to blur those lines between mixing and mastering is dangerous for the art form. As a mastering engineer I might not put any EQ on it, I might run it through a valve processor and realize it doesn't need any EQ at all.

George Massenburg and Mandy Parnell emphasize that there are certain cases where the mix engineer might master their own music. George states:

> There are occasions where there's barely a budget for mixing, let alone mastering, so I'll do some EQ and a little bit of dynamics on the bus. But if I run into

a wall I go back and change the bass level or I change the kick if it's punching holes, I'll pull it back a tad. I think there is a case to be made for doing that with sensitivity and objectivity sometimes.

Parnell also states that:

> It works maybe for some kinds of electronic music and some metal which is very aggressive. I think there is a place for it, but most of the time I'd say it should be avoided.

Darcy Proper has practical advice for producers working on a tight budget, where the mixing and mastering must be conducted by the same engineer:

> My advice is don't mix and master in one step. Finish your mix, print your mix, then take that mix and master it as a separate process, not within the same DAW session. If you do have to master it yourself, take a break, take some time between the mixing and the mastering. It should be two separate steps.

One unique and regularly prevalent issue is that review mixes are rarely at the loudness of commercial music, so clients and artists struggle to evaluate the mix, given that it will be much quieter than mastered songs that are heard on the radio or in the charts. In this case, it is often necessary to supply 'listener copies' of the mix that have been raised in loudness with some form of compression or limiting. Listener copies made in this way cause much confusion for artists and engineers, so it is important to have a clear and justified method for creating the listener copies. In addition to Darcy's previous advice about mixing and mastering as separate steps, she adds:

> If I could say one thing to every student or young mixer out there, it would be "don't mix through any loudness software". It is understandable that in order to get client approval—as many lack the imagination to understand how a mix might sound when it is mastered—you can give them a loudness-enhanced version, but don't mix through the software that makes it loud. Make a good solid mix and then right at the end put the loudness tool across it and keep it as close to the intention of your mix as possible. That way you can give the mastering engineer both the reference version that you have sent to the client and you can also provide a finished mix that doesn't have that extra loudness maximization on it.

Mandy Parnell also discusses the issue of mixing with limiters on the mix bus:

> There is a problem with this listening copy model, that mix engineers are being pushed to use limiters because artists and record labels want to hear loud mixes for review. The problem with mixing into limiters is that the final mix sent for mastering then has no relation to the listening copy, which is distorted and noisy. But that's what everyone has signed off from, and that is our starting point. Many times I have to go back and ask them to put a bit of the listener limiting back on in order to get the sound they are after, but still it's too loud, and I'm starting from such a wrong place to master the record. I'd like to see us getting away from that, but it's about education and people have been working like this for maybe ten years now.

Darcy Proper also adds that any mix bus processing can actually mask issues that will still need to be resolved at the mastering stage, whereas they could have been treated better during mixing:

> If you mix through a loudness maximizer, or any kind of multiband compression, it does things through individual frequency bins, so you could potentially have too much 4 kHz content in your mix, which the maximizer is keeping under control. But even though the copy the client approves sounds good, the mix still has an issue that needs to be rectified. Build the mix first, make that a good solid thing and then add the loudness afterwards.

Mastering Engineer as Mix Consultant

The mastering engineer acts as a quality control expert for the final stage of a project, but can also act as an invaluable mix consultant, too. George Massenburg explains that he would regularly seek a mastering engineer's opinion throughout the mixing process:

> Doug Sax had a unique knack for, through mastering, moving the irrational to the practical. For many years he was our great advisor who we would talk to in the middle of the project, take him rough mixes and he would have no problem telling us if it sucked, or to do nothing.

Mandy Parnell explains that it is in her interests to have as good a mix as possible. It therefore saves her time by helping out at the mix stage, or by requesting further modifications before mastering begins:

> Mastering engineers really need to advise producers and mixers more nowadays. I want to get a really great mix in, so it's in my interest to get a good mix first than to try to work miracles.

> Something I've started doing in the last five years, if they have the capability of going back and mixing in the box, I will send them back to remix rather than try to polish it in mastering. I give them a lot of direction and listen to lots of references with them and explain what they need to try and do with the mix, where it's going wrong.

Darcy Proper emphasizes that the mastering engineer's first listen is a valuable part of the music production process, the importance of which shouldn't be underestimated:

> We can be objective and have perspective and bring a fresh set of ears. When I hear a song for the first time I'm hearing it as a listener would be hearing it for the first time, and I know I've got it right when I go from not knowing the song and not caring about it or having any feeling for it, to having dialed into the emotional content to the point where I have the goosebumps and I'm emotionally connected to it.

Mandy Parnell agrees:

> My first perspective when a client comes in is the most important thing that they are paying for. My first listen, my first response, my first notes. The feedback they get from that point is the most important I can give, especially if the mix is wrong.

Mandy also emphasizes the importance of building up a relationship with the mastering engineer and continuing to collaborate on a number of projects to consolidate ideas and working practices:

> I say always attend the mastering session and build a relationship. I tell young producers to pick a mastering engineer and stick with them for a couple of years or projects. Use them as a sounding board so you can develop your mixing and build up an understanding.

Conclusions

Clearly, the boundary between mixing and mastering has become gradually porous in recent years, with digital technologies and the availability of home production systems creating an interesting circumstance which allows anyone to effectively take the role of mixing or mastering engineer, or both, at any given time in the production process. These advancements have in some way enabled us to cut corners with the production process and reduce the number of expert ears that might be involved in the recording, mixing and mastering. The opportunity to put off decision-making to a later stage in the production is enabled by technology and the possibility to master from stems, for example. However, this trend actually appears to lead to greater indecision and insecurity in artists and engineers. Mastering engineers are wise and experienced listeners who can assist with mixing as well as mastering, and it is therefore beneficial to utilize them in any professional project. Ultimately, all studio engineers are here to assist the artists and to enable them to most accurately achieve their creative vision for a record, and any professional recording, mix or mastering engineer will be passionate about helping the artist realize that goal, as George Massenburg humbly concludes:

> At the end of the day, it's their record. Their picture is on the front and we are here to serve the artist and at the pleasure of the artist—and if we don't get it right we have every right to be fired. I've tried to do everything an artist ever asked me to do and sometimes I learned something and sometimes they learned something. We are here to chase down their vision.

18

Conclusion

Mixing as Part-History, Part-Present and Part-Future

Russ Hepworth-Sawyer

In the introduction to this book, we discussed our intention that this book, and indeed the *Perspectives on Music Production* series as a whole, should serve as a multi-perspectival investigation into the history, present and future of the production of music for auditory consumption.

Many of the articles in this book explore historical and present incarnations of the mixing process. For this, the concluding chapter, we felt it best to extract and consider the future implications of these contributions. We also felt that to indulge me in a few of my own observations, and some of my predictions, would be useful to conclude this introductory and exploratory volume.

History

To our forefathers in record production, many of the job titles we commonly use today would either baffle or mislead, at best. The term 'producer' has developed quite considerably since even the 1960s, for instance. An electronic musician working today can today mix and release music from their laptop in the spare bedroom without any contact with another living soul, and it would be acceptable if they called themselves 'producers'. What does 'paying your dues' look like in this world? The experience of the 'bedroom mix engineer', for lack of a better term, seems a far cry from that of those engineers who earned the mantle in decades past and likely began their careers sweeping floors and making tea.

In fact, it seems that the titles Sound Engineer and Audio Engineer have all but been eradicated from contemporary industry language, certainly in pop music production parlance. Instead, we use the rather more bland titles Recording Engineer or, more likely, simply Engineer. Alongside this, the term Assistant has changed considerably too, not perhaps in name, but in the work undertaken. Certainly, it has mopped up the humble Tape Operator. Having recently worked in Abbey Road Studios on a number of projects, their assistant engineers are a testament to how that role can been reconceived within the studio. These assistant engineers are all, in my opinion, extremely talented and knowledgeable, and once upon a time, that level of knowledge might have only been reserved for those with the title of engineer at a later stage in their careers. Their roles are, and continue to be, to develop and change in tandem with technological developments. In today's contemporary music studio, the assistant needs to understand an ever-widening range of equipment for those clients, such as me, who visit Abbey Road for the odd day without the tacit knowledge or experience of that particular room.

To further understand these changes in roles, it is worth us revisiting the changes to the equipment over the generations. Our rich heritage in audio engineering had its own pioneers who progressed our art forward with each new technique and tool. As almost each new development emerged, seized upon because of perceived benefits to the quality of the audio, new roles would often be created. For example, the creation of magnetic tape and its adoption within the studio environment meant that no longer were engineers capturing the performance that would effectively serve as the master for production (production here meaning manufacture of the listener's product). As soon as magnetic tape made an entrance, it was necessary for a new stage, a new task, whether undertaken by the same person as the recording, whereby the audio material needed to be transferred or 'cut' to vinyl.

Similarly, as compact disc (CD) became the new format for everyone to jump on to in the 1980s, new roles specializing in digitizing analog audio appeared in all major studios to ensure they can translate their past catalog masters to that medium. A new role emerged ensconced in Sony PCM1610's and U-Matic tape machines. The trend can be analyzed further with the adoption of Sonic Solutions to enable the first widespread non-linear system for CD compilation.

The continual metamorphosis of the audio engineer throughout the ages has led us to a huge diversity of engineering duties, yet concurrently one person could, once more as in the very beginning, fulfill all roles necessary. Burgess (2014) observes that in the early days of recording the "mix capability was limited to positioning the musicians, instructing them with respect to dynamics and tone control, and guiding the vocalists in their proximity to the collecting horn". The notion of the mix being 'complete' at the time of the recording is paralleled in the classical conducting of an orchestra. Terrell et al. (2014) note that an orchestral conductor has the equivalent role of the mix engineer: "A critical control parameter is acoustic intensity, which the conductor controls through instructions to the musicians and the mixing engineer controls through adjustments to his mixing interface".

'Mixing' as a term in itself would not have been considered so before the dawn of the discrete stage. 'Balance' would have been the more appropriate term. Recordists prior to any form of multichannel audio would have considered the blend of the instrumental performance and moved the musicians closer to the transducer to ensure the required outcome. The producer's role could be arguably a reinterpretation or extension of the word 'arranger' as the instruments were arranged around the room. This was a strategy that Phil Spector continued at times even as multitrack techniques were prevalent to great success.

The rise of the engineer as personality perhaps began with Joe Meek, who made a name for himself within IBC studios during the 1950s. Meek was perhaps one of the first engineers to have what Gary Bromham[1] notes as a 'sonic signature'—a sound desired by artists unique to that engineer. Bromham also coins the phrase 'sonic trend' to describe the marketability or 'brand' of an engineer. Meek became a highly respected and sought after engineer for achieving chart success and a unique sound that made his records stand out. Meek quickly and inquisitively assessed and understood the qualities that made a good song translate well over vinyl in jukeboxes and on radio. Meek in those days would have been expected, and he himself relished, the need to work on each aspect of the project from recording to the 'mix'. Meek's propensity to secrecy for fear of espionage of his techniques (from fellow engineers) made him less than collegiate within IBC. The sonic 'brand' of Meek, of course, even within IBC, became well known with his own built equipment, such as the spring reverb unit made from a fan heater. Meek also made pivotal limiting devices to improve the sound of the audio. Despite the multitrack dream, much larger track counts of eight or even sixteen was to be over a decade away. Joe Meek shone the way forward with his enormous and sustaining influence on sound production per se, shaping many of the mixing engineer's 'given' techniques that persist to this day. In many ways, Meek was one of the first superstar mix engineers.[2]

Meek was also a pioneer within the UK studio scene by being one of the first engineers to leave a major studio business to go it alone. Meek, whilst not technically freelance as the term would later be known, did work in his own studio in his flat in Holloway Road. It would be, again, a decade away before engineers could start to be independent of studio employment in the mainstream.

As multitrack recording started to become the norm for most professional recording studios in the 1960s, again new skill sets were emerging because of the technology. Engineers, certainly in my native England at that time, were most often employees of one of the major studios of the day, whether that be IBC, Abbey Road, Apple Corps or later Trident, for example. Engineering roles in those early days were more a rite of passage that saw their training start in the cutting or transfer room. Engineers (or even producers later in their careers) may all have started at this point. The belief was that recording engineers ought to understand the limitation of the medium to which they would record. It was common to move between studios and projects within the building as an employee also. So to record or mix any given project would not have necessarily been of concern. Artists and producers would not necessarily question the ability of any engineer in mixing or recording employed by the studio. This, in the most part, was accepted practice. So to assert that there was any consistent delineation between a recording and mix engineer to any great extent, certainly in the UK, would have been less than common.

As the 1960s progressed, the undeniable technical push The Beatles provided for Abbey Road and its engineers would not only shape what could be achieved by the equipment, but also by the personnel. By the late 1960s, engineers were developing a freelance culture of their own,[3] and it became more common for bands to ask for certain engineers for certain projects, including The Beatles towards the end.

The aforementioned delineation of the sound engineer's traditional role was slowly starting to develop away from the norm as freelance culture developed within the industry. Despite this, a traditional norm sustained itself in the most part throughout the early 1970s. The working models for engineers employed by a studio in those days, as Phill Brown (2010) expresses in his excellent book, could mean that you could track the record and a colleague could end up mixing it if the studio manager wanted you working with another band. In the most part people did not mind, especially if working within the same studio complex, whether it be Abbey Road, Trident and so on. Brown later describes how disappointing it could be, though, to be removed from a project before completion.

Technologically, during the 1970s there was steady development. It was, in many ways, the start of a consistent period, with the two inch 24 track becoming ubiquitous alongside the large-format mixing console and the rise of Neve & SSL in the 1970s as the soon-to-be benchmark standards.

The Mix Engineer as an independent title does not emerge as a frequently appearing album credit for many years to come. During the 1980s and 1990s, we see a growing practice and expectation that a specialist mix engineer would touch your single or album, making it a hit. Bob Clearmountain is one such engineer that is credited as one of the first to obtain this mantle, despite also being a successful producer and recording engineer in his own right.

Many have followed and led a path for others to specialize in the area. Names such as Chris Lord-Alge, Manny Marroquin, Serban Ghenea, Cenzo Townshend and Robert Orton occupy just a few of the many of today's 'present' mix engineer slots.

The Present

The mix has always been borne out of the need to achieve balance: to arrange order from chaos. Mix engineers today have, in many cases, a more complex role than they once had. Bob Clearmountain in the 1980s in many cases may have just received an Ampex 456 two-inch track tape with 24 tracks on which he'd need to mix. He'd ensure (where budget and deadline provided) that he'd have ample time mixing through the installed console in front of him and the outboard processing around him, mindful in the knowledge that the mix needed to be committed there and then. To recall the mix (either by memory or by the primitive total recall systems of the day) was many hours' work. Similarly, due to the relatively limited number of tracks (compared with today's near-unlimited capacity), producers and musicians made necessary decisions in pre-production and the arrangement of their work to ensure the system was used to the maximum. Bands such as Queen in the 1970s proved just how much could be achieved.

As Justin Paterson explores in his chapter, there has been a significant change of practice from the physical studio to that of the computer studio. Mixing has thus a past and present that are, in some ways, completely different to each other. While skeuomorphic graphical designs have ensured that the transition from hardware to software has been tempered as best as possible, significant practices have had to alter to achieve the results. The largest move has been, for many engineers, the move from touching the controls of the SSL or Neve to the mouse and keyboard of their Apple Mac. There have been many investigations as to how this has affected the process of mixing, some of which is explored in the chapters within this book.

As noted in Alastair Sims's chapter, there is now much more dependence on digital editing to hone and polish the performance prior to mixing. In many cases, 'order' has to be drawn from the almost unlimited number of sources and possibilities. Basic editing, alongside tuning, has become commonplace, replacing honed performance in many cases. Given the ability to make such changes, indecisiveness within the production process has also been alluded to in earlier chapters. Fellow mastering engineer Mandy Parnell, quoted within Rob Toulson's chapter, confirms the attitude whereby the mix is not even totally completed because it can even come in as stems to her mastering studio, thus providing opportunity for further refinement.

Mix engineering has, due to the Internet and working practices, become a remote art. Robert Orton explained that much of the work he receives is directly from the artist's management or the label (Hepworth-Sawyer and Golding, 2010). As such, the mix engineer can be detached from the process of production, whereas once he or she would have occupied a chair within the same room as the engineer. The audio engineer's role has again changed due to onset of technology. It is understood that labels send a mix to multiple mix engineers to get them to compete with each other, or so that they can select the best mixes.

Mixing is an art form that has enjoyed greater interest in the past decade, expanding in tandem with the number of specialist mix engineers. As noted by Alex Krotz in his interview chapter "the big engineers that people actually think about . . . for the most part . . . are mix engineers". In the 1970s and 1980s, it was important who the producer and the recording engineer (probably just the engineer) were, but Krotz notes, it's all about the mixer and the producer. Krotz also importantly alludes to the fact that the role of the mix engineer is likely to be yet further entrenched as label budgets decrease and more artists work on Pro Tools at home.

With the propensity of plug-ins for mixing available, Dean Nelson notes that potential mixers "can become sidetracked by the latest . . . plug in, or update". Nelson also shares the frustration felt when he too falls into the trap of the distracting plug-in search. Engineers of a previous generation used the equipment available to them on the whole: the EQ within the desk, the few compressors in the rack and the plate reverb or echo chamber. Records were made and achieved. In many ways innovation, and technological advantage, has slowed the music creation process.

The outcome or expectation of standard stereo mixing remains relatively unchanged to that of few decades before, despite key technological changes. Formats, however, have come and gone. Digital Audio Tape (DAT) once used to deliver two-track masters after open reel tape became suddenly considered 'old school'. CD-R later became an option for many as the players and media became more reliable and widespread. Today, mixes are delivered by audio file most often over the Internet. Two major areas of change have been that of surround sound mixing and high-resolution audio. Surround sound mixing has been on the agenda ever since Quadraphonic in the 1970s. Despite enjoying huge acceptance within the film community, encouraging music consumers to adopt surround sound has been difficult.[4] This is due to a whole raft of issues, but mostly with the playback. Consumers' living rooms were not always of an ample size and optimum shape to perfectly situate five full-range monitors based on ITU standards. Then, of course, there's the expense of this. Surround sound for music was not easily possible at the desired resolution until Super Audio Compact Disc (SACD), which was adopted slowly due to cost of both the machine and the discs.

Since the launch of the SACD into the mainstream, there has been a slow progression towards listening to music on headphones, due to the prevalence of the mp3 player and most significantly the iPod. This has meant that mixing has never been so far removed from the consumer mainstream possibility of surround sound, as headphones could not deliver true 5.1 surround.

As the iPod made significant changes, alongside the iTunes platform, engineers had to ensure their mixes and masters would translate both to the still-popular CD and also to the highly data-compressed mp3. Convenience had won over quality. While the humble CD has stuck in there, we're now faced with many format options when considering buying a new release. The CD may remain (although this format is often being missed off the checklist with current releases in my studio), mp3 will be a potential upload to an aggregator, but more commonly some form of high-definition file, most likely Mastered for iTunes (MFiT), will be uploaded at 96 kHz/24 bit, which is discussed more a little later.

However, the format to ensure is mentioned at this juncture is the humble LP, often released even when CD is not. Vinyl has made a significant comeback in the past few years and whilst not of immediate interest to the mix engineer, there are a number of reasons that the future of mixing might permit different working methods. The present has become somewhat uncertain as to what the future will hold.

Part-Future

In the recent past, one might be able to argue that there has been a techno-logical slowdown when it comes to music production development commer-cially. Computing power has become ample for large-scale, high-resolution, multitrack projects right from a notebook computer. On the face of it, few game-changing developments are commercially available for mixing.

Recent developments, however, do show signs of the future. Touch-screen technology, currently being championed by the Slate Raven, starts to demonstrate that mixing by mouse may become less prevalent and that there could be a return to console-style mixing. The relative cost of mix-ing using a touch surface such as the Raven is significantly lower than an Avid C|24 for the same Pro Tools functionality. It is likely that this area may develop further, meaning that more engineers will stop sitting at right angles to the studio's SSL or Neve that only has two channels up panned hard left and right.

The main game changer is highlighted within this book. Joshua Reiss' work at Queen Mary University will slowly pave the way to true auto-mation of many of the functions of mixing. It is unlikely that automated mixing using intelligent systems will ever take away the gloss an accom-plished mix engineer can apply, but it is expected that a number of the routine editing functions and application of processing on the channel strip could be automated. Not automated by selecting just the Logic Pro chan-nel preset, but actually the system appreciating what the sound is, how it sounds and responding as though you had made the decisions yourself. This could save considerable time and leave more time for mix refinement by the engineer. As this book was going to press, iZotope announced the development of Neutron, which starts to offer mix automation based on the material played into it. It is anticipated that these intelligent system developments, as with Reiss' initial involvement with LandR.com will slowly become more commonplace in the future of mixing music.

LandR currently only offers standard resolution masters at a time when the industry has the option to move to higher resolution. One of the major changes, intertwined with the development of technology, or the lowering cost of hard drives, will be a common system resolution above current CD/PCM standards of 44.1 kHz/16 bit. Many professionals record at 24 bit, but many have yet to make the move permanently to 96 kHz or above. Given MFiT, HDtracks.com and devices such as the PonoPlayer, there is a slow, but certain, move back to high fidelity playback. This has also been given the shove to the forefront by the masses of people returning (or young people flocking) to the turntable and vinyl.

While technically a mastering engineer's issue, equal loudness stan-dards are being adhered to in the broadcast industry much faster than ever thought would be the case. EBU's R-128 standard has been widely adopted and is playing its part in broadcast. Similar principles are being employed, but for the playback of material within software, whether that be a vol-ume-levelling system within iTunes or now YouTube. Mastering engineer

Ian Shepherd's post on YouTube's equal loudness system[5] is rather important alongside high-definition audio in predicting a change in the mix engineer's practice. Loudness wars could be a real thing of the past, paving the way for mix engineers, in time, to choose to employ higher dynamic ranges as they wish.

The future of mix engineering practice cannot be predicted as an island alone. As such, recordists will need to start increasing their resolutions to ensure they can provide the higher sample rates for the mix engineers, as the market, post-mastering, will demand that high-resolution material is deliverable. For example, there's likely to be a real 'hole' should the high-resolution audio marketplace take off. All the mixes made to ½" open reel tape will be of a high definition, of sorts, and can be recaptured for remastering at 96 kHz or higher. However, those mixes made to DAT tape will be restricted in the most part to 44.1 kHz/16 bit (apart from later 24 bit enabled machines). There's a perceived hole in the catalog of potential high definition throughout the mid-1980s and 1990s. Of course, it should be noted that the mixes are of high quality even at standard CD resolution and will continue to be enjoyed as such. However, moving forward, resolution ought to increase. It could be argued that we, as an audio engineering industry, have the possibility to put the quality back into the music: to put the high fidelity back into hi-fi.

Notes

1 Gary Bromham notes 'sonic trends' within his chapter for this book and expands this to the 'sonic signature', which is related in principle to 'watermarking' as discussed in Hepworth-Sawyer and Golding (2010).
2 It must be noted that Joe Meek was much more than just 'a mix engineer'. Meek was perhaps the first superstar producer.
3 Sir George Martin had already made moves away from EMI as a staff employee, acknowledging his potential as a freelance producer.
4 We acknowledge that we'd have wished for more representation of this within the book, but the contributions were not forthcoming.
5 Production Advice—http://productionadvice.co.uk/youtube-loudness/ accessed 29/04/2016 14:16.

Bibliography

Brown, P. (2010). *Are We Still Rolling?: Studios, Drugs and Rock 'n' Roll—One Man's Journey Recording Classic Albums*. n.p.: Tape Op Books.
Burgess, R.J. (2014). *The History of Music Production*. Oxford, Oxford University Press.
Clarke, E., Cook, N., Leech-Wilkinson, D. and Rink, J., eds. (2009). *The Cambridge Companion to Recorded Music*. Cambridge: Cambridge University Press.
Cleveland, B. (2015). *Joe Meek's Bold Techniques*. Nashville: ElevenEleven Publishing.
Day, T. (2000). *A Century of Recorded Music, Listening to Musical History*. London: Yale University Press.
Emerick, G. and Massey. H. (2006). *Here, There & Everywhere: My Life Recording The Beatles*. New York: Gotham Press.
Hepworth-Sawyer, R. and Golding, C. (2010). *What Is Music Production?* Boston, MA: Focal Press.

Hodgson, J. (2010). *Understanding Records: A Field Guide to Recording Practice*. New York: Continuum Books.

Horning, S. Schmidt (2013). *Chasing Sound: Technology, Culture & the Art of Studio Recording from Edison to the LP*. Baltimore, MD: John Hopkins University Press.

Kealy, E.R. (1979). From Craft to Art: The Case of Sound Mixers and Popular Music. *Work and Occupations* 6 (1): 3–29.

Millard, A. (2005). *America on Record: A History of Recorded Sound*, Second edition. Cambridge: Cambridge University Press.

Shepherd, I. (2016). *YouTube Just Put the Final Nail in the Loudness War's Coffin*. Available at: http://productionadvice.co.uk/youtube-loudness/ [Accessed: 29 April 2016, 14:16].

Terrell, M., Simpson A. and Sandler, M. (2014). The Mathematics of Mixing. *Journal of the Audio Engineering Society* 62 (1/2): 4–13.

Index

Note: Italicized page numbers indicate a figure on the corresponding page. Page numbers in bold indicate a table on the corresponding page.

Lightning Source UK Ltd.
Milton Keynes UK
UKHW030335190419
341292UK00014B/352/P